AF887064

TECHNICAL-VOCATIONAL MATHEMATICS

PRENTICE-HALL SERIES IN TECHNICAL MATHEMATICS
Frank L. Juszli, *Editor*

APPLIED MATHEMATICS FOR ELECTRONICS, *Westlake and Noden*
APPLIED MATHEMATICS FOR ENGINEERING AND SCIENCE, *Shere and Love*
APPLIED MATHEMATICS FOR TECHNICIANS, *Moore, Griffin, and Polhamus*
BASIC MATHEMATICS FOR TECHNICAL OCCUPATIONS, *Bohuslov*
CALCULUS WITH ANALYTIC GEOMETRY, *Niles and Haborak*
CONTEMPORARY TECHNICAL MATHEMATICS, *Paul and Shaevel*
CONTEMPORARY TECHNICAL MATHEMATICS WITH CALCULUS, *Paul and Shaevel*
ELEMENTARY TECHNICAL MATHEMATICS, 2ND ED., *Juszli and Rodgers*
ELEMENTARY TECHNICAL MATHEMATICS WITH CALCULUS, *Juszli and Rodgers*
ESSENTIALS OF TECHNICAL MATHEMATICS, *Paul and Shaevel*
MATHEMATICS FOR APPLIED ENGINEERING, *Cairns*
MATHEMATICS FOR INDUSTRIAL TECHNICIANS, *Pachucki*
MATHEMATICS FOR TECHNICAL EDUCATION, *Ewen and Topper*
MATHEMATICS FOR TECHNICIANS, *Tronaas*
PRECALCULUS MATHEMATICS FOR TECHNICAL STUDENTS, *Mavrommatis and Reichmeider*
TECHNICAL CALCULUS, *Ewen and Topper*
TECHNICAL MATHEMATICS WITH CALCULUS, *Placek*
TECHNICAL-VOCATIONAL MATHEMATICS, *Mrachek and Komschlies*

TECHNICAL-VOCATIONAL MATHEMATICS

Leonard Mrachek
Charles G. Komschlies

Hennepin Technical Centers
Minneapolis, Minnesota

Prentice-Hall, Inc., Englewood Cliffs, New Jersey 07632

Library of Congress Cataloging in Publication Data

MRACHEK, LEONARD, (date)
Technical-vocational mathematics.

(Prentice-Hall series in technical mathematics)
Includes index.
1.–Mathematics—1961 I.–Komschlies,
Charles G., (date) joint author. II.–Title.
QA39.2.M7 512'.1 76-48917
ISBN 0-13-898569-3

©1978 by PRENTICE-HALL, INC.
Englewood Cliffs, N.J. 07632

All rights reserved. No part of this book may be reproduced in any form or by any means without permission in writing from the publisher.

Printed in the United States of America

20 19 18 17 16 15 14 13 12 11

PRENTICE-HALL INTERNATIONAL, INC., *London*
PRENTICE-HALL OF AUSTRALIA PTY. LIMITED, *Sydney*
PRENTICE-HALL OF CANADA, LTD., *Toronto*
PRENTICE-HALL OF INDIA PRIVATE LIMITED, *New Delhi*
PRENTICE-HALL OF JAPAN, INC., *Tokyo*
PRENTICE-HALL OF SOUTHEAST ASIA PTE. LTD., *Singapore*
WHITEHALL BOOKS LIMITED, *Wellington, New Zealand*

CONTENTS

SECTION FOUR: **TRIGONOMETRY**

SECTION FIVE: **METRICATION**

PREFACE

Since the beginning of recorded history, mathematics has broadened the outreach of every civilization and generation. Today, as the total knowledge of mankind doubles every seven years, the role of mathematics in the life of man cannot be overestimated. Mathematics has become essential to life. Any vocation you might choose to enter will require you to have a practical knowledge of mathematics. Our primary objective is to help you gain the mathematical skills you will need in your chosen vocation. It is also our desire that you be able to make use of these skills in the totality of your life.

To help you attain these skills, we have been careful to present all mathematical concepts in words and examples which are simple to understand and easy to remember. Constant use is made of mathematical problems which daily confront the technician and tradesman. Consequently, this book is written for you as a student as you are about to enter industry. It will help you learn the things you need to know—not what might be nice to know in a rare situation.

This book may be used in several ways:

1. *Self-study:* Using the book as a guide for individualized instruction, you will be able to set your own goals and schedules. Thus you will be able to move through the text as rapidly or as slowly as you wish. Since human nature often causes one to put things off until tomorrow, it would be wise for you to determine which chapters you want to study and when you plan to complete them. Remember, all successful people set goals and then find ways to meet them.
2. *Review:* You may use the book to brush up on mathematical skills you learned some time ago. The mathematical rules are written concisely and the problems are trade-oriented to provide you with a good, effective tool for review.
3. *Partial study:* Since some vocations and trades require only certain mathematical skills, you can choose to study only those portions of the book which are of value to you. This will enable you to learn what you need to learn in a short period of time. The remainder of the text can be studied when needed.
4. *Assigned textbook:* If you are using this book as a part of your course of study in a technical-vocational school or community college, you may be assigned specific areas of study by your instructor. Hopefully, you will be able to work with your instructor in the determination of your needs and interests.

It is important for you to know *how* to use this book. This will serve as a distinct asset to you in your study. Read through the following description of chapter sections, become familiar with their purpose and use, and think of ways you may want to use each section.

1. *Chapter Title:* The titles will give you a glimpse of the contents of the chapter. They will enable you to rapidly select your area of study.

2. *Objectives:* The objectives tell you what you will know and what you will be able to do when you complete the chapter.
3. *Self-test:* Use the self-test to measure your need to study the chapter. If you are able to do well with the self-test, you can be somewhat certain that you have a good comprehension of the materials presented in the chapter. Thus, if you should so desire, you may move on to the next self-test at the beginning of the next chapter.
4. *Introduction:* The introduction will bring the contents of the chapter into perspective and enable you to catch a glimpse of some of the concepts you will be learning within the chapter.
5. *Definitions:* Careful study of the definitions will familiarize you with the mathematical terms and concepts which are used in the chapter.
6. *Content:* This is the meat of the chapter. Here you will be introduced to the mathematical skills and knowledge you will need to master in order to meet the chapter objectives. Numerous examples and illustrations will help you gain these skills and knowledge.
7. *Exercises:* These problems will give you practice in the skills you have learned. Most often, the exercises will be drawn from the trades and everyday experiences.
8. *Think Time:* This is a time when you will want to think of ways you can put to use what you have learned. Perhaps you can think of problems you now face on the job that can be handled more effectively with what you have learned.
9. *Procedures to Remember:* Mathematical procedures and operations you will need to remember are written concisely and to the point. You will find that these procedures will be of great help to you in solving problems quickly and effectively.
10. *Summary:* The summary serves as composite of all that has been presented within the chapter. It includes those concepts which must be learned for further development of your mathematical skills.
11. *Chapter Test:* The test at the end of each chapter will evaluate your ability to do the mathematical operations taught within the chapter. Wrong answers and/or procedures for solutions will indicate precise areas where mathematical skills have not been properly or adequately developed. Correct answers and procedures will indicate properly developed skills and your readiness to begin work on the skills taught in the next chapter.

LEONARD MRACHEK
CHARLES G. KOMSCHLIES

Section One

ARITHMETIC

CHAPTER

1

FUNDAMENTAL OPERATIONS OF ARITHMETIC

OBJECTIVES

1. To review the fundamental concepts of arithmetic as a basis for further mathematical studies.
2. To provide an opportunity to improve skills and speed in computation, reason, and logic.

SELF-TEST

This test will determine whether or not you need to study this chapter. if you are able to compute your answers accurately, you are ready to move on to the next chapter and the next self-test. Be sure to show your work. This will enable you or someone else to check over your work to determine where you made your mistakes and what changes you need to make so that you will not make the same mistakes again.

S-1-1.

$$\begin{array}{r} 474 \\ 439 \\ 56 \\ +822 \\ \hline \end{array}$$

S-1-2.

$$\begin{array}{r} \$\,15.98 \\ .39 \\ 8.88 \\ +\quad 1.99 \\ \hline \end{array}$$

S-1-3.

$$\begin{array}{r} 789 \\ -454 \\ \hline \end{array}$$

S-1-4.

$$\begin{array}{r} \$\,12.39 \\ -\quad 4.44 \\ \hline \end{array}$$

S-1-5.

$$\begin{array}{r} 47{,}463 \\ -16{,}587 \\ \hline \end{array}$$

S-1-6.

$$\begin{array}{r} \$12.89 \\ \times\quad 12 \\ \hline \end{array}$$

S-1-7.

$$\begin{array}{r} 539 \\ \times\ 321 \\ \hline \end{array}$$

S-1-8. $7\overline{)287}$

S-1-9. $4\overline{)\$\,5.56}$

S-1-10. $408{,}060 \div 20 =$

S-1-11. An electrician worked 48 hours, 42 hours, 44 hours, and 50 hours. How many hours did he work altogether?

S-1-12. Find the total yards of concrete in the following loads (all are given in cubic yards): 8.3, 7.9, 12.4, 14.6, and 11.6.

S-1-13. Jim Lund received checks for $5.86, $14.29, $22.46, $4.29, and $18.42 as a carry-out boy at the supermarket. How much did he make?

S-1-14. A painter used 28 gallons of paint from a 55-gallon drum. How much paint was left?

S-1-15. A water meter read 2412 cubic feet on September 1 and 3901 cubic feet on October 1. Find the number of cubic feet of water used during the month of September.

S-1-16. A boat and trailer together weigh 1235 pounds. The trailer weighs 359 pounds. How much does the boat weigh?

S-1-17. Jack Burke's car odometer read 93,842.4 at the beginning of the summer and 99,326.7 at the end of the summer. How many miles did he drive during the summer?

S-1-18. If there are 5280 feet in 1 mile, how many feet are there in 12 miles?

S-1-19. If gasoline costs 76.9¢ per gallon, find the cost of 12.5 gallons.

S-1-20. A test tube weighs 10.4 grams. Find the weight of a gross of test tubes. (144 = 1 gross.)

S-1-21. A trucking company hauling gravel from a river bed estimates that 8% of each load is water. How many pounds of water are there in a load weighing 3000 pounds?

S-1-22. How long would a steel bar need to be to make 18 pieces 17 inches long? (Disregard the amount of material lost in cutting.)

S-1-23. How much wire would be needed to wrap 14 coils if 2528 inches of wire are needed for each coil?

S-1-24. The "Front Four" of a football team are listed in the program as weighing 280, 265, 260, and 275 pounds. What is their average weight?

S-1-25. Find the value of D in the diagram below.

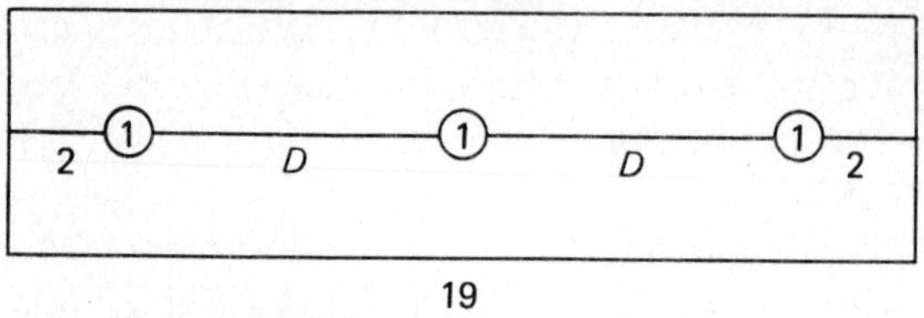

INTRODUCTION

Kevin wanted to cut down the cost of operating his car and at the same time eliminate unnecessary gasoline air pollutants. He decided to run a test to determine which type of gasoline, leaded or unleaded, would provide the greater cost saving to him. He knew that the price of unleaded gasoline was generally about 2¢ per gallon more than the price of leaded gasoline and that it could be used in place of leaded gasoline in his car. The choice between the two types of gasolines would be determined by the actual cost of the gasoline per driving mile.

Kevin knew that more research would be needed if he were to consider the total performance of his car and the amount of pollutants emitted from his car. But, for the moment, he was primarily concerned with the cost of gasoline per mile. In order to carry out his tests as accurately as possible, Kevin needed to have knowledge and skills in arithmetic. But what things must he know and be able to do? What skills will help him solve his problem?

First, Kevin must have an understanding of the terms and procedures used in arithmetic. One of the major causes of failure in the use of mathematics is a lack of knowledge of just what to do. Second, Kevin must not only know what to do, but he must be able to follow these procedures correctly. In the following sections you will become familiar with the things Kevin needed to know in order to solve his problem. These things can be of benefit to you, too, if you will learn them.

DEFINITIONS

As a small child, you were taught to count. At first you could count accurately to perhaps four or five. Soon you increased your ability to count to ten, then twenty, and finally to hundreds, thousands, and millions. The numbers you used are called *integers*. An *integer* is a whole or natural number used in counting. Each integer is made up of one or more *digits*. Thus the number 8 has one digit, the number 23 has two digits (2 and 3), and the number 4146 has four digits (4, 1, 4, and 6). An *even* integer is a number whose last digit (reading from left to right) is divisible by 2. Examples of even numbers are 2, 4, 6, 12, and 96. An *odd* integer is a number whose last digit is not divisible by 2. Thus 3, 5, 7, 17, and 29 are examples of odd numbers. A *prime number* is a number that can be divided only by itself and by the number 1 (unity). The first eight prime numbers are: 2, 3, 5, 7, 11, 13, 17, and 19. The number 1 may or may not be considered a prime number.

Numbers which are combined or added to give a total or sum are called *addends*. Addition is indicated by the plus (+) sign, which is like a road sign along the highway. It tells you what you are to do with the numbers. In mathematics this kind of symbol is called a *sign of operation*. The plus sign of operation tells you to combine (add) the given numbers to form a total. The result of combining the addends is called the *sum*.

$$\begin{array}{rl} 12 & \text{(addend)} \\ +\ 25 & \text{(addend)} \\ \hline 37 & \text{(sum)} \end{array}$$

The purpose of *subtraction* is to find out how much larger one number is than another. Generally, the larger number is placed on top of the smaller number. The top or larger number is called the *minuend*, and the smaller number is called the *subtrahend*. The result is called the *difference*. The sign of operation of subtraction is the minus (−) sign. Subtraction is the opposite of addition.

$$\begin{array}{rl} 25 & \text{(minuend)} \\ -\ 12 & \text{(subtrahend)} \\ \hline 13 & \text{(difference)} \end{array}$$

There are times when you may wish to add the same number several times. For example, if you wanted to find out how many inches there are in 7 feet, you could add seven 12's together to arrive at the answer of 84.

$$\begin{array}{r} 12 \\ 12 \\ 12 \\ 12 \\ 12 \\ 12 \\ +\ 12 \\ \hline 84 \end{array} \quad \text{or} \quad \begin{array}{rl} 12 & \text{(multiplicand)} \\ \times\ 7 & \text{(multiplier)} \\ \hline 84 & \text{(product)} \end{array}$$

You could also solve this problem by multiplying 12 by 7 and arrive at the same answer of 84. This second process, called *multiplication*, is often considered a form of rapid addition. It saves time and decreases the chances of making mistakes. When two numbers are multiplied, the resulting answer is called the *product*. The top number is called the *multiplicand*, and the bottom number is called the *multiplier*. The sign of operation of multiplication is the times (×) sign. In other words, the bottom number tells you how many times to combine the top number with itself.

There are times when you may wish to divide a quantity into two or more parts. For example, if you wanted to find out how many dozen rolls you would need for 72 people, you could keep on subtracting 12's from 72 until you reached a number that was less than 12. You would then count the number of 12's you had subtracted to find out how many dozen rolls you would need.

$$\begin{array}{r} 72 \\ -\ 12 \\ \hline 60 \end{array} \quad \begin{array}{r} 60 \\ -\ 12 \\ \hline 48 \end{array} \quad \begin{array}{r} 48 \\ -\ 12 \\ \hline 36 \end{array} \quad \begin{array}{r} 36 \\ -\ 12 \\ \hline 24 \end{array} \quad \begin{array}{r} 24 \\ -\ 12 \\ \hline 12 \end{array} \quad \begin{array}{r} 12 \\ -\ 12 \\ \hline 0 \end{array}$$

You could also solve this problem by dividing 72 by 12 to arrive at the same answer, 6.

$$\text{(divisor)}\ 12 \overset{\displaystyle 6}{\overline{\smash{)}72}}\ \text{(dividend)} \qquad 6 \text{ (quotient)}$$

This process, called *division*, may also be considered as rapid subtraction. When one number is divided by another, the resulting answer is called the *quotient*. The number which is used to divide is called the *divisor*, and the number that is divided is called the *dividend*. The sign of operation of division is the divide (÷) sign. When this sign appears in a problem, it tells you to divide the second number into the first number. For example, 72 ÷ 12 tells you to divide 72 by 12. When you attempt to solve the problem through the process of division, you often rewrite the problem so that you can more easily arrive at the correct answer. Thus, you could rewrite 72 ÷ 12 as $12\overline{\smash{)}72}$.

CONTENT

The four basic operations of mathematics are addition, subtraction, multiplication, and division. They are the essential tools for every technician or tradesman.

Let us return to Kevin and his problem. The following examples will illustrate how Kevin used the basic operations of mathematics to determine whether it cost less to use leaded or unleaded gasoline even though he could use either gasoline in his car.

Example 1-1

PROBLEM: Kevin drove 169 miles the first week, 96 miles the second week, 174 miles the third week, and 121 miles the fourth week. During these four weeks he used unleaded gasoline. How many miles did he drive using unleaded gasoline?

SOLUTION: Addition (add from top to bottom)

$$\begin{array}{rl} 169 & \text{(addend)} \\ 96 & \text{(addend)} \\ 174 & \text{(addend)} \\ +\ 121 & \text{(addend)} \\ \hline 560 & \text{(sum)} \end{array}$$

CHECK: Re-add from bottom to top.

Example 1-2

PROBLEM: Kevin bought 35 gallons of unleaded gasoline to drive 560 miles. How many miles did he travel on each gallon of unleaded gasoline?

SOLUTION: Division. Divide the total number of miles driven (560) by the total number of gallons of unleaded gasoline (35).

$$\begin{array}{lrl} & \text{16 miles per gallon} & \text{(quotient)} \\ \text{(divisor)}\ \text{35 gallons} & \overline{\smash{)}560}\ \text{miles driven} & \text{(dividend)} \\ & 35 & \\ & \overline{210} & \\ & 210 & \\ & \overline{} & \end{array}$$

CHECK: Multiplication

	35	(multiplicand)
×	16	(multiplier)
	210	
	35	
	560	(product)

Example 1-3

PROBLEM: Kevin paid $.78 for each gallon of unleaded gasoline. How much did he pay for the 35 gallons of unleaded gasoline?

SOLUTION: Multiplication. Multiply the total number of gallons by the price per gallon.

$	.78	price per gallon	(multiplicand)
×	35	gallons of unleaded	(multiplier)
	390		
	234		
	$27.30	total cost for unleaded gasoline	(product)

CHECK: Division

	.78	(quotient)
(divisor)	35)27.30	(dividend)
	24 5	
	2 80	
	2 80	

Example 1-4

PROBLEM: Kevin found that he used 40 gallons of leaded gasoline to drive 560 miles. How many miles did he travel on each gallon of leaded gasoline?

SOLUTION: Division. Divide the total number of miles driven (560) by the total number of gallons of leaded gasoline (40).

	14 miles per gallon	(quotient)
(divisor)	40 gallons)560 miles driven	(dividend)
	40	
	160	
	160	

CHECK: Multiplication

	40	(multiplicand)
×	14	(multiplier)
	160	
	40	
	560	(product)

Example 1-5

PROBLEM: Kevin paid $.76 for each gallon of leaded gasoline. How much did Kevin pay for the 40 gallons of leaded gasoline?

SOLUTION: Multiplication. Multiply the total number of gallons by the price per gallon.

$	.76	price per gallon	(multiplicand)
×	40	gallons of leaded	(multiplier)
	$30.40	total cost for leaded gasoline	(product)

CHECK: Division

	.76	(quotient)
(divisor)	40)30.40	(dividend)
	28.0	
	2 40	
	2 40	

Example 1-6

PROBLEM: Kevin wanted to find out whether it cost him less to use leaded or unleaded gasoline. It cost Kevin $30.40 for leaded gasoline to drive the 560 miles and $27.30 for unleaded to drive the same distance. Which gasoline cost him less to drive the 560 miles?

SOLUTION: Subtraction. Subtract the cost of unleaded ($27.30) from the cost of leaded ($30.40).

	$30.40	cost for leaded gasoline	(minuend)
−	27.30	cost for unleaded gasoline	(subtrahend)
	$ 3.10	less using unleaded gasoline	(difference)

CHECK: Addition

	$27.30	(addend)
+	3.10	(addend)
	$30.40	(sum)

Example 1-7

PROBLEM: As an afterthought, Kevin wondered just how many miles he drove on an average each week. He decided to use the total miles driven each week for the previous 8 weeks. The miles driven each week were 169, 96, 174, 121, 84, 137, 211, and 128. How many miles did Kevin drive on an average each week?

SOLUTION: Addition (step 1) and Division (step 2). First, add the total number of miles driven each week.

169	(addend)
96	(addend)
174	(addend)
121	(addend)
84	(addend)
137	(addend)
211	(addend)
+ 128	(addend)
1120	(sum)

Second, divide the sum by the total number of weeks.

```
                         average miles
                    140  driven per week   (quotient)
(divisor) 8 weeks )1120  miles driven      (dividend)
                     8
                    32
                    32
```

CHECK: First, re-add from bottom to top. Second, multiply the average mileage by the number of weeks.

140	(multiplicand)
× 8	(multiplier)
1120	(product)

EXERCISE 1-1

Do the following problems using the basic arithmetic operations. Be sure to check your results.

1-1.
$9.98
4.49
+ 3.29

1-2.
10.8 gallons
5.7 gallons
9.4 gallons
+ 8.0 gallons

1-3.
$199.95
59.95
+ 14.95

1-4.
386 miles
513 miles
121 miles
+ 86 miles

1-5.
4.25 hours
8.75 hours
10.25 hours
12.50 hours
+ 4.25 hours

1-6. The air distance from Atlanta to Boston is approximately 837 miles. From Atlanta to New Orleans it is approximately 393 miles. How far is it from Boston to New Orleans via Atlanta?

1-7. A wholesaler sells 504 fishing reels to one dealer, 103 to another, 97 to another, and still has 321 remaining. How many did he have at the beginning?

1-8. The area in square miles of the following states is: Alaska 586,400; Alabama 51,609; Arizona 113,909; and Arkansas 53,104. What is the total area of all these states?

1-9. Find the total area (in square miles) of the five largest countries in the world. They are: Soviet Union 8,599,300; Canada 3,851,809; China 3,691,500; United States 3,675,633; and Brazil 3,286,478.

1-10. Dayton's Service Department did the following labor and parts business in 1 hour:

	Labor	*Parts*
Mr. Hanson	$22.50	$19.50
Mr. Dunbar	22.75	27.00
Mr. Benson	25.00	22.00
Mr. Freeberg	20.50	19.85

(a) Find the total income from labor.
(b) Find the total income from parts.
(c) Find the total income from labor and parts for the 1 hour.

1-11.
$$\begin{array}{r} \$14.95 \\ -\ \ 9.98 \\ \hline \end{array}$$

1-12.
$$\begin{array}{rl} 28.8 & \text{gallons} \\ -\ \ 8.9 & \text{gallons} \\ \hline \end{array}$$

1-13.
$$\begin{array}{rl} 17{,}001 & \text{miles} \\ -\ \ 9{,}314 & \text{miles} \\ \hline \end{array}$$

1-14.
$$\begin{array}{rl} 50{,}000 & \text{miles} \\ -\ 19{,}876 & \text{miles} \\ \hline \end{array}$$

1-15.
$$\begin{array}{rl} 69{,}014 & \text{pounds} \\ -\ \ 8{,}789 & \text{pounds} \\ \hline \end{array}$$

1-16. How much greater is 12,000 than 9249?

1-17. The warranty on a new car is 50,000 miles. The odometer shows 39,875 miles. How many miles are left on the warranty?

1-18. A doctor recommended a dosage of 450 cubic centimeters per day and then changed it to 750 cubic centimeters. How much did he increase the dosage?

1-19. The fuel capacity of a 1976 Chevrolet is listed at 22.0 gallons. If it takes 19.7 gallons of gasoline to fill the tank, how much fuel was left in the tank?

1-20. A television set was priced at $398.99. Then it was marked down to $375.98. Later it was further marked down, to $339.49. What was the total discount?

1-21.
$$\begin{array}{r} 98 \\ \times\ 8 \\ \hline \end{array}$$

1-22.
$$\begin{array}{r} 2406 \\ \times\ \ \ 39 \\ \hline \end{array}$$

1-23.
$$\begin{array}{r} 5280 \\ \times\ 707 \\ \hline \end{array}$$

1-24. $\begin{array}{r} \$14.98 \\ \times \quad 8 \\ \hline \end{array}$

1-25. $\begin{array}{r} \$3.89 \\ \times \quad 14 \\ \hline \end{array}$

1-26. Office space at a new office complex rents for $4.85 per square foot. What would the rent be for a 1800-square-foot office space?

1-27. If a truck averages 51.4 miles per hour, how many miles will the truck cover in an 8-hour day?

1-28. If it takes 2.5 hours of flying time between Billings, Montana and Minneapolis, Minnesota, how much flying time would it take to make three round trips?

1-29. Ed Spiegel bought three records at $3.98 each and two records at $4.97 each. How much did he spend for the records?

1-30. What is the tax on a $199.95 tape deck at 4%?

1-31. $6\overline{)528}$

1-32. $17\overline{)663}$

1-33. $22.50 ÷ 6 =

1-34. 5412 ÷ 451 =

1-35. $16\overline{)\$2239.68}$

1-36. Tim Lause spent $15.18 for 20 gallons of gasoline. How much did he pay per gallon?

1-37. The starting salary at the Medical Center for a nurse is $13,004 per year. How much is this per month?

1-38. If you traveled 496 miles on 31 gallons of gas, what was your average mileage per gallon?

1-39. A lot in a suburb sold for $21,367.50. If the lot contained 12,210 square feet, what was the cost per square foot?

1-40. Bobby Levy spent $14.50 for gas on a 116-mile trip with his old army truck. What was his cost for gas per mile?

1-41. What is the average of 100, 220, 440, and 880?

1-42. What is the average age of a family whose father is 35; mother, 31; daughter, 7; daughter, 6; son, 3; and daughter, 2?

1-43. What is the average resistance in ohms of a group of resistors whose values are 570, 450, 630, and 710?

1-44. What is the average dosage per patient if the following dosages (in cubic centimeters) are prescribed: 420, 180, 100, 200, and 225?

1-45. The weekly salaries at the supermarket are as follows: Velda $204, Jake $189, Thelma $173, Fred $210, Harold $165, and George $193. What is the average *annual* salary? (52 weeks per year.)

1-46. Find the missing distance.

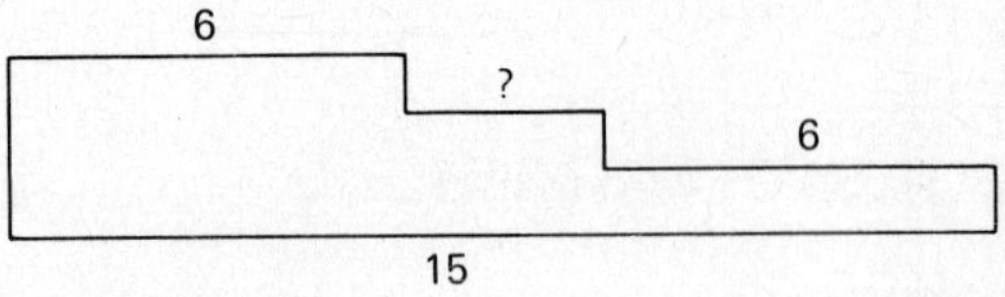

1-47. Find the missing dimension.

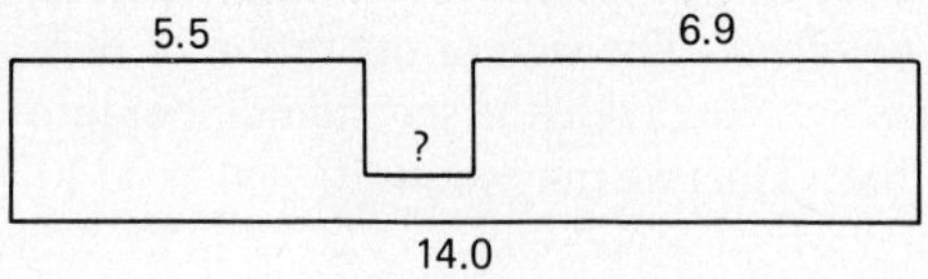

1-48. What is the thickness of this pipe?

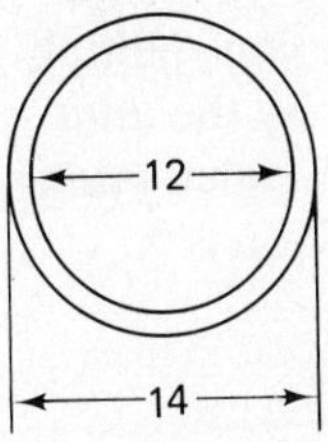

1-49. Find the missing dimension.

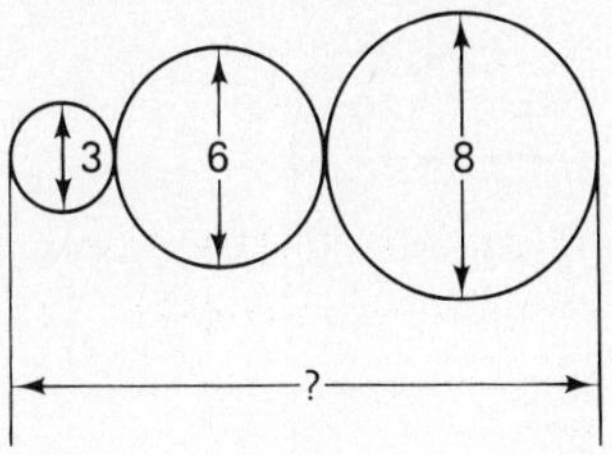

1-50. Find this missing dimension.

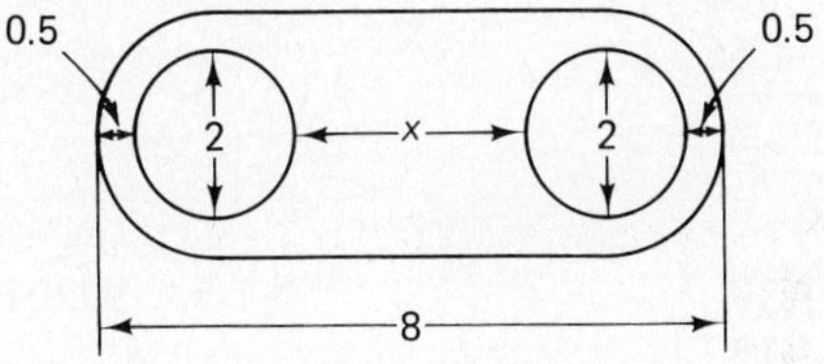

THINK TIME

As you worked through the examples and exercises of this chapter, did you recall using any of the four basic operations (addition, subtraction, multiplication, and division) within the last three days? Write some of these uses on a sheet of paper. What things have you learned in this chapter which will be of benefit to you on the job? What things aren't clear to you at this time? If you have an opportunity to discuss some of these questions with a teacher or a friend, it would be to your advantage.

PROCEDURES TO REMEMBER

The basic operations of mathematics are:

1. Read the signs of operation (road signs) correctly.
2. Perform the arithmetic operations by the most efficient method and show all necessary work.
3. Check your computations to make sure they are reasonable and correct. Methods of checking are as follows:

   Addition: Re-add from bottom to top.

   Subtraction: Add the difference to the subtrahend to obtain the minuend.

   Multiplication: Divide the product by the multiplier to obtain the multiplicand.

   Division: Multiply the quotient by the divisor to obtain the dividend.

CHAPTER SUMMARY

1. An *integer* is a whole or natural number used in counting.
2. An *even number* is a number which is divisible by 2.
3. An *odd number* is a number which is *not* divisible by 2.
4. A *prime number* is a number that can be divided only by itself and 1 (unity).
5. The numbers which are added to give a *sum* or *total* are called *addends.* The plus sign (+) indicates addition. This is one of the *signs of operation* of mathematics.
6. *Subtraction* is the arithmetic operation of finding the *difference* between the *minuend* (top number) and the *subtrahend* (bottom

number). The sign of operation is the minus or negative sign (−).

7. The *product* is obtained when you multiply the *multiplicand* (top number) by the *multiplier* (bottom number). This is called *multiplication* and the sign of operation is ×.

8. The *quotient* is obtained when you divide one number, the *divisor,* into another number, the *dividend.* The sign of operation of division is ÷. We divide the second number into the first. Thus we may show 20 ÷ 4 as $4\overline{)20}$.

CHAPTER TEST

Solve the following problems using the fundamental operations of arithmetic. Show all necessary work.

T-1-1.

$$\begin{array}{r} \$15.29 \\ 1.64 \\ 8.56 \\ +\ \ 4.66 \\ \hline \end{array}$$

T-1-2.

$$\begin{array}{r} \$183.98 \\ -\ \ 84.53 \\ \hline \end{array}$$

T-1-3.

$$\begin{array}{r} \$38.29 \\ 5.46 \\ .29 \\ 4.98 \\ +\ \ .39 \\ \hline \end{array}$$

T-1-4.

$$\begin{array}{r} \$1.39 \\ \times\ \ 15 \\ \hline \end{array}$$

T-1-5. $9\overline{)2016}$

T-1-6. If a dozen golf balls costs $14.76, what is the price of one ball?

T-1-7. A new electric typewriter is priced at $189.95. How much will it cost if the sales tax is 4¢ per dollar?

T-1-8. If a truck driver earns $70.00 a day, how much would he make in 250 working days?

T-1-9. Connie Denham's temperature dropped from 101.4° to the normal temperature of 98.6°. How much change was there in her temperature?

T-1-10. An electrician used the following lengths of number 14 wire: 63 inches, 132 inches, 86 inches, 168 inches, and 1024 inches. How much wire did he use?

T-1-11.

$$\begin{array}{r} 8877 \\ 2345 \\ 618 \\ 6824 \\ +\ \ 36 \\ \hline \end{array}$$

T-1-12.

```
  $20,382.00
      527.50
      322.46
+   1,296.78
```

T-1-13.

```
  23,273 miles
− 16,439 miles
```

T-1-14.

```
   999
×   99
```

T-1-15. 408)34,272

T-1-16. A washing machine in a bottling plant washes 1512 bottles in 12 hours. How many bottles does it wash per hour?

T-1-17. A halfback carried the ball 22 times in one game for a total yardage of 198. What was the average yardage he gained per carry?

T-1-18. A garment worker makes 368 golf shirts in an 8-hour day. How many shirts are made per hour?

T-1-19. An electrical generator produces 2700 watts of power. How many 75-watt light bulbs will this light?

T-1-20. The shipping weight of each of 15 electronic testing scopes is 219 pounds including the carton. What is the total shipping weight?

T-1-21. A farmer wants to paint his barn, which has 14,950 square feet. How many gallons of paint will be needed if each gallon covers 650 square feet?

T-1-22. Vic Goswick, a cabinet maker, uses 22 board feet of birch at 65¢ per foot and 8 board feet of walnut at 96¢ per foot. How much did the wood cost?

T-1-23. Some duck hunters purchased a raft at $153 and a tent at $95. How many hunters were involved if each paid an equal amount of $62?

T-1-24. Some chemical technicians compared the distances they drove to work. The distances were 19.3 miles, 2.4 miles, 8.6 miles, 11.3 miles, and 8.4 miles. What was the average distance driven?

T-1-25. Find *T*.

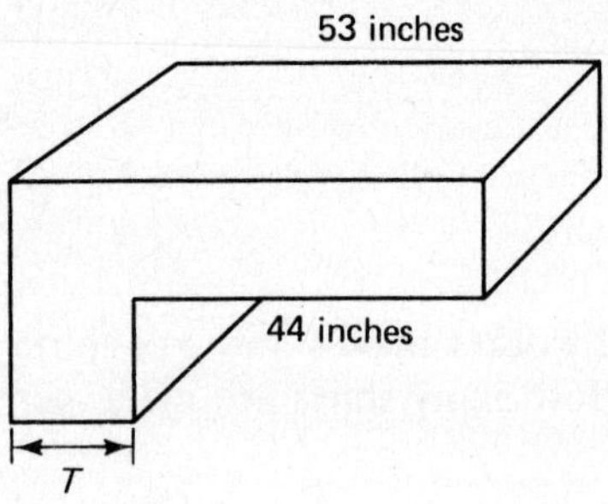

CHAPTER

2

COMMON FRACTIONS

OBJECTIVES

1. To review the basic concepts and computational skills used in working with common fractions.
2. To provide the opportunity to develop your skill and ability to work with common fractions.

SELF-TEST

This test will determine your need to develop your skills in working with common fractions. If you are able to work these problems successfully, you will have shown you are competent in working with common fractions and thus may move on to the study of decimal fractions. As you work each problem, show how you worked the problem. This will indicate areas in which you may need to develop your skills.

S-2-1. $\frac{5}{6} + \frac{3}{4} + \frac{1}{3}$

S-2-2. $24\frac{1}{3} + 10\frac{5}{6}$

S-2-3. $6\frac{3}{4} + 7\frac{1}{2} + 2\frac{2}{3}$

S-2-4. $\frac{4}{5} - \frac{1}{3} =$

S-2-5. $15\frac{2}{5} - 5\frac{2}{3}$

S-2-6. $34\frac{1}{3} - 29\frac{5}{6}$

S-2-7. $\frac{2}{3} \times \frac{6}{18} =$

S-2-8. $3\frac{1}{2} \times 1\frac{3}{4} =$

S-2-9. $24\frac{1}{2} \div 18\frac{5}{6} =$

S-2-10. $12\frac{1}{2} \div 18\frac{5}{6} =$

S-2-11. $\frac{3}{4} \div 12 =$

S-2-12. $\frac{7}{8} \times \frac{7}{16} \div \frac{21}{32} =$

S-2-13. $12\frac{1}{2} + 4\frac{2}{3} \div 8\frac{3}{8} =$

S-2-14. $3\frac{2}{3} \times 1\frac{3}{5} \times 2\frac{1}{12} =$

S-2-15. $\dfrac{\frac{7}{8} - \frac{1}{2}}{4} =$

S-2-16. $\dfrac{3\frac{1}{2} + \frac{3}{4}}{3\frac{1}{2} \times 2\frac{1}{4}} =$

S-2-17. $\dfrac{1\frac{1}{2} - \frac{3}{4}}{1\frac{5}{6} \div 6}$

S-2-18. Determine the dimension of *A*.

S-2-19. Determine the dimension of *B*.

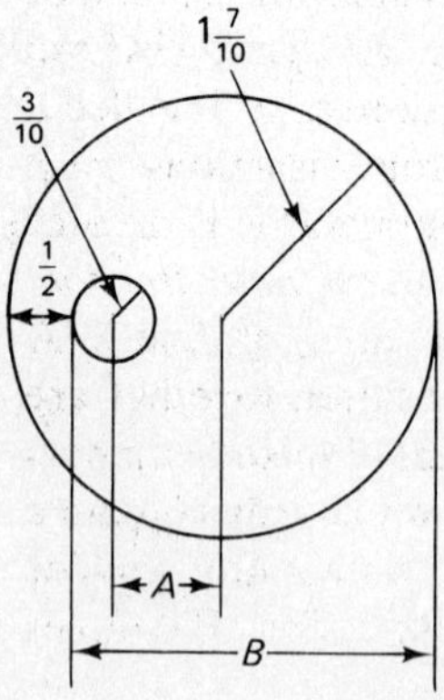

S-2-20. An oil stock sold for $11\frac{1}{2}$ a share when the stock market opened. The stock closed at $12\frac{1}{8}$. How much did the stock gain?

S-2-21. Steak sells for $3 per pound. What is the total cost for steaks weighing: $1\frac{1}{8}$ pounds, $1\frac{1}{4}$ pounds, $1\frac{3}{8}$ pounds, and $1\frac{5}{16}$ pounds.

S-2-22. Total resistance is the sum of the individual resistances. What is the resistance of three motors connected in series if their resistances are: $1\frac{1}{25}$, $2\frac{3}{50}$, and $5\frac{37}{150}$.

S-2-23. The gas tank of a car shows $\frac{3}{4}$ full. If the full tank contains 22 gallons and gasoline costs $75\frac{9}{10}$¢ per gallon, how much is the gas worth that is in the tank?

S-2-24. A truck travels $12\frac{4}{10}$ miles on a round trip from a plant to a construction site. On the eighth trip, the truck stops at the construction site. How many miles has the truck been driven?

S-2-25. A master painter can paint a garage in 6 hours. An apprentice can paint the same garage in 9 hours. How much of the garage can they paint in 1 hour working together?

INTRODUCTION

Common fractions are important to our day-to-day living. We use fractions to indicate: (1) time—"I will meet you in a half hour"; (2) quantity—"You have only a quarter of a tank of gasoline left"; (3) distance—"The runners have passed the three-quarter mark." Fractions are a part of the language of industry and technology. Terms such as $\frac{3}{32}$ welding rods, eighth-inch plywood, three-quarters of a cup of flour, $\frac{9}{64}$ drill, three-quarter ton truck, and one third off are used thousands of times each day across the country. Consequently, it is essential for most men and women to have skills in working with fractions in order to do their jobs.

DEFINITIONS

A *fraction* is the division of one number by another number. Examples of fractions are $\frac{1}{8}$, $\frac{2}{3}$, and $\frac{6}{7}$. The number above the line is called the *numerator* and the number below the line is called the *denominator.* The numerator and denominator are the *terms* for a fraction. Use the name, North Dakota, and its initials, N.D., to help you remember numerator over denominator. A fraction whose denominator is larger than its numerator is called a *proper fraction.* Examples of proper fractions are $\frac{3}{4}$, $\frac{5}{8}$, and $\frac{3}{16}$. A fraction whose numerator is greater than its denominator is called an *improper fraction.* Examples of improper fractions are $\frac{5}{4}$, $\frac{32}{9}$, and $\frac{8}{5}$. A whole number and a fraction written together are called *mixed numbers.* Examples of mixed numbers are $4\frac{3}{4}$, $5\frac{3}{16}$, and $6\frac{7}{8}$. A fraction which may contain a fraction or fractions in the numerator and/or denominator is called a *complex fraction.* Examples of complex fractions are:

$$\frac{\frac{6}{7}}{\frac{3}{5}}, \quad \frac{\frac{2}{3}}{\frac{1}{2}}, \text{ and } \frac{\frac{3}{1}}{\frac{7}{8}}$$

To *reduce* a fraction to lowest terms, divide the numerator and the denominator by the same number.

CONTENT

To be a competent technician or skilled craftsman, you will need to be able to perform fundamental operations with fractions. These fundamental operations, as you recall, are addition, subtraction, multiplication, and division. We will begin with the multiplication of fractions which is perhaps the easiest operation.

Multiplication of Fractions

Example 2-1

PROBLEM: $\frac{2}{3} \times \frac{1}{5} =$

SOLUTION: Multiply the numerator (2) by the numerator (1) and the denominator (3) by the denominator (5).

$$\frac{2}{3} \times \frac{1}{5} = \frac{2 \times 1}{3 \times 5} = \frac{2}{15}$$

Example 2-2

PROBLEM: $\frac{3}{4} \times \frac{7}{15} =$

SOLUTION: Before multiplying fractions, reduce the fractions to the lowest terms. Reducing fractions before multiplying will make the multiplication process easier. Reduction may be done through two different methods.

METHOD 1: Divide the numerator (3) and the denominator (15) by the same number (3). The selection of a number which will divide evenly into the numerator and the demoninator is done through observation. Practice will refine your observation and selection ability.

$$\frac{\overset{1}{\cancel{3}}}{4} \times \frac{7}{\underset{5}{\cancel{15}}} = \frac{1 \times 7}{4 \times 5} =$$

METHOD 2: Divide the *top* (numerator) of one fraction (3) into the bottom (denominator) of another fraction (15). Thus the numerator (3) divides into the denominator (15) five times and into itself (3) once. A simple rule to remember is "*top* into *bottom*" and "*bottom* into *top*." After reducing the fractions to lowest terms, multiply the numerator (1) by the numerator (7) and the denominator (4) by the denominator (5).

$$\frac{1 \times 7}{4 \times 5} = \frac{7}{20}$$

Example 2-3

PROBLEM: $\frac{39}{9} \times \frac{45}{13} \times \frac{1}{10} =$

SOLUTION: Reduce fractions to lowest terms. Remember, there may be several possible reductions. Make certain you have made all possible reductions before you multiply the fractions. Move from left to right as you reduce fractions.

METHOD 1: Divide the numerator (39) and the denominator (13) by the same number (13). Divide the denominator (9) and the numerator (45) by the same number (9). Divide the new numerator (5) and the denominator (10) by the same number (5).

$$\frac{\overset{3}{\cancel{39}}}{\underset{1}{\cancel{9}}} \times \frac{\overset{\overset{1}{\cancel{5}}}{\cancel{45}}}{\underset{1}{\cancel{13}}} \times \frac{1}{\underset{2}{\cancel{10}}} = \frac{3 \times 1 \times 1}{1 \times 1 \times 2} =$$

METHOD 2: Divide the *bottom* (denominator) (9) into the *top* (numerator) (45) and into itself (9). Divide the *bottom* (denominator) (13) into the *top* (numerator) (39) and into itself (13). Divide the new *top* (numerator) (5) into the *bottom* (denominator) (10) and into itself (5). After reducing fractions, multiply the numerators (3, 1, 1) together and the denominators (1, 2, 2) together.

$$\frac{3 \times 1 \times 1}{1 \times 1 \times 2} = \frac{3}{2}$$

Reduce the improper fraction to lowest terms. Divide the numerator (3) by the denominator (2). The remainder (1) becomes the new numerator.

$$\frac{3}{2} = 1\tfrac{1}{2}$$

Division of Fractions

CONTENT

The procedures for dividing fractions are similar to the procedures used in multiplying fractions except for one added procedure. The first step in dividing fractions calls for the *inversion* of the second fraction or the fraction on the right. To *invert* means to place the denominator where the numerator was and the numerator where the denominator was. In the inversion process, the second fraction is literally turned upside down. After inverting the second fraction, multiply the fractions.

Example 2-4

PROBLEM: $\frac{4}{5} \div \frac{1}{3} =$

SOLUTION: Invert the second fraction and multiply. Reduce the resulting solution to lowest terms.

$$\frac{4}{5} \div \frac{1}{3} = \frac{4}{5} \times \frac{3}{1} = \frac{12}{5} = 2\tfrac{2}{5}$$

Example 2-5

PROBLEM: $\frac{5}{8} \div \frac{3}{16} =$

SOLUTION: After inverting the fraction, it may be necessary to reduce the fractions to lowest terms. Remember that the multiplication is done *after* the inversion, not before.

$$\frac{5}{8} \div \frac{3}{16} = \frac{5}{\underset{1}{\cancel{8}}} \times \frac{\overset{2}{\cancel{16}}}{3} = \frac{10}{3} = 3\tfrac{1}{3}$$

EXERCISE 2-1

Do the following problems which will give you practice in multiplying and dividing fractions.

2-1. $\frac{5}{9} \times \frac{2}{7} =$

2-2. $\frac{3}{4} \times \frac{7}{8} =$

$\frac{4}{1} \times \frac{3}{2} =$

Example 2-6

PROBLEM: $14 \div \frac{7}{8} =$

SOLUTION: In this problem, you will note that there is no denominator for the number 14. When a number is without a denominator, it is understood that the denominator is 1. When a denominator is 1, the numerator may be written without the denominator. After inverting the fraction on the right, reduce to lowest terms and multiply.

$$14 \div \frac{7}{8} = \frac{\overset{2}{\cancel{14}}}{1} \times \frac{8}{\underset{1}{\cancel{7}}} = \frac{2 \times 8}{1 \times 1} = \frac{16}{1} = 16$$

Example 2-7

PROBLEM: If half a pizza is to be divided equally among four people, how much will each person receive?

SOLUTION: $\frac{1}{2}$ pizza $\div$ 4 people $= \frac{1}{2} \div \frac{4}{1} =$

$$\frac{1}{2} \times \frac{1}{4} = \frac{1}{8}$$

Note that each person will receive $\frac{1}{8}$ of the *whole* pizza. See figure below.

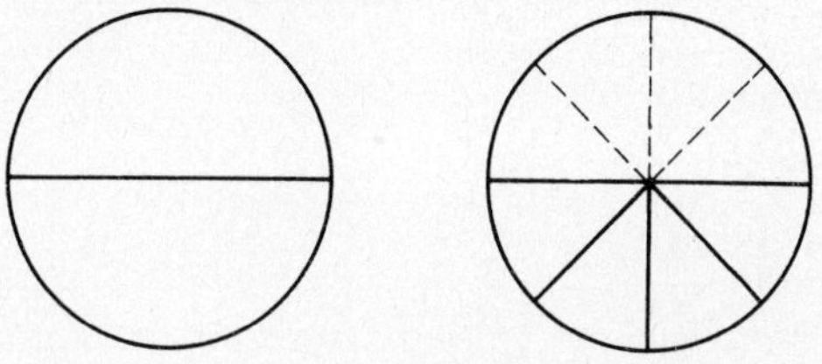

2-4. $\frac{4}{7} \times \frac{2}{3} =$

2-5. $\frac{3}{4} \times \frac{1}{2} =$

2-6. $\frac{5}{6} \times \frac{12}{11} =$

2-7. $\frac{3}{4} \times \frac{8}{7} \times \frac{14}{21} =$

2-8. $\frac{3}{4} \times \frac{5}{16} \times \frac{7}{8} =$

2-9. $\frac{3}{4} \div \frac{2}{3} =$

2-10. $\frac{1}{2} \div \frac{1}{3} =$

2-11. $6 \div \frac{2}{3} =$

2-12. $\frac{1}{10} \div \frac{3}{5} =$

2-13. $\frac{16}{5} \div \frac{18}{5} =$

2-14. $\frac{4}{3} \div \frac{1}{3} =$

2-15. $13 \div \frac{169}{15} =$

Addition of Fractions

Example 2-8

PROBLEM: $\frac{1}{2} + \frac{1}{3} =$

SOLUTION: In the addition of fractions, each fraction must have the same denominator. Thus the first step is to determine the lowest common denominator. This means that a number is to be selected which is a multiple of each denominator in the problem. In this problem we must find a number which is a multiple of 2 and 3. One way to do this is to multiply the denominators together. However, this method does not necessarily give the *lowest* common denominator. In our problem, we can multiply the two denominators (2 × 3) and arrive at the common denominator 6, which is a multiple of both 2 and 3. Once a common denominator has been selected, it is important to determine if it is the lowest common denominator. In other words, is there a number less than 6 which is a multiple of both 2 and 3? There is none, so 6 is the lowest common denominator for this problem.

The second step is to multiply the numerators and denominators of both fractions equally by this number. The only way to do this is by using unity or a fraction that is equal to 1. In this problem $\frac{6}{6}$ will be used. Note that the outline of the number 1, representing unity, is superimposed over the fraction for illustrative purposes. This is only used to emphasize the importance of making certain the fraction is equal to 1.

$$\frac{1}{\cancel{2}_{1}} \times \frac{\cancel{6}^{3}}{6} = \frac{3}{6}$$

$$+\frac{1}{\cancel{3}_{1}} \times \frac{\cancel{6}^{2}}{6} = \frac{2}{6}$$

The third step is to add the numerators. Note that *only* the numerators are added and the denominators remain the same.

$$\begin{array}{r} \frac{3}{6} \\ +\frac{2}{6} \\ \hline \frac{5}{6} \end{array}$$

Example 2-9

PROBLEM:

$$\begin{array}{r} \frac{2}{5} \\ \frac{5}{8} \\ +\frac{3}{10} \\ \hline \end{array}$$

SOLUTION: First, determine the lowest common denominator for the denominators 5, 8, and 10. By multiplying the denominators 5 and 8, we obtain a possible common denominator of 40. If 40 is to be a common denominator, it must also be a multiple of the denominator 10. Dividing the possible common denominator, 40, by 10, we obtain a quotient of 4. Thus 40 is a common denominator. The only other possible common denominators for 10 are 20 or 30. In neither case does the denominator 8 have a whole-number multiple which would give us an answer of 20 or 30. Thus 40 is the lowest common denominator.

Second, multiply each fraction by the unity number $\frac{40}{40}$.

$$\frac{2}{\not{5}_1} \times \frac{\not{40}^{\,8}}{40} = \frac{16}{40}$$

$$\frac{5}{\not{8}_1} \times \frac{\not{40}^{\,5}}{40} = \frac{25}{40}$$

$$+\frac{3}{\not{10}_1} \times \frac{\not{40}^{\,4}}{40} = \frac{12}{40}$$

An *alternative* second step is to multiply each fraction by a form of unity which will produce the common denominator 40. In order to do this, take the first fraction, $\frac{2}{5}$. Ask yourself the question, "5 times what number equals 40?" With the answer of 8, we can now set up the unity number $\frac{8}{8}$. Following the same steps for the fraction $\frac{5}{8}$, we obtain the unity number $\frac{5}{5}$. The unity number for $\frac{3}{10}$ is $\frac{4}{4}$. This method eliminates the need for cancellation.

$$\frac{2}{5} \times \frac{8}{8} = \frac{16}{40}$$

$$\frac{5}{8} \times \frac{5}{5} = \frac{25}{40}$$

$$+\frac{3}{10} \times \frac{4}{4} = \frac{12}{40}$$

Third, add the fractions. Remember to add only the numerators and place the sum over the common denominator, 40.

$$\begin{array}{r} \frac{16}{40} \\ \frac{25}{40} \\ +\frac{12}{40} \\ \hline \frac{53}{40} \end{array}$$

Fourth, reduce the common fraction $\frac{53}{40}$ to a mixed fraction by dividing the denominator (40) into the numerator (53) and rewriting the remainder as a fraction.

$$\begin{array}{r} 1\frac{13}{40} \\ 40\overline{)53} \\ 40 \\ \hline 13 \end{array}$$

Example 2-10

PROBLEM:

$$\begin{array}{r} 6\frac{1}{9} \\ 1\frac{9}{14} \\ +\ 4\frac{5}{21} \\ \hline \end{array}$$

SOLUTION: First, determine the lowest common denominator for 9, 14, and 21. This could be done by multiplying 9 × 14 × 21. The resulting common denominator, 2646, could then be reduced to a lowest common denominator through trial-and-error division. This process is time-consuming. Another way to find the lowest common denominator is to identify the *prime factors* of each denominator. Prime factors are those numbers which, when multiplied together, will be equal to the denominator. In this problem, the prime factors of the denominators are:

$$9 = 3 \times 3$$
$$14 = 2 \times 7$$
$$21 = 3 \times 7$$

The next step is to select those prime factors which will determine the lowest common denominator. Select the prime factors of the first fraction, 3×3, and then select additional prime factors which are different from 3×3. Use these additional prime factors only once. Thus the prime factors $3 \times 3 \times 7 \times 2$ are selected. The lowest common denominator is 126.

Second, multiply each fraction by a unity number. Method 1 uses $\frac{126}{126}$. Method 2 eliminates cancellation by selecting a unity number which will produce the common denominator 126 through multiplication. Study both examples given below. In order to prevent mistakes, remember to show all whole numbers as you work.

METHOD 1

$$6\frac{1}{\cancel{9}_{1}} \times \frac{\cancel{126}^{14}}{126} = 6\frac{14}{126}$$
$$1\frac{9}{\cancel{14}_{1}} \times \frac{\cancel{126}^{9}}{126} = 1\frac{81}{126}$$
$$+4\frac{5}{\cancel{21}_{1}} \times \frac{\cancel{126}^{6}}{126} = 4\frac{30}{126}$$

METHOD 2

$$6\frac{1}{9} \times \frac{14}{14} = 6\frac{14}{126}$$
$$1\frac{9}{14} \times \frac{9}{9} = 1\frac{81}{126}$$
$$+\ 4\frac{5}{21} \times \frac{6}{6} = 4\frac{30}{126}$$

Third, add the mixed fractions. Add the whole numbers first and then the fractions.

$$\begin{array}{r} 6\frac{14}{126} \\ 1\frac{81}{126} \\ +\ 4\frac{30}{126} \\ \hline 11\frac{125}{126} \end{array}$$

Subtraction of Fractions

Example 2-11

PROBLEM:

$$\begin{array}{r} 12\frac{7}{8} \\ -\ 4\frac{3}{16} \\ \hline \end{array}$$

SOLUTION: First, as in addition, determine the lowest common denominator for 8 and 16. The lowest common denominator, easily determined through simple observation, is 16. Second, multiply each fraction by a unity number.

METHOD 1

$$12\frac{7}{\cancel{8}_{1}} \times \frac{\cancel{16}^{2}}{16} = 12\frac{14}{16}$$
$$-4\frac{3}{\cancel{16}_{1}} \times \frac{\cancel{16}^{1}}{16} = 4\frac{3}{16}$$

METHOD 2

$$12\frac{7}{8} \times \frac{2}{2} = 12\frac{14}{16}$$
$$-\ 4\frac{3}{16} \times \frac{1}{1} = 4\frac{3}{16}$$

Third, subtract the fractions. Be sure to subtract the subtrahend (bottom) numerator from the minuend (top) numerator. Denominators are *not* subtracted.

$$\begin{array}{r} 12\frac{14}{16} \\ -\ 4\frac{3}{16} \\ \hline 8\frac{11}{16} \end{array}$$

EXERCISE 2-2

Solve the following problems which involve addition and subtraction of fractions. Remember that you must have the same denominator when adding or subtracting fractions. You may wish to indicate the unity fraction with $\square$.

2-16. $\begin{array}{r} \frac{2}{7} \\ +\ \frac{3}{7} \\ \hline \end{array}$

2-17. $\begin{array}{r} \frac{3}{5} \\ +\ \frac{4}{10} \\ \hline \end{array}$

2-18. $\begin{array}{r} \frac{5}{20} \\ +\ \frac{3}{5} \\ \hline \end{array}$

2-19. $\frac{5}{9} + \frac{3}{18}$

2-20. $1\frac{1}{2} + 1\frac{3}{4} + 1\frac{1}{8}$

2-21. $\frac{1}{3} + \frac{1}{4}$

2-22. $7\frac{1}{2} + 8$

2-23. $\frac{11}{16} + \frac{1}{2}$

2-24. $\frac{2}{3} - \frac{1}{3}$

2-25. $\frac{2}{3} - \frac{1}{6}$

2-26. $3\frac{3}{4} - 2\frac{1}{2}$

2-27. $16 - 12\frac{5}{8}$

2-28. $15\frac{31}{32} - 9$

2-29. $\frac{3}{8} + \frac{13}{32} + \frac{5}{64}$

2-30. $\frac{3}{20} + \frac{2}{3} + 7\frac{11}{12}$

Simplifying Complex Fractions

CONTENT

In working with fraction, you have learned how to: (1) change fractions from one form to another to solve problems, (2) perform the fundamental operations of multiplication, division, addition, and subtraction with fractions, and (3) simplify proper and improper fractions by reducing them to lowest terms. Simplifying *complex fractions* will require you to use all of these skills.

Example 2-12

PROBLEM: Reduce the following fraction to lowest terms:

$$\frac{\frac{3}{4}}{\frac{1}{3}}$$

SOLUTION: Since the line between the two fractions indicates division, restate the fraction in normal division of fraction terms. Then solve the problem.

$$\frac{\frac{3}{4}}{\frac{1}{3}} = \frac{3}{4} \div \frac{1}{3} = \frac{3}{4} \times \frac{3}{1} = \frac{9}{4} = 2\tfrac{1}{4}$$

Example 2-13

PROBLEM: $\dfrac{1}{\frac{1}{2}+\frac{1}{3}+\frac{1}{5}} =$

SOLUTION: This complex fraction contains three fractions in the denominator. When reducing complex fractions, perform the operations indicated in the numerator and denominator first. Note that the lowest common denominator for the fractions in the denominator is 30. Thus, unity $\frac{30}{30}$ is used to determine the common denominator for each fraction.

$$\frac{1}{\frac{1}{2}+\frac{1}{3}+\frac{1}{5}} = \frac{1}{\frac{1}{\cancel{2}_1} \times \frac{\cancel{30}^{15}}{30} + \frac{1}{\cancel{3}_1} \times \frac{\cancel{30}^{10}}{30} + \frac{1}{\cancel{5}_1} \times \frac{\cancel{30}^{6}}{30}}$$

$$= \frac{1}{\frac{15}{30}+\frac{10}{30}+\frac{6}{30}}$$

Second, add the fractions in the denominator.

$$\frac{1}{\frac{15+10+6}{30}} = \frac{1}{\frac{31}{30}}$$

Third, restate the fraction in normal division of fraction terms and divide.

$$\frac{1}{\frac{31}{30}} = \frac{1}{1} \div \frac{31}{30} = \frac{1}{1} \times \frac{30}{31} = \frac{30}{31}$$

EXERCISE 2-3

Solve the following problems which involve simplifying complex fractions. Remember to use all of the skills you have previously learned to simplify the complex fractions.

2-31. $\dfrac{\frac{3}{16}}{\frac{1}{4}} =$

2-32. $\dfrac{\frac{1}{2}}{\frac{7}{8}} =$

2-33. $\dfrac{5}{\frac{3}{8}} =$

2-34. $\dfrac{\frac{7}{8}}{28} =$

2-35. $\dfrac{1\frac{1}{16}}{\frac{3}{16}} =$

2-36. $\dfrac{\frac{1}{7}}{3\frac{1}{7}} =$

2-37. $\dfrac{\frac{1}{2} + \frac{5}{8}}{\frac{5}{8} - \frac{1}{2}} =$

2-38. $\dfrac{\frac{5}{7} + \frac{3}{2}}{\frac{5}{3} \times 4} =$

2-39. $\dfrac{12\frac{1}{2} \times 4}{\frac{3}{4} \div \frac{6}{16}} =$

2-40. $\dfrac{\frac{23}{8} \div \frac{1}{8}}{69 \times \frac{1}{5}} =$

2-41. $\dfrac{8\frac{1}{3} \times 2\frac{1}{2}}{8\frac{1}{3} \div 2\frac{1}{2}} =$

2-42. $\dfrac{1}{\frac{1}{2} + \frac{3}{4} + \frac{5}{8} + \frac{7}{16}} =$

2-43. $\dfrac{5 \div \frac{1}{10}}{\frac{1}{2} + \frac{1}{5} + \frac{1}{10}} =$

2-44. $\dfrac{\frac{9}{6} \times \frac{2}{11}}{14\frac{3}{8} - 6\frac{17}{24}} =$

Blueprint and Specification Fractions

Example 2-14

PROBLEM: Determine the unknown quantities, A and B, in the blueprint.

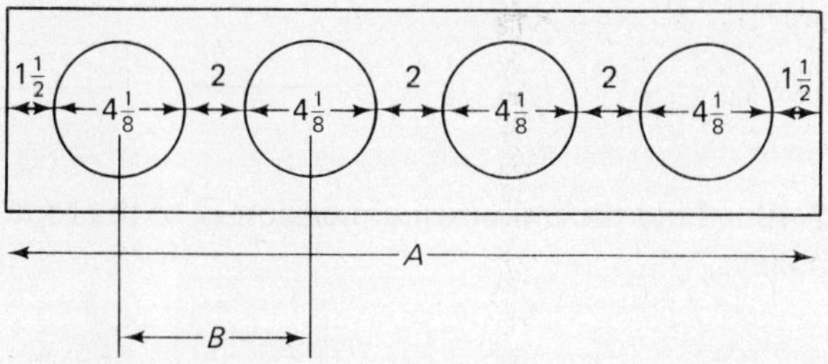

SOLUTION: Determine the length of A. The length of A is equal to the sum of all the distances. Thus all the fractions should be added together.

$1\frac{1}{2} + 4\frac{1}{8} + 2 + 4\frac{1}{8} + 2 + 4\frac{1}{8} + 2 + 4\frac{1}{8} + 1\frac{1}{2} = A$

METHOD 1: Determine the lowest common denominator. This number is 8. Multiply each fraction requiring change to the lowest common denominator by unity $\boxed{\frac{8}{8}}$. Then add the fractions and reduce to lowest terms.

$$
\begin{aligned}
1\tfrac{1}{2} \times \boxed{\tfrac{8}{8}} &= 1\tfrac{4}{8} \\
4\tfrac{1}{8} &= 4\tfrac{1}{8} \\
2 &= 2 \\
4\tfrac{1}{8} &= 4\tfrac{1}{8} \\
2 &= 2 \\
4\tfrac{1}{8} &= 4\tfrac{1}{8} \\
2 &= 2 \\
4\tfrac{1}{8} &= 4\tfrac{1}{8} \\
+1\tfrac{1}{2} \times \boxed{\tfrac{8}{8}} &= 1\tfrac{4}{8} \\
\hline
24\tfrac{12}{8} = 25\tfrac{4}{8} &= 25\tfrac{1}{2} = A
\end{aligned}
$$

METHOD 2: Another way to solve the problem is to combine like dimensions through multiplication, reduce to lowest terms, and then add.

$$2 \times 1\tfrac{1}{2} = \frac{2}{1} \times \frac{3}{2} = \frac{6}{2} = 3$$

$$4 \times 4\tfrac{1}{8} = \frac{\overset{1}{\cancel{4}}}{1} \times \frac{33}{\underset{2}{\cancel{8}}} = \frac{33}{2} = 16\tfrac{1}{2}$$

$$+3 \times 2 = 6 \quad = 6 = 6$$

$$25\tfrac{1}{2} = A$$

SOLUTION: Determine the length of B. B is a center-to-center length.

B is equal to $\frac{1}{2}$ of $4\frac{1}{8} + 2 + \frac{1}{2}$ of $4\frac{1}{8}$.

Remember that "of" always means multiply.

$$\frac{1}{2} \times 4\tfrac{1}{8} + 2 + \frac{1}{2} \times 4\tfrac{1}{8} = B$$

Complete the multiplication operation first. Convert mixed fractions to improper fractions, multiply, reduce to lowest terms, add, and then reduce to lowest terms once again.

$$\frac{1}{2} \times 4\tfrac{1}{8} = \frac{1}{2} \times \frac{33}{8} = \frac{33}{16} = 2\tfrac{1}{16}$$

$$2 \quad = 2 \quad = 2 = 2$$

$$+\frac{1}{2} \times 4\tfrac{1}{8} = \frac{1}{2} \times \frac{33}{8} = \frac{33}{16} = 2\tfrac{1}{16}$$

$$6\tfrac{2}{16} = 6\tfrac{1}{8} = B$$

EXERCISE 2-4

Solve the following problems concerning the reading of blueprints. Use the operations you have learned. Be sure to select the most efficient method or procedure.

2-45. Find the unknown dimension A of the washer.

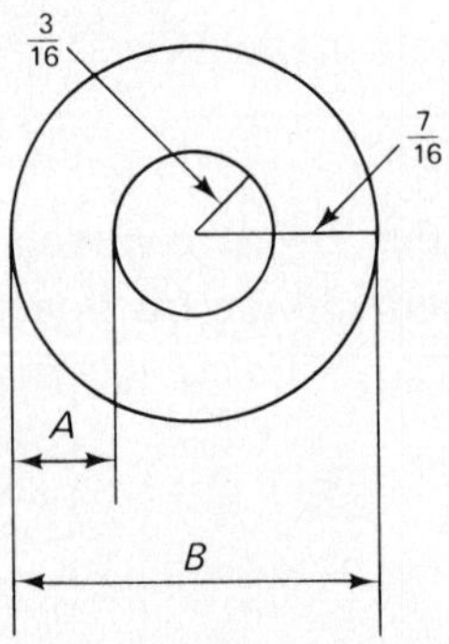

2-46. Find the unknown dimension B of the washer.

2-47. Find the unknown dimension C of the cone pulleys.

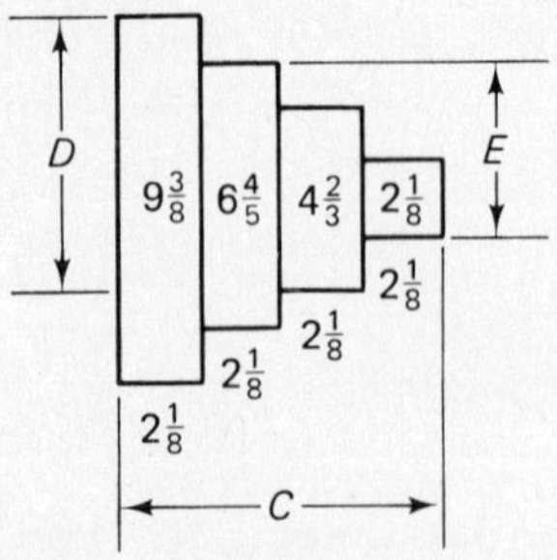

2-48. Find the unknown dimension D of the cone pulleys.

2-49. Find the unknown dimension E of the cone pulleys.

2-50. Find the missing dimension F on the concentric circles.

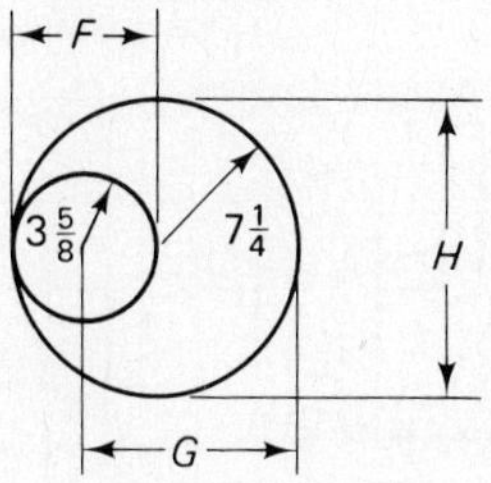

2-51. Find the missing dimension G on the concentric circles.

2-52. Find the missing dimension H on the concentric circles.

2-53. Find the missing dimension I of the eyebolt.

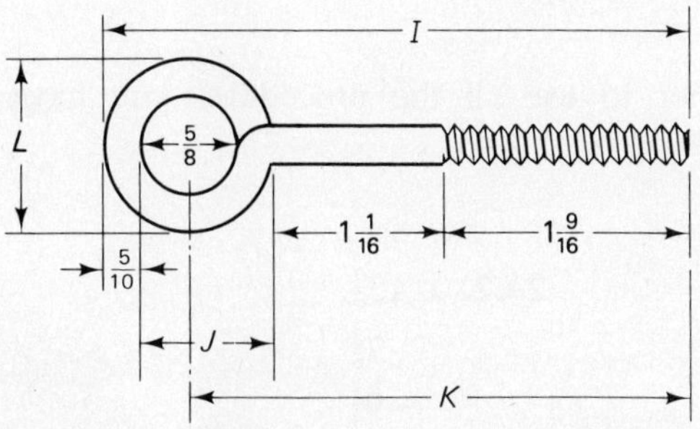

2-54. Find the missing dimension J of the eyebolt.

2-55. Find the missing dimension K of the eyebolt.

2-56. Find the missing dimension L of the eyebolt.

2-57. Find the missing dimension M on the crankshaft.

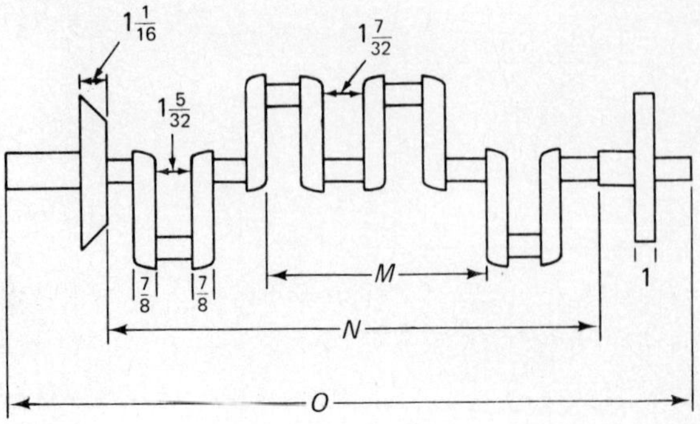

2-58. Find the missing dimension N on the crankshaft.

2-59. Find the missing dimension O on the crankshaft.

EXERCISE 2-5

Solve the following problems which involve operations with fractions. There are also some problems with practical applications of fractions. Remember to use all the procedures you have learned.

2-60. $\frac{1}{4} + \frac{1}{2} + \frac{5}{12}$

2-61. $5\frac{1}{6} + 7\frac{3}{4}$

2-62. $13\frac{2}{3} + 7\frac{3}{8} + 9\frac{1}{4}$

2-63. How many pounds of corn were produced if 5 plants produced the following amounts: $6\frac{1}{2}$, $4\frac{3}{4}$, $3\frac{7}{8}$, $4\frac{7}{32}$, and $5\frac{3}{16}$ pounds?

2-64. How much trim would you need for a window which is $6\frac{1}{2} \times 8$ feet?

2-65. $\dfrac{7}{12}$
$-\dfrac{1}{3}$

2-66. $17\frac{1}{16}$
$-\ 14\frac{3}{8}$

2-67. $16\frac{3}{8}$
$-\ 9\frac{15}{16}$

2-68. What is the difference in thickness between $\frac{5}{8}$-inch and $\frac{1}{2}$-inch sheet rock?

2-69. How much of an 8-foot heater hose is left if you cut off a piece $3\frac{3}{4}$ feet long?

2-70. $32 \times 4\frac{1}{4} =$

2-71. $\dfrac{1}{3} \times \dfrac{3}{8} \times \dfrac{64}{72} \times \dfrac{16}{54} =$

2-72. $2\frac{1}{3} \times 4\frac{3}{5} \times 2\frac{5}{30} =$

2-73. What would the total length of 8 studs be if each is $7\frac{3}{4}$ feet long?

2-74. If the travel time each way to work is $\frac{3}{4}$ hour, how many hours are spent in transit during a 5-day workweek?

2-75. $\dfrac{5}{6} \div \dfrac{3}{16} =$

2-76. $7\frac{1}{5} \div 3\frac{3}{5} =$

2-77. $7\frac{1}{2} \div 1\frac{1}{4} =$

2-78. If $3\frac{1}{2}$ pizzas are to be divided among 12 persons, how much will each person receive?

2-79. A room $15\frac{1}{2}$ feet long is to have three electrical outlets equally spaced. How far apart will they be?

2-80. $\dfrac{\frac{3}{5}}{2\frac{1}{8}} =$

2-81. $\dfrac{3\frac{1}{4} \div 4\frac{1}{8}}{4\frac{1}{4} \times 2\frac{7}{16}} =$

2-82. $\dfrac{1\frac{1}{3} \div 3}{2\frac{1}{16} + 1\frac{13}{32}} =$

2-83. What is the thickness of a pipe whose outside diameter is $3\frac{3}{8}$ inches and whose inside diameter is $2\frac{3}{4}$ inches?

2-84. A 2×4 is actually $1\frac{5}{8}$ inches $\times$ $3\frac{5}{8}$ inches. How much is planed off *each* side of the $3\frac{5}{8}$ inch sides?

2-85. A piece of copper tubing is $14\frac{1}{2}$ inches long. How much is left after cutting off pieces $2\frac{1}{2}$ inches, $1\frac{1}{16}$ inches, and $3\frac{5}{32}$ inches and allowing $\frac{1}{16}$ inch for each cut?

2-86. A taper has a diameter of $2\frac{5}{16}$ inches on one end and $1\frac{5}{8}$ inches at the other end. Find the difference between the diameters.

2-87. The volume of a rectangular solid (block) is determined by multiplying the length times the width times the height. Find the volume of a block of concrete $11\frac{3}{4}$ inches $\times$ 8 inches $\times$ $6\frac{1}{2}$ inches.

2-88. A cubic foot contains approximately $7\frac{1}{2}$ gallons. How many gallons of gasoline would go into a $2\frac{3}{4}$-cubic foot gas tank?

2-89. Find the missing dimension A on the blueprint:

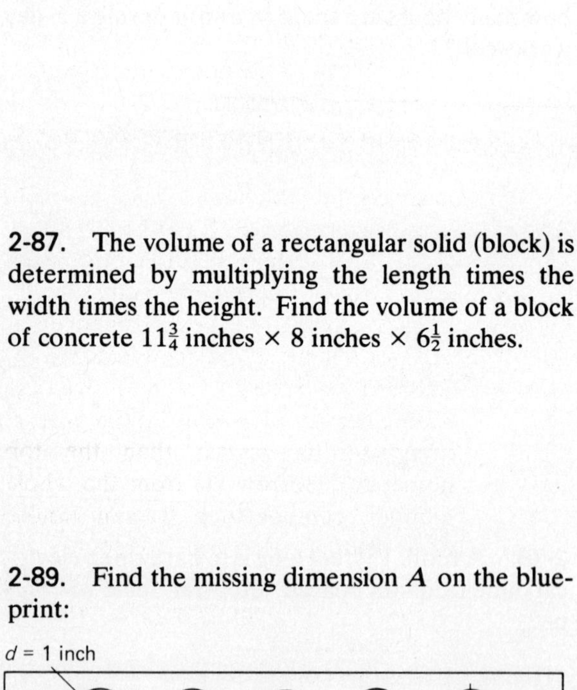

THINK TIME

There are many different methods used to determine the lowest common denominator of a group of fractions. Several of these have already been presented. Perhaps you can recall other methods you have been taught, or maybe you can create new ones. Practice the new method on this problem: Find the lowest common denominator for the fractions $\frac{1}{24} + \frac{3}{28} + \frac{7}{15}$. If you cannot think of a different method, you might want to try the one presented below. It looks rather complicated, but after a little practice, you may want to use it quite often.

PROBLEM: Find the lowest common denominator for the fractions: $\frac{1}{24} + \frac{3}{28} + \frac{7}{15}$.

SOLUTION: Place the denominators on an inverted division symbol as shown below. Divide each of the denominators by a prime number (2). Place the divided numbers (12, 14) as well as the number which could not be divided (15) on a new inverted division symbol. Repeat the process until no more divisions are possible.

2	~~24~~	~~28~~	15
2	~~12~~	~~14~~	15
2	~~6~~	7	15
3	~~3~~	7	~~15~~
	1	7	5

Multiply the divisors and remaining numbers to arrive at the lowest common denominator.

$$2 \times 2 \times 2 \times 3 \times 1 \times 7 \times 5 = 840$$

PROCEDURES TO REMEMBER

1. To reduce a fraction, divide the numerator and denominator by an equal quantity. Example: $\frac{2}{4}$ divided by $\frac{2}{2} = \frac{1}{2}$.
2. To convert an improper fraction to a mixed number, divide the denominator into the numerator, write the quotient, and place the remainder over the denominator. Example: $\frac{3}{2} = 1\frac{1}{2}$.
3. To convert a whole number or a mixed number to an improper fraction, multiply the denominator (or the denominator desired) by the whole number and add the original numerator to create the new numerator. Examples: $2 = \frac{8}{4}$; $1\frac{7}{8} = \frac{15}{8}$.
4. When multiplying fractions:
   (a) Write all fractions to be multiplied.
   (b) Convert mixed numbers to improper fractions.
   (c) Reduce fractions. This can be done by dividing the numerator of one fraction and the denominator of another fraction by the same number, or by dividing the numerator into the denominator and the denominator into the numerator. (Top into bottom and bottom into top.)
   (d) Multiply numerators times numerators and denominators times denominators.
   (e) Reduce the fraction to a common fraction or to a mixed number.
5. When dividing fractions:
   (a) Write all fractions to be multiplied.
   (b) Convert mixed numbers to improper fractions.
   (c) Invert the fraction to the right of the division symbol.
   (d) Reduce fractions.
   (e) Multiply numerators times numerators and denominators times denominators.
   (f) Reduce the fraction to a common fraction or to a mixed number.
6. When adding fractions:
   (a) Write the fractions to be added.
   (b) If they do not have a common denominator, find a common denominator.
   (c) Change all fractions to equivalent fractions with a common denominator.
   (d) Add all numerators.
   (e) Place the sum of the numerators over the common denominator.
   (f) Reduce to a common fraction or mixed number.
7. When subtracting fractions:
   (a) Write the fractions to be subtracted.
   (b) If they do not have a common denominator, find a common denominator.
   (c) Change all fractions to equivalent fractions with a common denominator.
   (d) Subtract the bottom numerator from the top numerator. However, if the bottom numerator is greater than the top numerator, borrow (1) from the whole number, convert to a fraction whose denominator is the common denominator, add to the numerator, and subtract the bottom numerator from the top numerator.
   (e) Place the difference over the common denominator.
   (f) Reduce to a common fraction or mixed number.
8. To find the lowest common denominator:
   (a) Determine the prime factors of each denominator.

(b) Using the prime factors of the first denominator and the prime factors from other denominators not found in the first denominator, multiply these together.

(c) Check to make certain each denominator divides evenly into the common denominator.

CHAPTER SUMMARY

A *fraction* expresses mathematically a portion of a whole. Fractions are written in such a way as to show the relationship of the portion to the whole. An example of a fraction is $\frac{5}{9}$. The number below the line, 9, is called the *denominator* and indicates the number of parts into which the whole has been divided. The number above the line, 5, is called the *numerator* and indicates the number of parts of the whole given.

The numerator and denominator are called the *terms* of the fraction. A fraction whose numerator is smaller than its denominator is called a *proper* fraction. A fraction whose numerator is larger than its denominator is called an *improper* fraction.

A whole number and a fraction written together is called a *mixed number.* A fraction which contains a fraction or fractions in the numerator or denominator is called a *complex* fraction. A *common denominator* is a number which is a multiple of all denominators in a set of fractions. The *lowest common denominator* is the smallest number which is a multiple of all the denominators in a set of fractions. To *reduce* a fraction to its lowest terms, divide the numerator and the denominator by the same quantity. *Prime factors* are those numbers which, when multiplied together, will be equal to the composite number.

CHAPTER TEST

Find answers to the following problems.

T-2-1. $\frac{3}{4} + \frac{1}{2} + \frac{5}{16}$

T-2-2. $2\frac{2}{7} + 3\frac{1}{8}$

T-2-3. $15\frac{1}{2} + 6\frac{1}{3} + 4\frac{3}{8}$

T-2-4. $\frac{15}{16} - \frac{5}{8}$

T-2-5. $2\frac{3}{4} - 1\frac{5}{8}$

T-2-6. $12\frac{1}{32} - 9\frac{7}{8}$

T-2-7. $\frac{5}{9} \times \frac{3}{10} =$

T-2-8. $7\frac{1}{3} \times 2\frac{1}{2} =$

T-2-9. $4\frac{5}{16} \times 5\frac{1}{3} =$

T-2-10. $\frac{1}{8} \div \frac{3}{4} =$

T-2-11. $\frac{25}{36} \div \frac{5}{9} =$

T-2-12. $5\frac{3}{8} \div 3\frac{1}{4} =$

T-2-13. $\frac{9}{3\frac{2}{3}} =$

T-2-14. $\dfrac{\frac{5}{8} \times \frac{16}{45}}{\frac{3}{4} \div \frac{4}{3}} =$

T-2-15. $\dfrac{\frac{2}{3} \times 7\frac{3}{4}}{\frac{1}{2} \times 37\frac{1}{16}} =$

T-2-16. $18\frac{2}{3} \times 8\frac{1}{4} \div 72 =$

T-2-17. $6\frac{1}{4} \times 8\frac{1}{2} \times 5\frac{1}{2} \div 36 =$

T-2-18. Determine the missing dimension *A* on a set of eight steps.

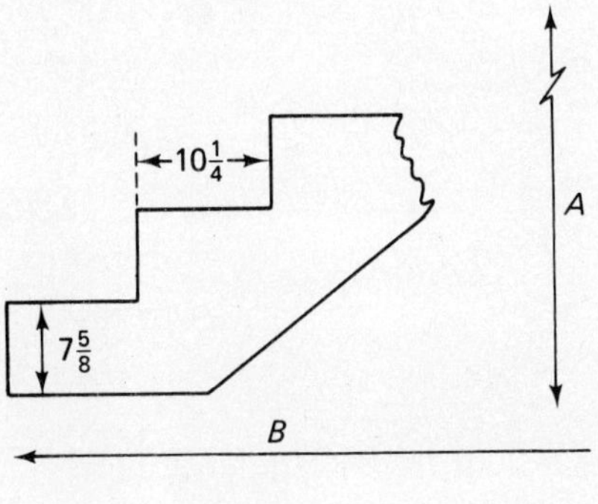

T-2-19. Determine the missing dimension *B* on a set of eight steps.

T-2-20. A truck is loaded with bales of scrap iron, each weighing $\frac{1}{20}$ ton. If the total weight of the bales is $5\frac{1}{2}$ tons, how many bales are on the truck?

T-2-21. Eight inches of plastic tubing weighs $\frac{3}{4}$ ounce. How much will 2 inches of the tubing weigh?

T-2-22. Twenty-four strips, each $1\frac{7}{8}$ inches wide, are to be ripped from a sheet of paneling. Assuming 24 cuts are necessary and $\frac{1}{16}$ inch is lost on each cut, how wide must the paneling be?

T-2-23. A stonemason spends the following hours on different jobs: $4\frac{1}{2}$, $8\frac{1}{2}$, $1\frac{3}{4}$, $2\frac{1}{4}$, and $5\frac{3}{4}$. At \$$10\frac{1}{2}$ per hour, what will the charge for labor be?

T-2-24. Partners Marge, John, and Steve agreed to share gains and losses in proportion to their respective investments. Marge invested \$21,000; John, \$28,000; and Steve, \$35,000. If they are to share a gain of \$21,000, how much should each receive?

T-2-25. A freshly molded plastic boat bumper weighed $14\frac{1}{2}$ kilograms. After being shaped, it weighed $12\frac{5}{8}$ kilograms. The shaping cost is \$2.00 per kilogram, and the plastic removed in shaping can be sold as scrap at \$.75 per kilogram. Find the net cost of shaping the boat bumper.

CHAPTER

3

DECIMAL FRACTIONS

OBJECTIVES

1. To review the basic concepts and computational skills used in working with decimal fractions.
2. To provide the opportunity to develop your skill, ability, and speed in working with decimal fractions.
3. To use decimal fractions to solve problems related to your occupation.

SELF-TEST

This test will help you evaluate your ability to work with decimal fractions. If you are successful in solving these problems, you do not need to work through this chapter and thus may move on to developing your skills in working with percentages. Be sure to show your work. This will help you identify skills which you need to develop.

Read the following numbers then write them as figures.

S-3-1. Twelve and three tenths

S-3-2. Seventy-nine thousandths

S-3-3. One hundred and one hundredth

S-3-4. Three and thirty-five thousandths

S-3-5. Two hundred sixty-five thousandths

S-3-6. Four hundred five and sixty thousandths

S-3-7. Forty-nine hundred-thousandths

S-3-8. Four and twenty-four millionths

Convert the following common fractions to decimal fractions.

S-3-9. $\frac{3}{5}$

S-3-10. $\frac{3}{16}$

S-3-11. $\frac{7}{11}$

S-3-12. $\frac{1}{50}$

S-3-13. $\frac{11}{28}$

S-3-14. $1\frac{8}{9}$

S-3-15. $\frac{230}{6}$

Convert the following decimal fractions to common fractions.

S-3-16. .75

S-3-17. .375

S-3-18. .9375 (to the nearest 16th)

S-3-19. $.34\frac{2}{3}$

S-3-20. What is the difference between $\frac{3}{4}$ and $\frac{5}{8}$ expressed as a decimal fraction?

S-3-21. There are seven steps in a set of stairs. If each step is $8\frac{1}{4}$ inches high, what is the distance between the floors? Express your answer as a mixed decimal fraction.

S-3-22. The wall of a building consists of cement blocks $9\frac{5}{8}$ inches thick, with plaster on one side $\frac{3}{4}$ inch thick. What is the total thickness of the wall? Express your answer as a mixed decimal fraction.

S-3-23. An airline stock is selling for $25\frac{5}{8}$. What will be the cost of 40 shares?

S-3-24. Number 12 wire is .081 inch in diameter. What is its diameter to the nearest 64th of an inch?

S-3-25. A piece of steel was $27\frac{7}{8}$ inches long. Four pieces $4\frac{5}{8}$ inches long were cut off. How long was the remaining portion if the saw blade was .012 inch wide?

INTRODUCTION

Decimal fractions are used every time you make a purchase in a store. When you check over the deductions on your pay check stub to determine if the amounts are correct, several decimal fractions are used. The dollars and cents of everyday living require you to have a basic knowledge of decimal fractions.

As the metric system becomes more and more a fact of life in our society, you will need to develop the skills of decimal fractions to work effectively within this system.

As more and more information is fed into computers to assist you in your occupation and life, you will need to know the mathematical language of the computer—decimal fractions. Even though you may not work directly with computers, your occupation will rely heavily on the use of the computer for time and money-saving results.

The decimal fraction is the language of your future.

DEFINITIONS

A *decimal fraction* is a fraction whose denominator is 10 or some multiple of 10 such as 10, 100, 1000, etc. Examples of decimal fractions are: $\frac{7}{10}$, $\frac{35}{100}$, $\frac{783}{1000}$, and $\frac{5,281}{10,000}$. In writing decimal fractions, the denominator is not placed below the numerator as it is in common fractions. We indicate the denominator by placing a *decimal point* (.) in the numerator. The decimal point is placed at the point where there are as many figures to the right of the decimal point as there are zeros in the denominator. Thus $\frac{47}{100}$ is written .47, $\frac{783}{1000}$ is written .783, and $\frac{532}{100}$ is written 5.32. The basis for this simple method may be best illustrated by dividing the denominator into the numerator. Thus:

$$\frac{47}{100} = 100\overline{)47.00}\quad\begin{array}{r} .47 \\ \hline 47.00 \\ 40\,0 \\ \hline 7\,00 \\ 7\,00 \\ \hline 0 \end{array}$$

You will find that $\frac{47}{100}$ is referred to as the *fractional form* and .47 is called the *decimal form*. Occasionally a zero is placed to the left of the decimal point when there is no whole number. Examples of this are 0.47 instead of .47, and 0.783 instead of .783. This is done to emphasize that the number is a decimal fraction.

When decimal fractions are read out loud, the decimal point may be simply called *point* or *and*. Thus 3.5 may be read "three point five" or "three and five tenths."

CONTENT

The word forms for decimal fractions are directly related to the decimal fraction denominator 10 or some form or power of 10 × 10. Thus .3 is read "three tenths" and .74 is read "seventy-four hundredths." Names are assigned to each *place* to the left and to the right of the decimal point. Study the chart entitled "Place Values For Decimals." The two columns on the left show the number and its word form (name). The third column shows how the multiplication of tens or tenths produces the number which appears in the first column on the left. The multiplication process shows the *powers of ten.* The fourth column indicates how the powers of ten are written in a simpler form called the *exponential form.* The two columns on the right will be of value to you when you develop skills in solving algebraic equations.

PLACE VALUES FOR DECIMALS

Number	*Word Form*	*Powers-of-Ten Form*	*Exponential Form*
1,000,000	One Million	10 × 10 × 10 × 10 × 10 × 10	10^6
100,000	One Hundred Thousand	10 × 10 × 10 × 10 × 10	10^5
10,000	Ten Thousand	10 × 10 × 10 × 10	10^4
1,000	One Thousand	10 × 10 × 10	10^3
100	One Hundred	10 × 10	10^2
10	Ten	10	10^1
1	One	1	10^0
.1	One Tenth	$\frac{1}{10}$	$\left(\frac{1}{10}\right)^1$ or 10^{-1}
.01	One Hundredth	$\frac{1}{10} \times \frac{1}{10}$	$\left(\frac{1}{10}\right)^2$ or 10^{-2}
.001	One Thousandth	$\frac{1}{10} \times \frac{1}{10} \times \frac{1}{10}$	$\left(\frac{1}{10}\right)^3$ or 10^{-3}
.0001	One Ten-Thousandth	$\frac{1}{10} \times \frac{1}{10} \times \frac{1}{10} \times \frac{1}{10}$	$\left(\frac{1}{10}\right)^4$ or 10^{-4}
.00001	One Hundred-Thousandth	$\frac{1}{10} \times \frac{1}{10} \times \frac{1}{10} \times \frac{1}{10} \times \frac{1}{10}$	$\left(\frac{1}{10}\right)^5$ or 10^{-5}
.000001	One Millionth	$\frac{1}{10} \times \frac{1}{10} \times \frac{1}{10} \times \frac{1}{10} \times \frac{1}{10} \times \frac{1}{10}$	$\left(\frac{1}{10}\right)^6$ or 10^{-6}

Thus the number 69373.56354 is read "sixty-nine thousand, three hundred seventy-three and fifty-six thousand, three hundred fifty-four one hundred thousandths."

If no decimal point is given, it is assumed that the decimal point is to the right of the last digit. Thus 12 = 12.0 and 528 = 528.0.

EXERCISE 3-1

Read, and then write the following decimal fractions.

3-1. .09

3-2. .085

3-3. 521.14

3-4. 16.141

Convert the following numbers to decimal fractions.

3-5. Forty hundredths.

3-6. Eight tenths

3-7. Seven and four tenths

3-8. Five and thirty-two thousandths

3-9. Two and two hundredths

3-10. Thirty-six and twenty-four thousandths

Read, and then write in words the following decimal fractions:

3-11. .909

3-12. 3.005

3-13. 5.0055

3-14. 168.09

3-15. 16,000.00016

Convert the following numbers to decimal fractions.

3-16. Thirty-three and three tenths

3-17. Five hundred and four hundred-thousandths

3-18. Three and five hundred thirty-five ten-thousandths

3-19. Seventy-seven hundred-thousandths

3-20. Nine hundred thirty-four thousand and six ten-thousandths

CONTENT

A common fraction may be converted to a decimal fraction by dividing the denominator into the numerator.

Example 3-1

PROBLEM: Convert the fraction $\frac{3}{4}$ to decimal form.

SOLUTION: Divide the denominator (4) into the numerator (3).

$$\begin{array}{r} .75 \\ 4\,\overline{)3.00} \\ 2\,8 \\ \hline 20 \\ 20 \end{array}$$

Therefore the common fraction for $\frac{3}{4}$ equals the decimal fraction .75.

Example 3-2

PROBLEM: Convert the fraction $\frac{9}{16}$ to a decimal form.

SOLUTION: Divide the denominator (16) into the numerator (9).

$$\begin{array}{r} .5625 \\ 16\overline{)9.0000} \\ \underline{8\,0} \\ 1\,00 \\ \underline{96} \\ 40 \\ \underline{32} \\ 80 \\ \underline{80} \end{array}$$

Example 3-3

PROBLEM: Convert the mixed number $15\frac{5}{6}$ to decimal form.

SOLUTION: Since 15 is a whole number, it may be placed to the left of the decimal point when the division of the numerator (5) by the denominator (6) has been completed.

$$\begin{array}{r} .8333 \\ 6\overline{)5.0000} \\ \underline{4\,8} \\ 20 \\ \underline{18} \\ 20 \\ \underline{18} \\ 20 \\ \underline{18} \\ 2 \end{array}$$

Note that, regardless of how long you continue to divide 5 by 6, you will always obtain a 3. This result is called a *repeating decimal.* It may be indicated in your answer in one of three ways: (1) by placing a dot above the 3: $\dot{3}$; (2) by placing a short line above the 3: $\bar{3}$; or (3) by placing three dots immediately after the last 3: 3 . . . . Thus

$$15\tfrac{5}{6} = 15.833\dot{3} = 15.833\bar{3} = 15.833\ldots$$

EXERCISE 3-2

Convert the following common fractions to decimal form.

3-21. $\frac{3}{10}$

3-22. $\frac{5}{8}$

3-23. $\frac{5}{16}$

3-24. $\frac{49}{100}$

3-25. $\frac{1}{9}$

3-26. $\frac{2}{32}$

3-27. $\frac{21}{32}$

3-28. $\frac{37}{8}$

3-29. $\frac{31}{64}$

3-30. Write 5¢ as a decimal part of a dollar.

CONTENT

It is often necessary in industry to change decimal fractions to common fractions. One of the more common needs is to change decimal fractions of an inch to fourths, eighths, sixteenths, thirty-seconds, and sixty-fourths of an inch. Because these changes must be made as rapidly as possible, many companies post a *table of decimal equivalents* in their shops. The exact value is frequently not on the table. Consequently, the decimal must be rounded off to the nearest fraction, depending of course on the purpose of the conversion. These conversions will become more and more important as industry adopts the metric system as its basic form of measurement.

To convert any decimal fraction to a common fraction, simply write the decimal as a fraction. For example .5 may be written as $\frac{5}{10}$; .79 as $\frac{79}{100}$; and .329 as $\frac{329}{1000}$. These fractions may be reduced to lowest terms by dividing the numerator and the denominator by the same number. In decimal fractions, the denominator is always divisible by the numbers 2, 5, and 10.

Example 3-4

PROBLEM: Convert .65 to a common fraction and reduce to lowest terms.

SOLUTION: First, convert the decimal fraction to its common fraction form.

$$.65 = \frac{65}{100}$$

Second, reduce the fraction by dividing the numerator and denominator by the number 5.

$$\frac{\overset{13}{\cancel{65}}}{\underset{20}{\cancel{100}}} = \frac{13}{20}$$

Thus $.65 = \frac{13}{20}$.

Example 3-5

PROBLEM: Convert 3.875 inches to a mixed fraction and reduce to lowest terms.

SOLUTION: First, convert the decimal fraction to its common fraction form.

$$3.875 = 3\frac{875}{1000}$$

Second, reduce the fraction by dividing the numerator and denominator by the number 25.

$$3\frac{\overset{35}{\cancel{875}}}{\underset{40}{\cancel{1000}}} = 3\frac{35}{40}$$

Third, reduce the fraction once again by dividing the numerator and denominator by the number 5.

$$3\frac{\overset{7}{\cancel{35}}}{\underset{8}{\cancel{40}}} = 3\frac{7}{8}$$

Thus $3.875 = 3\frac{7}{8}$. Remember that you may take as many steps as necessary to reduce a fraction to its lowest form.

Example 3-6

PROBLEM: Convert .316 to the nearest sixteenth.

SOLUTION: First, convert the decimal fraction to its common fraction form.

$$.316 = \frac{316}{1000}$$

Second, divide the numerator and denominator by 1000. This will make the problem easier to work.

$$\frac{316}{1000} = \frac{.316}{1}$$

(You may want to write $\frac{.316}{1}$ immediately.)

Third, multiply the numerator and denominator by the number 16.

$$\frac{.316}{1} \times \left\{\frac{16}{16}\right\} = \frac{.316 \times 16}{1 \times 16} = \frac{5.056}{16}$$

Fourth, round to the nearest sixteenth.

$$\frac{5.056}{16} = \frac{5}{16}$$

Thus $.316 = \frac{5}{16}$ to the nearest sixteenth.

Example 3-7

PROBLEM: Convert $.83\frac{1}{3}$ to a common fraction.

SOLUTION: First, convert the decimal fraction to its common fraction form.

$$.83\tfrac{1}{3} = \frac{83\frac{1}{3}}{100}$$

Second, convert the numerator to an improper fraction, by multiplying the numerator, 83, by the denominator, 3.

$$\frac{83\frac{1}{3}}{100} = \frac{\frac{250}{3}}{\frac{100}{1}}$$

Third, perform the indicated division.

$$\frac{\frac{250}{3}}{\frac{100}{1}} = \frac{250}{3} \div \frac{100}{1} = \frac{\overset{5}{\cancel{250}}}{3} \times \frac{1}{\underset{2}{\cancel{100}}} = \frac{5}{6}$$

Thus $.83\frac{1}{3} = \frac{5}{6}$.

EXERCISE 3-3

Convert the following decimal fractions to common fractions and reduce to lowest terms:

3-31. .7

3-32. .16

3-33. .625

3-34. .375

3-35. $.72\frac{8}{11}$

3-36. .0125

3-37. Convert .34 to the nearest eighth.

3-38. Convert .367 to the nearest sixteenth.

3-39. Convert 8.096 to the nearest thirty-second.

3-40. Convert 12.5326 to the nearest sixty-fourth.

3-41. Convert $\frac{7}{8}$ inch to a decimal form.

3-42. Convert $40\frac{1}{2}$ inches to decimal form in feet to the nearest thousandth.

3-43. What is the inside diameter, in decimal form, of a pipe whose outside diameter is $6\frac{1}{2}$ inches and which is $\frac{3}{16}$ inch thick?

3-44. A board 6 inches wide is to be finished to seven eighths of its original width. Express the new width in decimal form.

3-45. What is the dollar value of stock that has a market value of $75\frac{3}{8}$?

3-46. A machine produced 495 outboard motor shear pins in 8.25 minutes. What is the average number of shear pins it can produce per minute?

3-47. A piece of marine plywood is made up of one piece of $\frac{1}{4}$-inch wood, two pieces of $\frac{1}{8}$-inch wood, and two pieces of $\frac{1}{16}$-inch wood. What is the total thickness of the plywood in decimal form?

3-48. A steel shaft is $1\frac{5}{8}$ inches in diameter. How deep a cut must be made to bring the diameter down to 1.594 inches?

3-49. A mechanic desires to set the spark plug gap at .035. Express the gap as a common fraction to the nearest sixty-fourth.

3-50. A blueprint calls for a .8365-inch hole or slightly larger. The available drills are sized in thirty-seconds of an inch. What drill should be chosen?

THINK TIME

It is important for tradespeople to know decimal and fractional equivalents. Below you will find a mixture of decimals and fractions. Connect the pairs of equivalents with straight lines. When you have completed your work, you will find a perfect five-pointed star in the midst of the lines. Then think of other ways in which you can review the decimal and fractional equivalents which are important to you in your trade.

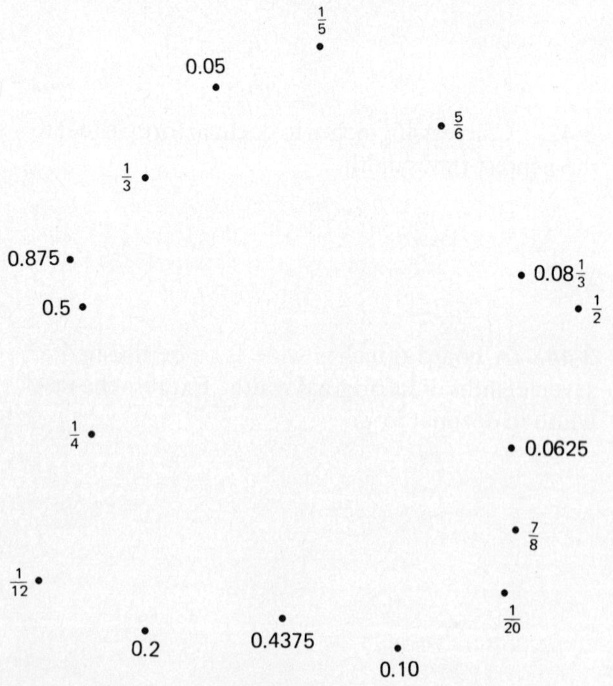

PROCEDURES TO REMEMBER

1. To change a common fraction to a decimal fraction:
   (a) Set up the division of the numerator by the denominator.
   (b) Locate the decimal point.
   (c) Divide, carrying the division as many places as conditions require.
2. To change a decimal fraction to a common fraction form:
   (a) Write the decimal fraction in its common fraction form.
   (b) Reduce the fraction to its lowest terms.
3. To change a decimal fraction to a common fraction with a specific denominator (eighth, sixteenth, etc.):
   (a) Write the decimal fraction in its common fraction form.
   (b) Divide the numerator and denominator by the denominator so that the new

denominator will be 1 and the numerator will be in decimal form.

(c) Multiply the numerator and denominator by the specific denominator (eighth, sixteenth, etc.).

CHAPTER SUMMARY

A *decimal fraction* is a fraction whose denominator is 10 or some form or power of 10 ×

10. The *decimal point* (.) separates the whole digits from the decimal part of a number. The word *point* or *and* is used when reading decimals. A value which continues to result when one quantity is divided into another is called a *repeating decimal.* Example: $\frac{1}{3} = .\bar{3}$ or $.\dot{3}$ or .333 . . . . A repeating decimal is indicated by a bar or dot over the repeating digit or by three dots after the repeating digit. Fractions may be written in two ways: (1) a *fractional form* ($\frac{37}{100}$) and (2) a *decimal form* (.37).

CHAPTER TEST

Read the following numbers, and then write them as figures.

T-3-1. Seven and five hundredths

T-3-2. Eighty-four thousandths

T-3-3. One thousand and one thousandth

T-3-4. Seven and sixty-six thousandths

T-3-5. Three hundred and thirty-three thousandths

T-3-6. Three thousand and six thousandths

T-3-7. Fifty-five ten-thousandths

T-3-8. Three and three thousand one hundred twenty-five hundred thousandths

Convert the following common fractions to decimal fractions:

T-3-9. $\frac{4}{5}$

T-3-10. $\frac{11}{32}$

T-3-11. $\frac{7}{9}$

T-3-12. $\frac{2}{25}$

T-3-13. $\frac{26}{27}$

T-3-14. $3\frac{3}{11}$

T-3-15. $\frac{25}{16}$

Convert the following decimal fractions to common fractions.

T-3-16. .15

T-3-17. .1875

T-3-18. Convert .844 to the nearest thirty-second.

T-3-19. Convert $.44\frac{4}{9}$ to a common fraction.

T-3-20. What is the difference between $\frac{7}{8}$ and $\frac{5}{6}$ expressed as a decimal fraction?

T-3-21. A machinist turned a $\frac{27}{64}$-inch-diameter shaft from a piece of stock $\frac{1}{2}$ inch in diameter. Express the difference in diameter as a decimal.

T-3-22. A plumber made a run of pipe consisting of three pieces of pipe whose lengths were: $10\frac{3}{8}$ inches, $15\frac{7}{16}$ inches, and $18\frac{7}{8}$ inches. Express the total length of pipe as a mixed decimal.

T-3-23. A mining stock is selling for $48\frac{7}{8}$. What will 55 shares of the stock cost?

T-3-24. A blueprint calls for a .84375-inch opening. What size drill should be used to make the opening to the nearest thirty-second?

T-3-25. When plastic cures, it reduces in size by $\frac{1}{250}$. What will the length of a 30-foot plastic pipe be after curing. Express the cured length as a decimal fraction.

CHAPTER

4

PERCENTAGE

OBJECTIVE

1. To review the skills used to:
   (a) Convert common fractions to decimal fractions.
   (b) Convert decimal fractions to common fractions.
2. To develop the skills needed to:
   (a) Find the percent of a number.
   (b) Compare one number to another using percent.
   (c) Solve industry-related problems.

SELF-TEST

The skills you need to solve problems dealing with percentages are very similar to the skills you developed to work with decimals. Your results on the self-test may indicate that you have no need to work through this chapter. You can identify specific areas of needed skill development by having someone check over your work.

S-4-1. Convert $\frac{11}{16}$ to a decimal fraction.

S-4-2. Convert $5\frac{7}{8}$ to a decimal fraction.

S-4-3. Convert 35% to a common fraction.

S-4-4. Convert $37\frac{1}{2}\%$ to a common fraction.

S-4-5. Convert .37 to a percent.

S-4-6. Convert .08 to a percent.

S-4-7. Convert .005 to a percent.

S-4-8. Convert $\frac{1}{2}$% to a decimal fraction.

S-4-9. Convert $34\frac{2}{3}$% to a common fraction.

S-4-10. Convert .115 to a percent.

S-4-11. $63\% = \frac{?}{100}$

S-4-12. Convert $\frac{7}{10}$ to a percent.

S-4-13. $\frac{6}{25} = \frac{?}{100}$

S-4-14. What is 30% of 50?

S-4-15. What is 39% of 1974?

S-4-16. If $3\frac{1}{2}$% of the transistors manufactured are defective, how many would be defective if 5000 were produced?

S-4-17. If fertilizer is 12% nitrogen, how much nitrogen is in a 22-pound bag?

S-4-18. What is $6\frac{3}{4}$% of $2800?

S-4-19. 800 is 12% of what number?

S-4-20. What is the sales tax on a new automobile costing $3988.63 at a tax rate of 4%?

S-4-21. If 6.3% of your salary is deducted for the retirement fund, how much would be deducted from a gross weekly salary of $288?

S-4-22. What is the total interest you would pay on an automobile loan of $3500 if the interest rate was $9\frac{1}{4}$% per year for 3 years?

S-4-23. What is the total interest you would pay on a $400 loan on your life insurance at $5\frac{1}{4}$% each year for 12 years?

S-4-24. A transistor radio was marked down from $32.00 to $24.00. What was the percent of discount?

S-4-25. If the attendance at a local church was 2300 at the beginning of the year, what would it be at the end of the year if it increased $12\frac{1}{2}$%?

INTRODUCTION

Percentages are used to indicate the relationship between numbers. Most often we use percentage to show rate of change, interest, and tax. Concerns are voiced in percentages for increases or decreases in wages, crime, cost of living, production, natural resources, etc. Wise consumers carefully note the percentage rates for interest on loans, carrying charges, and mortgages. Many dollars are saved each year through the selection of lending agencies which offer the lowest rates of interest. The development of the skills of working with percentages will be of great value to you on the job or as a consumer.

DEFINITIONS

Percent is the comparison of a quantity or number to another number. Generally we compare a quantity or number by placing it as a numerator over the denominator, 100. The word "percent" is derived from the Latin and means "by the hundred." The symbol used for percent is %. Basically, there are only three types of percentage problems. Examples are as follows: (1) What is 20% of 540? (2) Thirty-seven is what percent of 97? (3) Sixty-three is 25% of what number? All problems dealing with percentages may be traced back to these fundamental examples.

Any percent problem involves three things: *the base, the rate of percent,* and *the percentage.* These concepts can best be explained through an example. A man earns $200 a week and saves 30% of what he earns. Thus he saves $60 a week. In this problem, $200 is the base, 30% is the rate of percent, and $60 is the percentage. To determine the percentage or amount saved, we take 30% of $200 which is $60. Thus 30% of $200 means .30 × $200 = $60.

The following formulas are to be used to determine percentage, base, and rate.

1. To find the percentage, use the formula:

   Percentage = base times rate ($p = b \times r$)

2. To find the base, use the formula:

   Base = percentage divided by rate $\left(b = \frac{p}{r}\right)$

3. To find the rate, use the formula:

   Rate = percentage divided by base $\left(r = \frac{p}{b}\right)$

At times it may be difficult to determine if a number represents the rate, base, or percentage.

Some helpful hints for the identification of rate, base, and percentage are as follows:

1. The rate always has a percent sign or the word "percent" with it.
2. The base is the number which follows the word "of" in a word problem and refers to the entire quantity being discussed.
3. The percentage is the result of determining the rate of a given base.

CONTENT

The rate of percent is often converted to a decimal fraction to make computation easier. To convert a percent to a decimal fraction, remove the percent sign and move the decimal point two places to the left. This is equivalent to dividing the percent by 100. To convert a decimal to a percent, move the decimal point two places to the right and write the percent sign, %, after the number. This is equivalent to multiplying the decimal by 100. To change a fraction to a percent, convert the fraction into a decimal and then convert the decimal to a percent. To change a percent to a fraction, remove the percent sign, place the number over 100, and then reduce to lowest terms. These processes can be better understood by working through the following examples:

Example 4-1

PROBLEM: Change 15% to a decimal.

SOLUTION: Remove the percent sign and move the decimal point two places to the left.

$$15\% = .15$$

Example 4-2

PROBLEM: Change .075 to a percent.

SOLUTION: Add the percent sign and move the decimal point two places to the right.

$$.075 = 7.5\% \quad \text{or} \quad 7\tfrac{1}{2}\%$$

Example 4-3

PROBLEM: Change $\frac{7}{8}$ to a percent.

SOLUTION: First, complete the division indicated below:

$$\begin{array}{r} .875 \\ 8\overline{)7.000} \\ \underline{6\,4} \\ 60 \\ \underline{56} \\ 40 \\ \underline{40} \\ 0 \end{array}$$

Second, add the percent sign and move the decimal point two places to the right.

$$.875 = 87.5\% \quad \text{or} \quad 87\tfrac{1}{2}\%$$

Example 4-4

PROBLEM: Change $62\frac{1}{2}\%$ to a fraction.

SOLUTION: First, remove the percent sign and place the number over 100.

$$62\tfrac{1}{2}\% = \frac{62\frac{1}{2}}{100} \quad \text{or} \quad \frac{62.5}{100} \quad \text{or} \quad \frac{625}{1000}$$

Second, reduce to lowest terms.

$$\frac{\overset{25}{\cancel{625}}}{\underset{40}{\cancel{1000}}} = \frac{\overset{5}{\cancel{25}}}{\underset{8}{\cancel{40}}} = \frac{5}{8}$$

Example 4-5

PROBLEM: Change $16\frac{2}{3}\%$ to a fraction.

SOLUTION: First, remove the percent sign and place the number over 100.

$$16\tfrac{2}{3}\% = \frac{16\frac{2}{3}}{100}$$

Second, change the numerator and denominator to improper fractions.

$$\frac{16\frac{2}{3}}{100} = \frac{\frac{50}{3}}{\frac{100}{1}}$$

Third, rewrite the complex fraction as a standard division-of-fractions problem.

$$\frac{\frac{50}{3}}{\frac{100}{1}} = \frac{50}{3} \div \frac{100}{1}$$

Fourth, invert and multiply

$$\frac{50}{3} \div \frac{100}{1} = \frac{\overset{1}{\cancel{50}}}{3} \times \frac{1}{\underset{2}{\cancel{100}}} = \frac{1}{6}$$

Thus, $16\frac{2}{3}\% = \frac{1}{6}$.

EXERCISE 4-1

Convert the following percents, decimals, and fractions into the indicated forms.

4-1. Convert 37% to a decimal.

4-2. Convert .865 to a percent.

4-3. Convert 16.5 to a percent.

4-4. Convert .173 to a percent.

4-5. $\frac{1}{8}$ equals what percent?

4-6. Convert .3% to a fraction.

4-7. Convert 25.5% to a decimal.

4-8. Convert $33.\bar{3}\%$ to a fraction.

4-9. Convert 37.5% to a fraction.

4-10. Convert .08 to a percent.

4-11. .0091 equals what percent?

4-12. Convert $\frac{4}{7}$ to a percent.

4-13. Convert $\frac{3}{8}$ to a percent.

4-14. Convert $28\frac{4}{7}\%$ to a fraction.

4-15. Convert .036 to a percent.

4-16. Convert $\frac{3}{10}$ to a decimal.

4-17. $12\frac{1}{2}\%$ equals what decimal value?

4-18. Convert $66\frac{2}{3}\%$ to a common fraction.

4-19. Convert $\frac{5}{6}$ to a percent.

4-20. Convert $\frac{5}{12}$ to a percent.

CONTENT

It is important for you to be able to apply your skills in working with percent to the everyday problems that face you. The next group of examples will give you an opportunity to do this and to further develop your skills in working with percents.

Example 4-6

PROBLEM: Find 30% of 360 days.

SOLUTION: First, convert 30% to a decimal.

30% = .30

Second, since "of" means multiply, carry out this operation:

```
   360   days
×  .30
------
   000
  1080
------
108.00   days
```

Thus, 30% of 360 days is 108 days.

Example 4-7

PROBLEM: Find 121% of $1500.

SOLUTION: First, convert 121% to a decimal.

121% = 1.21

Second, carry out the indicated multiplication.

```
    $1500
×    1.21
---------
     1500
    3000
   1500
---------
 $1815.00
```

Example 4-8

PROBLEM: Determine the selling price of a motorcycle if the cost to the dealer is $1224.00 and he would like to make a profit of 30%.

SOLUTION: First, change 30% to a decimal.

30% = .30

Second, multiply .30 × $1224.00 to determine the actual profit.

```
   1224.00
×      .30
----------
    000000
   267200
----------
  $3672000
```

Third, add the profit to the dealer's cost to determine the selling price.

```
  $1224.00   cost
+   367.20   profit
----------
  $1591.20   selling price
```

ALTERNATE SOLUTION: Since the selling price is 100% of the dealer's cost plus 30% profit, we could add these percents together and then multiply the dealers cost by the resulting percent. Follow these procedures to solve the problem. First, add 100% and 30%.

```
  100%
+  30%
------
  130%
```

Second, change 130% to a decimal

130% = 1.30

Third, multiply the decimal (1.3) and the dealer's cost ($1224) to determine the selling price.

```
    $1224.00   cost
×       1.30
------------
      000000
     367200
    122400
------------
  $1591.2000   selling price
```

Example 4-9

PROBLEM: Find the cost of a pair of snow tires which were priced at $56.00 but were reduced 20%.

SOLUTION: Reductions are given as a percent off the original price. To calculate them we multiply the original price ($56.00) by the rate of reduction (20%) and then subtract the reduction ($11.20) from the original selling price ($56.00). To determine the reduced price ($44.80), follow these procedures to solve the problems. First, change 20% to a decimal.

20% = .20

Second, multiply the original selling price ($56.00) by the decimal (.20).

```
     $56.00   original price
×       .20
    -------
      0000
     11200
    -------
  $11.2000   reduction
```

Third, subtract the reduction ($11.20) from the original price ($56.00).

```
   $56.00   original price
 −  11.20   reduction
   ------
   $44.80   reduced price
```

ALTERNATE SOLUTION: Since the reduced price is 100% of the original price minus 20% reduction, we could subtract 20% from 100% and then multiply the original price by the resulting percent to obtain the reduced price. Follow these procedures to solve the problem. First, subtract 20% from 100%.

```
   100%
 −  20%
 ------
    80%
```

Second, change 80% to a decimal.

80% = .80

Third, multiply the decimal (.80) and the original price ($56.00) to determine the reduced price.

```
     $56.00   original price
×       .80
    -------
      0000
     44800
    -------
  $44.8000   reduced price
```

EXERCISE 4-2

Do the following problems which will give you an opportunity to apply your skills in working with percents.

4-21. How much is 42% of 536?

4-22. What is 9% of 15?

4-23. What number is 275% of 828?

4-24. Find 343% of 606.

4-25. What is 9% of 180?

4-26. Find 123% of 41.

4-27. Find 76% of $584.

4-28. What is .4% of 5280?

4-29. What is 325% of 60?

4-30. What is .55% of $18?

4-31. The profit on an electric calculator is 55%. What is the selling price if the cost is $58?

4-32. What is the state sales tax on a new car priced at $4995.00 if the tax rate is 4%?

4-33. What number is 6% of $580?

4-34. A dress costing $27 is reduced 15%. What is the reduced price?

4-35. A football team looses $12\frac{1}{2}$% of its games. How many games did the team loose if it played 16 games?

4-36. A grain agent bought wheat at $5.40 a bushel and sold it at an 8% loss. What was the selling price of the wheat?

4-37. The 4% sales tax on a pontoon boat amounted to $66. What was the cost of the boat?

4-38. The city sales tax on a color television set was $24 for an $800 set. Determine the sales tax rate.

4-39. A mixture for concrete consists of 3 parts sand, 2 parts gravel, and 1 part cement. What is the percent of cement in the mixture?

4-40. A small business building is insured for 90% of its value at an insurance rate of $\frac{1}{4}$% per year. The building is valued at $140,000. Determine the annual premium.

CONTENT

Interest is money paid for the use of money. It is determined by a percent of the amount of money being loaned. Generally, interest is determined on the basis of one year. Banks and other lending agencies pay interest to those who let them use their money, and the rate of interest is the percent per year. The base is called the *principal* or amount of money loaned. The *percentage* is the interest paid. We determine interest by calculating the rate *times* the principal *times* the amount of time the loan is in effect. The formula we use to determine interest is $i = prt$. Sales taxes and commissions are calculated in the same manner.

Example 4-10

PROBLEM: Find the interest on \$2500 for 1 year at $7\frac{1}{2}\%$ per year.

SOLUTION: $i = prt$ or Interest = principal (\$2500) × rate ($7\frac{1}{2}\%$) × 1 year.

First, change $7\frac{1}{2}\%$ to a decimal.

$$7\tfrac{1}{2}\% = .075$$

Second, multiply the principal (\$2500) by the decimal (.075).

	\$2500.00	principal
×	.075	rate
	1250000	
	1750000	
	\$187.50000	interest

Example 4-11

PROBLEM: On a savings account, interest of \$120 is paid at the rate of 5% per year. How much money (principal) is in the account?

SOLUTION: The percentage or interest (\$120.00) and the rate or percent (5%) are known. Only the base, or principal, must be determined. To find the base, use the formula:

$$\text{base} = \frac{\text{percentage}}{\text{rate}} \quad \text{or} \quad b = \frac{p}{r}$$

First, identify the parts of the formula which are already known.

$$b = \frac{p}{r} = \frac{\$120}{5\%}$$

Second, change 5% to a decimal.

$$5\% = .05$$

Third. divide the percentage (\$120) by the rate (5%) to determine the base.

```
       $24 00.  base
 05.)$120 00.
      10
      --
       20
       20
       --
        0
```

Thus, there is a \$2400 principal in the savings account.

EXERCISE 4-3

Solve the following problems which will give you an opportunity to work with practical problems of percentage.

4-41. Find the interest on \$580 at 8% for 1 year.

4-42. Find the interest on \$1250 at $9\frac{1}{2}\%$ for 1 year.

4-43. Find the interest on \$17,500 at $7\frac{1}{4}\%$ for 20 years.

4-44. Find the interest on \$4050 at $9\frac{1}{2}\%$ for $3\frac{1}{2}$ years.

4-45. Determine the monthly payments on a \$3500 automobile at $10\frac{1}{2}\%$ for 3 years.

4-46. An AM-FM radio is priced at \$50.00. It was reduced to \$40.00 for quick sale. What is the rate of reduction (discount)?

4-47. Find the interest on a \$250.00 savings account for 5 years 9 months at $7\frac{1}{2}\%$.

4-48. If an electric bill is paid within 15 days, a discount of $1\frac{1}{2}\%$ is given. What would a customer pay on a \$33.00 electric bill if it was paid within 15 days?

4-49. A dealer allowed \$2875 for a used car. He then sold the car for \$3000. What was his percent of gain?

4-50. Find the rate if the principal is \$8000 and the interest for 1 year is \$400.

4-51. What is the rate of interest if the principal is \$3300 and the interest received after 8 years is \$924?

4-52. A pair of skis was priced at \$122.40. The price was first reduced $33\frac{1}{3}\%$ and then an additional 25%. Determine the final selling price.

4-53. Kristin bought two dresses at a 30%-off sale and saved \$41 off the marked price. What was the original price of the dresses?

4-54. A grain dealer bought peanut hearts for \$8.70 per bag and sold them for \$11.60. What was his percent of profit?

4-55. How long must \$30,000 be invested at $7\frac{1}{2}\%$ to yield \$6750?

4-56. What would be the sale price of an outboard motor which was marked 22% off a list price of \$589.00?

4-57. How much would you save if you received a $3\frac{1}{2}$% discount for paying cash for a $595.00 TV set?

4-58. A lumber company allows an 8% contractor discount. If a contractor bought lumber worth $8256, how much would he pay for the lumber after his discount?

4-59. A hardware store sold a radial arm saw for $392.50. The store bought the saw for 65% of the selling price. What did the store pay for the saw?

4-60. A land dealer bought two lots for $12,000 and $9000. He sold the first lot at a profit of 8% but lost 10% on the second. How much did he gain or lose on these deals?

CONTENT

Generally companies give substantial discounts for volume buying. An additional small discount is also given for cash payment or payment within a short period of time. Concern is also expressed for determining percent of increase or decrease, profit or loss, and the percent of change. These kinds of problems will be dealt with in the next two examples.

Example 4-12

PROBLEM: Determine the price paid for $240.00 worth of wool fabrics if discounts of 20% and 2% are given.

SOLUTION: First, determine the 20% discount by multiplying the original price ($240.00) by the rate of discount (.20).

$240.00	original price
× .20	discount rate
00000	
48000	
$48.0000	discount

Second, subtract the first discount from the original price to determine the discount price.

$240.00	original price
− 48.00	first discount
$192.00	discount price

Third, determine the 2% cash discount by multiplying the discount price ($192.00) by the rate of discount (.02).

$192.00	discount price
× .02	discount rate
$3.8400	cash discount

Fourth, subtract the cash discount from the discount price.

$192.00	discount price
− $3.84	cash discount
$188.16	price paid

ALTERNATE SOLUTION: Since the discounted price is 100% of the original price minus a 20% discount, we could subtract 20% from 100% and then multiply the original price by the resulting percent to obtain the discounted price.

First, subtract 20% from 100% and change to a decimal.

$$\begin{array}{r} 100\% \\ -\ 20\% \\ \hline 80\% \end{array} = .80$$

Second, multiply the decimal (.80) and the original price ($240.00) to determine the discounted price.

$240.00	original price
× .80	rate
00000	
192000	
$192.0000	discounted price

Third, subtract 2% from 100% and change to a decimal.

$$\begin{array}{r} 100\% \\ -\quad 2\% \\ \hline 98\% \end{array} = .98$$

Fourth, multiply the decimal (.98) and the discounted price ($192.00) to determine the price paid.

$$\begin{array}{rl} \$192.00 & \text{discounted price} \\ \times \quad .98 & \text{rate} \\ \hline 153600 & \\ 172800 & \\ \hline \$188.1600 & \text{price paid} \end{array}$$

Example 4-13

PROBLEM: Determine the percent of increase on the yield of soybeans through the application of fertilizer, if the yield increases from 48 bushels to 64 bushels.

SOLUTION: First, determine the amount of change.

$$\begin{array}{rl} 64 & \text{bushel yield with fertilizer} \\ -\,48 & \text{bushel yield without fertilizer} \\ 16 & \text{bushel increase} \end{array}$$

Second, divide the amount of change (16) by the yield without fertilizer (48) to determine the percent of change.

$$\frac{16}{48} = \frac{\text{change}}{\text{original}} = \text{percent of change}$$

Note: $\frac{16}{48}$ could be immediately reduced to $\frac{1}{3}$ or $33\frac{1}{3}\%$.

$$\begin{array}{r} .3333 \\ 48\,\overline{)16.00} \\ 14\,4 \\ \hline 1\,60 \\ 1\,44 \\ \hline 160 \\ 144 \\ \hline 160 \\ 144 \\ \hline 16 \end{array}$$

Third, change the decimal to a percent.

$$.333 = 33\tfrac{1}{3}\%$$

Example 4-14

PROBLEM: A basketball player makes 10 out of 13 shots. What percent of the shots does he make?

SOLUTION: The percentage (10) and the base (13) are known. Only the rate or percent must be determined. To find the rate, use the formula:

$$\text{rate} = \frac{\text{percentage}}{\text{base}} \quad \text{or} \quad r = \frac{p}{b}$$

First, identify the parts of the formula which are already known.

$$r = \frac{p}{b} = \frac{10}{13}$$

Second, divide the percentage (10) by the base (13) to determine the rate.

$$\begin{array}{r} .7692 \\ 13\,\overline{)10.00} \\ 9\,1 \\ \hline 90 \\ 78 \\ \hline 120 \\ 117 \\ \hline 30 \\ 26 \\ \hline 4 \end{array}$$

Third, change the decimal to the nearest tenth of a percent.

$$.7692 = 76.9\%$$

Thus, the basketball player made 76.9% of his shots. In the world of basketball, we would say his shooting average is 76.9%.

EXERCISE 4-4

Solve the following problems which will give you an opportunity to work more difficult problems dealing with percents.

4-61. A soap is advertised as $99\frac{44}{100}$% pure. How many grams of impurities would there be in 5000 grams?

4-62. A service charge of $1\frac{1}{2}$% is charged on the unpaid balance of charge accounts. Determine the interest on a bill for $59.95.

4-63. A real estate salesman receives 3% commission on the sale of a house. How much commission would he receive for selling a $44,500 house?

4-64. A trucking firm charges 3% commission for hauling and collects another $1\frac{1}{2}$% for insurance. What are the fees for a $12,000 cargo?

4-65. A certain type of crude oil is 14.875% sulfur. How many gallons of sulfur would there be in 8000 barrels of crude oil if each barrel contained 31.5 gallons?

4-66. At a baseball game, 37.42% of the people bought hot dogs. How much money was collected for hot dogs if 27,753 people attended the game and the hot dogs sold for 75¢?

4-67. If 37.42% is a constant for hot-dog buyers at baseball games, determine the profit on hot dogs for the season. Seventy-five games are played and average attendance is 24,276. Hot dogs sell for 75¢ each and the profit on the hot dogs is 62%.

4-68. A tolerance of ±.18% is allowable for a car axle. How much variation is allowable on a $\frac{7}{8}$-inch axle?

4-69. What are the upper and lower limits of tolerance of a 9100-ohm resistor if the tolerance factor is ±.5%?

4-70. What is the efficiency of an electric motor that draws 54 watts from an electrical supply and delivers 46.6 watts to its load?

4-71. Minnesota iron ore is 18.1% pure. How much iron can be refined from 150 tons of iron ore?

4-72. The list price of an office chair is $229.89. The decorator's discount is 40%. What does the decorator pay for the chair if the state sales tax is 4%?

4-73. A length of nylon rope 2 meters long was given a tensile test and stretched to 2.23 meters before it broke. Determine the percent of the original length it was stretched.

4-74. A drill-press operator was to drill a hole .34375 in diameter. The hole drilled measured .34345. Find the percent of error.

4-75. Marshall bought a lakeshore cabin for \$23,600 and later sold it for 25% more than he paid for it. The realtor charged a 6% fee on the selling price. What profit did Marshall make on his home?

4-76. A certain type of fertilizer is 22% nitrogen. If a 50-pound bag costs \$2.98, what is the price per pound of the nitrogen?

4-77. If $1\frac{1}{4}$ pounds of salt is dissolved in 5.5 gallons of water that weighs 45.8 pounds, what is the percent of salt in the solution?

4-78. How much would you need to invest if the interest rate was $7\frac{1}{2}$% and you wanted an income of \$1200.00 per month?

4-79. The cost of living rose 6.3%, 8.1%, and 5.8% over a 3-year period. The raise in pay amounted to 25% over 3 years. How much is a person ahead or behind at the end of 3 years if his initial yearly salary is \$12,500?

4-80. The depreciation of a new car is approximately 18% of the original price the first year, 12.5% of the original price the second year, 8% of the original price the third year. The car loan is for \$4200 at $8\frac{1}{2}$% per year for 3 years. Determine the depreciation and interest for 3 years.

THINK TIME

Many people have difficulty understanding the mathematical language involving percent. A free-spirited instructor, Scotty Jacobs, set up a system to translate percent language into simple arithmetic language. The system can best be explained through the use of examples.

PROBLEM: What is 20% of \$250?

JACOBS' LAW: ? = .20 × 250. Thus "what" becomes "?"; "is" becomes "="; "20%" becomes ".20"; and "of" becomes "times" or "×."

To complete the solution:

? = .20 × \$250

\$50 = .20 × \$250

PROBLEM: 52 is what percent of 260?

JACOBS' LAW: 52 = ? × 260. Thus "is" becomes "="; "what" becomes "?"; and "of" becomes "×."

To complete the solution, divide both sides by 260.

$$52 = ? \times 260$$

$$\frac{52}{260} = \frac{? \times 260}{260}$$

$$.20 = ?$$

$$20\% = ?$$

Check for correctness.

$$52 = 20\% \text{ of } 260$$
$$52 = .20 \times 260$$
$$52 = 52$$

PROBLEM: 105 is 75% of what number?

JACOBS' LAW: 105 = .75 × ? Thus "is" becomes "="; "75%" becomes ".75"; "of" becomes "×"; and "what" becomes "?."

To complete the solution, divide both sides by .75.

$$105 = .75 \times ?$$

$$\frac{105}{.75} = \frac{\not{.75} \times ?}{\not{.75}}$$

$$140 = ?$$

Check for correctness.

$$105 = .75 \times 140$$

$$105 = 105$$

PROCEDURES TO REMEMBER

1. To change a percent to a decimal, remove the percent sign and move the decimal point two places to the left.
2. To change a decimal to a percent, move the decimal point two places to the right and add the percent sign to the right of the number.
3. To change a fraction to a percent, complete the division indicated, move the decimal point two places to the right, and add the percent sign to the right of the number.
4. To change a percent to a common fraction, remove the percent sign, place the number over 100, and reduce to lowest terms.
5. To determine the percent of a number, convert the percent to a decimal and multiply.
6. To determine the selling price of an article based on a percentage of profit:

   METHOD 1
   (a) Convert the percent to a decimal.
   (b) Multiply the cost by the decimal.
   (c) Add the product to the cost to determine the selling price.

   METHOD 2
   (a) Consider the cost as 100%.
   (b) Add the percent of profit to the percent of cost (100%).
   (c) Multiply the cost by the sum of percents from step (b) to determine the selling price.
7. To determine the reduction of the selling price of an article:

   METHOD 1
   (a) Convert the percent to a decimal.
   (b) Multiply the selling price by the decimal.
   (c) Subtract the product from the selling price to determine the reduction price.

   METHOD 2
   (a) Consider the selling price as 100%.
   (b) Subtract the percent of reduction from the percent of the selling price (100%).
   (c) Convert the percent to a decimal.
   (d) Multiply the selling price by the decimal to determine the reduction price.
8. To determine interest, calculate the rate *times* the principal *times* the amount of time the loan is in effect ($i = prt$).
9. To determine the base or principal, convert the percent or rate of interest to a decimal and divide the interest paid by the rate of interest (decimal form).
10. To determine the rate of interest, divide the interest paid by the base or principal and change the quotient from a decimal to a percent form.

CHAPTER SUMMARY

1. *Percent* is the comparison of one number to another number.
2. The symbol for percent is %.
3. There are three types of percentage problems.
   (a) What is 5% of 265?
   (b) Sixteen is what percent of 64?
   (c) Thirty-nine is 20% of what number?

4. Any percent problem involves three things:
   (a) The base or principal.
   (b) The rate of percent or interest.
   (c) The percentage or the amount of the interest.
5. *Interest* is money paid for the use of money.
6. *Rate* of interest is the percent of interest per year.
7. *Base* is the principal or amount of money loaned.
8. Formulas:
   (a) *Percentage* equals base *times* rate ($p = b \times r$).
   (b) *Base* equals percentage *divided* by rate $\left(b = \frac{p}{r}\right)$.
   (c) *Rate* equals percentage *divided* by base $\left(r = \frac{p}{b}\right)$.
   (d) *Interest* equals principal *times* rate *times* time ($i = prt$).
   (e) *Principal* equals interest *divided* by rate *times* time $\left(p = \frac{i}{rt}\right)$.
   (f) *Rate of interest* equals interest *divided* by principal *times* time $\left(r = \frac{i}{pt}\right)$.

CHAPTER TEST

T-4-1. Change $\frac{7}{8}$ to a decimal.

T-4-2. Change $14\frac{2}{5}$ to a decimal.

T-4-3. Convert 45% to a common fraction.

T-4-4. Change $37\frac{1}{2}$% to a common fraction.

T-4-5 Convert .31 to a percent.

T-4-6. Convert 1.25 to a percent.

T-4-7. Convert .5% to a decimal.

T-4-8. Write $\frac{1}{4}$% as a decimal.

T-4-9. Convert .032% to a common fraction.

T-4-10. Convert $7\frac{3}{4}$% to a decimal.

T-4-11. $78\% = \dfrac{?}{100}$

T-4-12. Convert $\dfrac{7.5}{10}$ to a percent.

T-4-13. $\dfrac{3}{33\frac{1}{3}} = \dfrac{?}{100}$

T-4-14. Find 25% of 78.9.

T-4-15. Determine $7\frac{1}{4}\%$ of 5764.

T-4-16. A family sold a lot for $20,000 and gained 20% on the transaction. What did the lot cost the family?

T-4-17. What is the percent of increase from the application of fertilizer on a wheat field whose yield was 24 bushels before and 38 bushels after?

T-4-18. A man bought a house for $52,000 and 5 years later sold it for $87,000. Find the percent of gain.

T-4-19. An automobile depreciated 24% and was worth $3800 at the end of the year. What was it worth at the beginning of the year?

T-4-20. A lathe is set to operate at 4200 revolutions per minute. The belt slippage is 18%. What is the actual speed of the machine?

T-4-21. Dorothy works at $5.32 per hour for 22 hours per week. Determine her take-home pay if 27.2% is deducted for taxes, social security, and other benefits.

T-4-22. A set of golf clubs originally priced at $275.00 was discounted at 40%, 30%, and 10%. What was the final sale price?

T-4-23. The retail price of a radial tire is $75. If the tire store is allowed a 25% profit and the manufacturer earns 20% profit, what is the cost of making the tire?

T-4-24: How much money would you need to invest at $7\frac{1}{2}\%$ to have an income of $450 per month?

T-4-25. A plastic material loses 4.25% of its weight in drying. What was its original weight if its weight after drying is 134.05 grams?

CHAPTER

5

SQUARE ROOT

OBJECTIVES

1. To review the skills used to compute square root.
2. To develop the skills to use exponents and square root to solve industrial problems.

SELF-TEST

This test will determine your need to develop the skills taught in this chapter. Carefully work through each problem. Show your work as you proceed so that the difficulties you may encounter can easily be identified and needed help can be given.

S-5-1. $7^2 =$

S-5-2. $25^2 =$

S-5-3. Find the square of 37.

S-5-4. Find the square of $6\frac{1}{2}$.

S-5-5. Find the square of 1.732.

S-5-6. $\sqrt{625} =$

S-5-7. $\sqrt{289} =$

S-5-8. Find the square root of 3025.

S-5-9. $\sqrt{.1225} =$

S-5-10. Find the square root of $\frac{49}{81}$.

S-5-11. $21^2 =$

S-5-12. $.00494^2 =$

S-5-13. $6^3 =$

S-5-14. $\left(\frac{3}{8}\right)^3 =$

S-5-15. $3^6 =$

S-5-16. Calculate the square root of 141,000.

S-5-17. $6.35^3 =$

S-5-18. Calculate the square root of .120409.

S-5-19. Calculate the square root of 24,409 to the nearest thousandth.

S-5-20. Calculate the square root of 5280 to the nearest hundredth.

S-5-21. Find the diagonal of a square with 15-foot sides.

S-5-22. Find the hypotenuse of a right triangle if one leg is 12 units and the other leg is 5 units.

S-5-23. Find the cube root of 343.

S-5-24. Find the length, to the nearest tenth, of the side of a square whose diagonal is 16.97 meters.

S-5-25. Find the length, to the nearest tenth, of the diagonal of an $18\frac{1}{2}$-centimeter square.

INTRODUCTION

The calculation of square root is essential to many areas of industry. There are several ways to calculate square root, none of which are easy. To calculate the square root of a number means to determine what number, when multiplied by itself, will equal the original number. For example, the square root of 16 is 4, and 4 × 4 equals the original number, 16. Few numbers have even square roots. For example, the square root of 5 does not have an exact number, it can only be approximated. Possibly the easiest way to calculate square root is with an electronic calculator. Other simple ways are to use a table of square roots or a slide rule. Generally, craftsman and technicians use these simpler forms. However, there may be times when these tools are not available and it may be necessary to calculate square root by longhand. The remainder of this chapter will provide you with an opportunity to develop the skills for using the longhand method.

DEFINITIONS

To find the *square root* of a number means to find a number which, when multiplied by itself, will equal the original number. When several numbers are multiplied together to obtain another number, the numbers being multiplied are called *factors*. In the example $5 \times 3 \times 2 = 30$, the 5, 3, and 2 are factors. When a number is used as a factor more than once, it may be written with an exponent. An *exponent* is a number written above and to the right of another number and indicates how many times that number is to be used as a factor. In the example $5^2 = 5 \times 5 = 25$, the 2 is an exponent which indicates that the number 5 is used as a factor two times. When a factor is used two times, it may be referred to as *squared*. A number is *cubed* when it is used as a factor three times. In the example $2^3 = 2 \times 2 \times 2 = 8$, we say that the factor 2 is cubed. When a factor is used more than three times, it is generally called a *power*. In the example $3^4 = 3 \times 3 \times 3 \times 3 = 81$, the 4 indicates the use of 3 as a factor four times. This can also be read as 3 to the fourth power.

The square root of a number is one of two equal factors which, when multiplied together, equal the original number. Similarly, the cube root of a number is one of three equal factors, and the fourth root of a number is one of four equal factors.

The *radical sign*, $\sqrt[3]{8}$, indicates that a root is to be determined. The figure inside the radical sign (3) is called the *index of the root* and indicates what root is being determined. When no number appears, the number 2 or square root is understood. The number under the radical sign (8) is called the *radicand*. Numbers which have even square roots are called *perfect squares*. Examples of perfect squares are: 1, 4, 9, 16, 25, and 36.

CONTENT

The following examples illustrate the use of powers and factors. Refer to the definitions given above when necessary to recall the meanings of the terms.

Example 5-1

PROBLEM: Write $3 \times 3 \times 3 \times 3 \times 3$ in an abbreviated form and evaluate.

SOLUTION: Since 3 is used as a factor five times, it is written as 3^5. Multiplying through,

$$3^5 = 243.$$

Example 5-2

PROBLEM: Write $5 \times 5 \times 2 \times 2 \times 2$ in an abbreviated form and evaluate.

SOLUTION: Since 5 is used as a factor two times, it is written as 5^2; and since 2 is used as a factor three times, it is written as 2^3. Multiplying through, $5^2 \times 2^3 = 200$.

$$5 \times 5 = 25 \quad \text{and} \quad 2 \times 2 \times 2 = 8$$

$$25 \times 8 = 200$$

Example 5-3

PROBLEM: What does nine squared equal?

SOLUTION: Nine squared or 9^2 means to multiply 9×9, which equals 81.

Example 5-4

PROBLEM: What does eight cubed equal?

SOLUTION: Eight is used as a factor three times, which can be written as 8^3. Multiplying through, $8^3 = 512$.

$$8 \times 8 \times 8 = 512$$

EXERCISE 5-1

Solve the following problems which will give you practice in working with factors and exponents. Use the procedures shown in the examples.

5-1. Evaluate 5^3.

5-2. Evaluate $4^2 \times 4^2$.

5-3. Evaluate 4^5.

5-4. Evaluate 12^2.

5-5. Evaluate 150^2.

5-6. Write $2 \times 2 \times 2 \times 4 \times 4$ using exponents, and then evaluate.

5-7. Write $3 \times 3 \times 3 \times .3 \times .3 \times .3$ using exponents, and then evaluate.

5-8. Evaluate 12^3.

5-9. Evaluate $(\frac{1}{2})^4$.

5-10. Evaluate $\frac{1}{2} \times \frac{1}{2} \times \frac{1}{2} \times 4 \times 4 \times 2 \times 2 \times 2$.

5-11. Evaluate $.5^2$.

5-12. Evaluate $.22^2$.

5-13. Evaluate $.8^3$.

5-14. Evaluate $.102^2$.

5-15. Evaluate $.12^3$.

5-16. Evaluate 3.15^2.

5-17. Evaluate 10.1^3.

5-18. Evaluate $.4^4$.

5-19. Evaluate 2.5^5.

5-20. Evaluate $8^2 \times 8^2$.

CONTENT

As you develop the ability to use the longhand method of calculating square root as illustrated in the following examples, keep in mind that this is just one of the methods which may be used. You should also note that most of the calculations will determine only approximate numbers. The degree of accuracy will depend on the requirements of the individual problems.

Example 5-5

PROBLEM: Calculate the square root of 576.

SOLUTION: First, set up the square-root process. Place 576 under the radical symbol. Divide the digits into groups of two, starting from the decimal point and moving to the left.

$$\sqrt{5_\wedge 76.}$$

Second, take the first group (5) and find the largest perfect square smaller than 5. This is $2^2 = 4$. Place the 2 above the number 5 on top of the radical sign. Place the number 4 below the number 5. Subtract the number 4 from the number 5. Bring down the next group (76).

$$\begin{array}{l} \;\;2\;\;\;\;. \\ \sqrt{5_\wedge 76.} \\ \;\;4 \\ \hline \;\;1\;76 \end{array}$$

Third, multiply 20 times the number on top of the radical sign (2). Then divide the result of 20 × 2, or 40, into 176. This will give you the approximate quotient which is 4. Place the number 4 above the number 6 on top of the radical sign.

$$\begin{array}{r} 4 \\ 40\,\overline{)176.00} \\ 160 \\ \hline \end{array} \qquad \begin{array}{l} \;\;2\;\;\;4. \\ \sqrt{5_\wedge 76} \\ \;\;4 \\ \hline \;\;1\;76 \end{array}$$

Fourth, add the approximate quotient (4) to 40, with the result of 44. Then multiply 44 × 4 and place the result below the 176. Subtract the resulting number from 176.

$$\begin{array}{r} 4 \\ +\,40 \\ \hline 44 \end{array} \qquad \begin{array}{r} 44 \\ \times\;4 \\ \hline 176 \end{array} \qquad \begin{array}{l} \;\;2\;\;\;4. \\ \sqrt{5_\wedge 76} \\ \;\;4 \\ \hline \;\;1\;76 \\ \;\;1\;76 \\ \hline \;\;0\;00 \end{array}$$

Fifth, since the resulting subtraction of 176 from

176 equals zero, we have a perfect square of 24. Check your answer by multiplying 24 by 24.

$$\begin{array}{r} 24 \\ \times\ 24 \\ \hline 96 \\ 48 \\ \hline 576 \end{array}$$

Once you understand the procedure, the process can be condensed as follows:

$$\sqrt{5_\wedge 76.} \qquad \begin{array}{l} \phantom{\sqrt{}}2 \\ \sqrt{5_\wedge 76} \\ \phantom{\sqrt{}}4 \\ \hline \phantom{\sqrt{}}1\ 76 \end{array} \qquad \begin{array}{r} 20 \\ \times\ 2 \\ \hline 40 \end{array}$$

$$\begin{array}{r} 4 \\ 40\,\overline{)176} \\ 160 \\ \hline \end{array} \qquad \begin{array}{r} 40 \\ +\ 4 \\ \hline 44 \\ \times\ 4 \\ \hline 176 \end{array} \qquad \begin{array}{l} \phantom{\sqrt{}}2\ \ 4. \\ \sqrt{5_\wedge 76.} \\ \phantom{\sqrt{}}4 \\ \hline \phantom{\sqrt{}}1\ 76 \\ \phantom{\sqrt{}}1\ 76 \\ \hline \phantom{\sqrt{}}0\ 00 \end{array}$$

CHECK:

$$\begin{array}{r} 24 \\ \times\ 24 \\ \hline 96 \\ 48 \\ \hline 576 \end{array}$$

Example 5-6

PROBLEM: Calculate the square root of 4134.969.

SOLUTION: First, set up the square-root process. Place 4134.969 under the radical symbol. Divide the digits into groups of two, starting from the decimal point and moving in both directions. Since 9 is alone in the last group on the right, add a 0 to make the group complete.

$$\sqrt{{}_\wedge 41_\wedge 34.96_\wedge 90_\wedge}$$

Second, take the first group (41) and find the largest perfect square smaller than 41. This is $6^2 = 36$. Place the number 6 above the number 41 on top of the radical sign. Place the number 36 below the number 41. Subtract the number 36 from the number 41. Bring down the next group (34).

$$\begin{array}{l} \phantom{\sqrt{}}6 \\ \sqrt{41_\wedge 34.96_\wedge 90_\wedge} \\ \phantom{\sqrt{}}36 \\ \hline \phantom{\sqrt{}}5\ 34 \end{array}$$

Third, multiply 20 times the number on top of the radical sign (6). Then divide the result of 20 × 6, or 120, into 534 to obtain the approximate quotient of 4. Place the number 4 above the number 34 on top of the radical sign.

$$\begin{array}{r} 4. \\ 120\,\overline{)534.} \\ 480 \\ \hline \end{array} \qquad \begin{array}{l} \phantom{\sqrt{}}6\ \ 4. \\ \sqrt{41_\wedge 34.96_\wedge 90_\wedge} \\ \phantom{\sqrt{}}36 \\ \hline \phantom{\sqrt{}}5\ 34 \end{array}$$

Fourth, add the approximate quotient (4) to 120, with the result of 124. Then multiply 124 × 4 and place the result below the 534. Subtract the resulting number from 534. Bring down the next group (96).

$$\begin{array}{r} 4 \\ +\ 120 \\ \hline 124 \end{array} \qquad \begin{array}{r} 124 \\ \times\ 4 \\ \hline 496 \end{array} \qquad \begin{array}{l} \phantom{\sqrt{}}6\ \ 4. \\ \sqrt{41_\wedge 34.96_\wedge 90_\wedge} \\ \phantom{\sqrt{}}36 \\ \hline \phantom{\sqrt{}}5\ 34 \\ \phantom{\sqrt{}}4\ 96 \\ \hline \phantom{\sqrt{}}38\ 96 \end{array}$$

Note that if the result of multiplying 4 × 124 were larger than 534, the number 3 would have to be used instead of 4.

Fifth, multiply 20 times the number on top of the radical sign (64). Then divide the result of 20 × 64, or 1280, into 3896 to obtain the approximate quotient of 3. Place the number 3 above the number 96 on top of the radical sign.

$$\begin{array}{r} 3 \\ 1280\,\overline{)3896} \\ 3840 \\ \hline \end{array} \qquad \begin{array}{l} \phantom{\sqrt{}}6\ \ 4.\ 3 \\ \sqrt{41_\wedge 34.96_\wedge 90_\wedge} \\ \phantom{\sqrt{}}36 \\ \hline \phantom{\sqrt{}}5\ 34 \\ \phantom{\sqrt{}}4\ 96 \\ \hline \phantom{\sqrt{}}38\ 96 \end{array}$$

Sixth, add the approximate quotient (3) to 1280, with the result of 1283. Then multiply 1283 × 3 and place the result below 3896. Bring down the next group (90).

$$\begin{array}{r} 3 \\ +\ 1280 \\ \hline 1283 \end{array} \qquad \begin{array}{r} 1283 \\ \times\ 3 \\ \hline 3849 \end{array} \qquad \begin{array}{l} \phantom{\sqrt{}}6\ \ 4.\ 3 \\ \sqrt{41_\wedge 34.96_\wedge 90_\wedge} \\ \phantom{\sqrt{}}36 \\ \hline \phantom{\sqrt{}}5\ 34 \\ \phantom{\sqrt{}}4\ 96 \\ \hline \phantom{\sqrt{}}38\ 96 \\ \phantom{\sqrt{}}38\ 49 \\ \hline \phantom{\sqrt{}}47\ 90 \end{array}$$

Seventh, multiply 20 times the number on top of the radical sign (643). Note that the decimal point is disregarded in this process. Then divide the result of 20 × 643, or 12,860, into 4790 to obtain the approximate quotient. Since 12,860 is greater than 4790, it will not divide into 4790 with a whole number as a result. Place a 0 above the 90 and bring down the next group (00).

```
          0          6  4. 3  0
12,860 )4790      √41‸34.96‸90‸00‸
                   36
                    5 34
                    4 96
                      38 96
                      38 49
                         47 90 00
```

Eighth, multiply 20 times the number on top of the radical sign (6430). Then divide the result of 20 × 6430, or 128,600, into 479,000 to obtain the approximate quotient of 3. Place the number 3 above the number 00 on top of the radical sign.

```
              3        6  4. 3  0  3
128,000 )479,000    √41‸34.96‸90‸00‸
         384 000     36
                      5 34
                      4 96
                        38 96
                        38 49
                           47 90 00
```

Ninth, add the approximate quotient (3) to 128,600, with the result of 128,603. Then multiply 128,603 × 3 and place the result below 479,000. Subtract the resulting number from 479,000.

```
        3      128,603      6  4. 3  0  3
+ 128,600   ×        3   √41‸34.96‸90‸00
  128,603      385,809    36
                           5 34
                           4 96
                             38 96
                             38 49
                                47 90 00
                                38 58 09
                                 9 31 91
```

The square root of 4134.969 has now been calculated to three places. More places could be calculated if necessary. As a general rule, however, the square root of a quantity is calculated to the same number of significant digits as the original number. Before leaving this problem, it would be wise to check for accuracy. You will note that the result is close. The next three digits would bring the result very close to the original number.

```
       64.303
     × 64.303
       192909
       00000
     192909
    257212
   385818
  4134.875809
```

Example 5-7

PROBLEM: Calculate the square root of 0.003136.

SOLUTION: First, set up the square-root process. Place 0.003136 under the radical symbol. Divide the digits into groups of two, starting from the decimal point and moving to the right.

$$\sqrt{0.00_{\wedge}31_{\wedge}36_{\wedge}}$$

Second, since the first group contains two zeros, the largest perfect square is zero. Place a zero in the result and bring down the next group (31). Determine the largest perfect square smaller than 31. This is $5^2 = 25$. Place the number 5 above the number 31 on top of the radical sign. Place the number 25 below the number 31 and subtract the number 25 from the number 31. Bring down the next group (36).

```
  .0   5
√0.00‸31‸36
   00
   00 31
   00 25
       6 36
```

Third, multiply 20 times the number on top of the radical sign (5). Then divide the result of 20 × 5, or 100, into 636 to obtain the approximate quotient of 6. Place the number 6 above the number 36 on top of the radical sign.

```
      6.       . 0   5   6
100 )636.    √0.00‸31‸36
     600        00
                00 31
                00 25
                    6 36
```

Fourth, add the approximate quotient (6) to 100, with the result of 106. Then multiply 106 × 6 and place the result below the 636. Subtract the resulting number from 636.

```
   6           . 0  5  6
+ 100       √0.00∧31∧36
  106          00
               00 31
               00 25
                   6 36
                   6 36
                      0
```

Note that the subtraction ends in zero. Thus the square root of 0.003136 is .056. Check the result by squaring .056.

```
    .056
  × .056
     336
    280
 .003136
```

CONTENT

Skills used to determine square root can be used to determine an unknown distance or length. For example, carpenters can use square-root skills to determine the lengths of rafters. The formula for determining the length of a rafter is based on the formula: the square of the hypotenuse of a right triangle is equal to square of one side plus the square of the other side. Note the illustration below.

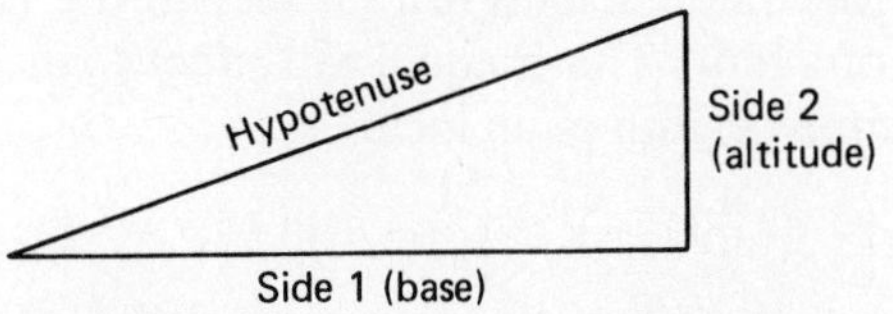

Likewise, the length of the hypotenuse is equal to the square root of side 1 squared plus side 2 squared. Study the other formulas given below. One of them will be used in the next example.

$$\text{Hypotenuse} = \sqrt{\text{base}^2 + \text{altitude}^2}$$

$$\text{Base} = \sqrt{\text{hypotenuse}^2 - \text{altitude}^2}$$

$$\text{Altitude} = \sqrt{\text{hypotenuse}^2 - \text{base}^2}$$

$$\text{Diagonal side} = \sqrt{(\text{side 1})^2 + (\text{side 2})^2}$$

$$\text{Side 1} = \sqrt{(\text{diagonal side})^2 - (\text{side 2})^2}$$

$$\text{Side 2} = \sqrt{(\text{diagonal side})^2 - (\text{side 1})^2}$$

Remember, these formulas are valid only for *right* triangles.

Example 5-8

PROBLEM: Find the length of a rafter if the rise is 6 feet and the run is 12 feet.

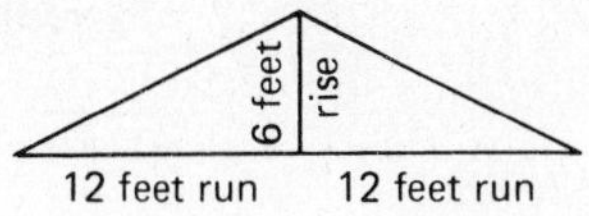

SOLUTION: To determine the length of a rafter, take the square root of the square of the rise plus the square of the run. First, set up the square-root process, then square each of the numbers (6, 12) and add the results together (36 + 144 = 180).

$$\text{Rafter length} = \sqrt{(\text{rise})^2 + (\text{run}^2)}$$

$$\text{Rafter length} = \sqrt{6^2 + 12^2}$$

$$\text{Rafter length} = \sqrt{36 + 144}$$

$$\text{Rafter length} = \sqrt{180}$$

Second, place the decimal point after the zero. Divide the digits into groups of two, starting from the decimal point and moving to the left.

```
√1∧80.
```

Third, take the first group (1) and find the largest perfect square. Obviously, the largest perfect square of 1 is 1. Place the number 1 above the number 1 on top of the radical sign. Place the number 1 below the number 1 and subtract. Bring down the next group (80).

```
  1
√1∧80.
  1
  0 80
```

Fourth, multiply 20 times the number on top of the radical sign (1). Then divide the result of 20 × 1, or 20, into 80 and obtain the approximate quotient of 4. Add the quotient (4) to 20, with the result of 24. Multiply 24 × 4, with the result of 96. Note that 96 is greater than 80; therefore, the approximate quotient should be 3. Place the number 3 above the number 80 on top of the radical sign.

```
    4                              1  3.
20 )80       4        24       √1∧80.00∧
    80    + 20      ×  4         1
            24        96         0 80
```

Fifth, add the approximate quotient (3) to 20, with the result of 23. Then multiply 23 × 3 and place the result below the 80. Subtract the resulting number from 80. Bring down the next group (00).

```
   3       23       1  3.
+ 20      × 3     √1∧80.00∧
  23       69      1
                   0 80
                     69
                   11 00
```

Sixth, multiply 20 times the number on top of the radical sign (13). Then divide the result of 20 × 13, or 260, into 1100 to obtain the approximate quotient of 4. Place the number 4 above the number 00 on top of the radical sign.

```
  20              4.       1  3. 4
× 13       260 )1100.    √1∧80.00∧00
  60             1040      1      ∧
 20                        0 80
 260                         69
                           11 00
```

Seventh, add the approximate quotient (4) to 260, with the result of 264. Then multiply 264 × 4 and place the result below the 1100. Subtract the resulting number from 1100. Bring down the next group (00).

```
    4       264       1  3. 4
+ 260     ×   4     √1∧80.00∧00
  264      1056       1     ∧
                      0 80
                      0 69
                        11 00
                        10 56
                           44 00
```

Eighth, multiply 20 times the number on top of the radical sign (134). Then divide the result of 20 × 134, or 2680, into 4400 to obtain the approximate quotient of 1. Place the number 1 above the number 00 on top of the radical sign.

```
 134                 1       1  3. 4  1
× 20       2680 )4400      √1∧80.00∧00∧
2680             2680        1
                             0 80
                             0 69
                               11 00
                               10 56
                                  44 00
```

Ninth, add the approximate quotient (1) to 2680, with the result of 2681. Then multiply 2681 × 1 and place the result below the 4400. Subtract the resulting number from 4400.

```
     1       2681       1  3. 4  1
+ 2680     ×    1     √1∧80.00∧00∧
  2681       2681       1      ∧
                        0 80
                        0 69
                          11 00
                          10 56
                             44 00
                             26 81
                             17 19
```

Thus, the rafter length is 13.4 feet to the nearest tenth of a foot. This is equal to 13 feet $4\frac{7}{8}$ inches to the nearest eighth of an inch.

EXERCISE 5-2

Work through the following problems which will help you to develop your skills in calculating square root.

Calculate the square root of the following problems to the nearest tenth.

5-21. $\sqrt{121}$

5-22. $\sqrt{256}$

5-23. $\sqrt{1024}$

5-24. $\sqrt{3136}$

5-25. $\sqrt{5476}$

5-26. $\sqrt{708964}$

5-27. $\sqrt{.0576}$

5-28. $\sqrt{38.4400}$

Calculate the square root to the nearest thousandth.

5-29. $\sqrt{.6561}$

5-30. $\sqrt{.002116}$

5-31. $\sqrt{38}$

5-32. $\sqrt{149}$

5-33. $\sqrt{85}$

5-34. $\sqrt{5280}$

5-35. $\sqrt{.1234}$

5-36. $\sqrt{10{,}000.}$

5-37. $\sqrt{30.603024}$

5-38. $\sqrt{.060516}$

5-39. $\sqrt{1976}$

5-40. $\sqrt{91.550451}$

5-41. Find the length of the diagonal (hypotenuse) of a 24-meter square to the nearest tenth.

5-42. Find the length of the diagonal (hypotenuse) of a 14.5-centimeter square to the nearest hundredth.

5-43. Find the length of the diagonal of the given rectangle.

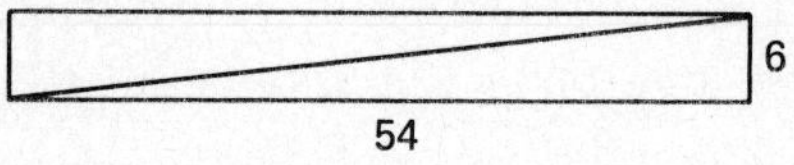

5-44. Find the altitude of the given triangle.

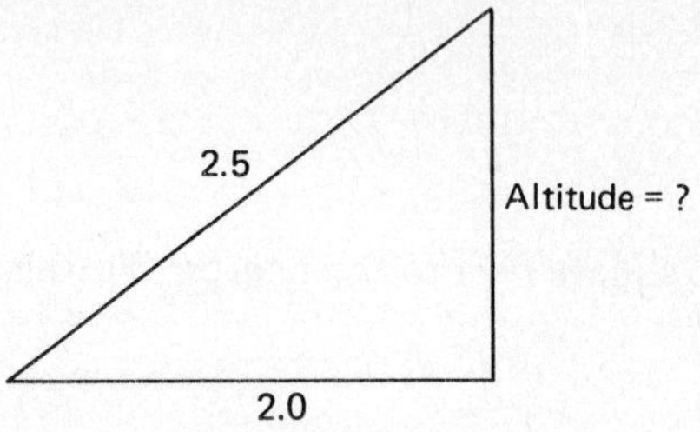

5-45. Find the hypotenuse of the given triangle to the nearest thousandth.

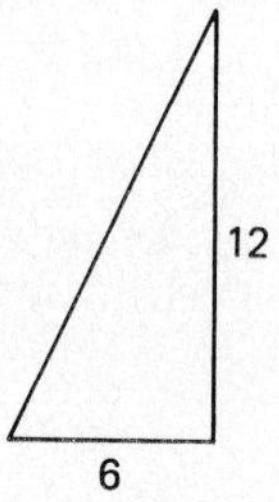

5-46. Find the altitude of a right triangle whose hypotenuse is 20 and whose base is 16.

5-47. Find the hypotenuse of the given triangle to the nearest tenth.

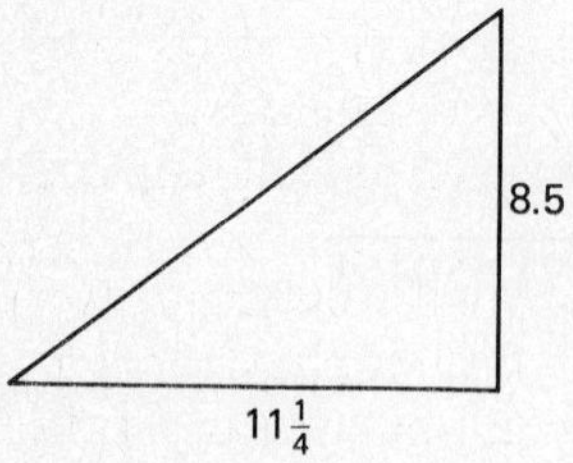

5-48. Find the rise if the rafter is 37.1 feet and the run is 36 feet.

5-49. Find the length of the rafter for the given roof to the nearest hundredth.

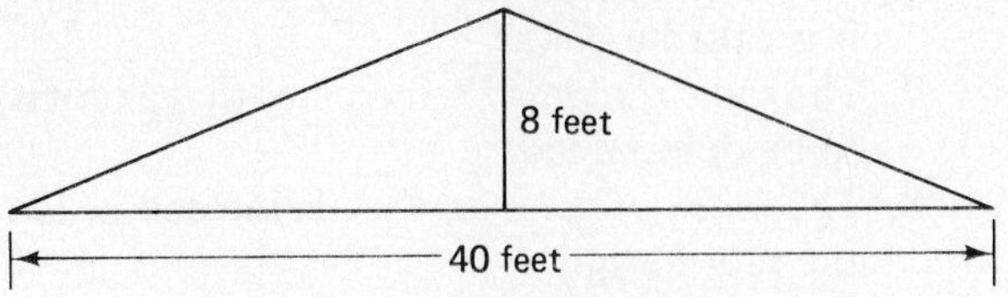

5-50. Good proportioning in printing requires the diagonal to be two times the width of the page. Find the length of the page if the width is 24 centimeters to the nearest hundredth.

THINK TIME

Many different ways are used to describe the approximate square root of a number. Generally, the process used is determined by the exactness of the approximation desired. A simple approximation process is illustrated below.

PROBLEM: Determine the approximate square root of 29.

SOLUTION: First, recall the perfect squares smaller and larger than 29.

$$5 \times 5 = 25 \quad \text{and} \quad 6 \times 6 = 36$$

Second, determine $\frac{1}{2}$ of these perfect squares combined.

$$\frac{1}{2}(5 + 6) = \frac{11}{2}$$

Third, square $\frac{11}{2}$ and note that this factor is too large.

$$\frac{11}{2} \times \frac{11}{2} = \frac{121}{4} = 30\tfrac{1}{4}$$

Fourth, divide 29 by $\frac{11}{2}$ and note that this factor is too small.

$$29 \div \frac{11}{2} = \frac{29}{1} \times \frac{2}{11} = \frac{58}{11}$$

$$\left(\frac{58}{11}\right)^2 = \frac{3364}{121} = 27.8$$

Fifth, take $\frac{1}{2}$ of the sum of the factors $\frac{11}{2}$ and $\frac{58}{11}$.

$$\frac{1}{2}\left(\frac{11}{2} + \frac{58}{11}\right)$$

$$\frac{1}{2}\left(\frac{121 + 116}{22}\right)$$

$$\begin{array}{r} 5.386 \\ 44\overline{)\,237.00} \\ \underline{220} \\ 17\,0 \\ \underline{13\,2} \\ 3\,80 \\ \underline{3\,52} \\ 280 \\ \underline{264} \\ 16 \end{array}$$

Sixth, square 5.386 and notice if it is close to 29.

$$\begin{array}{r} 5.386 \\ \underline{\times\ 5.386} \\ 32316 \\ 43088 \\ 16158 \\ \underline{26930} \\ 29.008996 \end{array}$$

PROCEDURES TO REMEMBER

To determine the square root of a number:

1. Place the number under the radical sign.
2. Divide the digits of the number into groups of two, starting from the decimal point and moving in both directions as necessary.
3. Begin with the group on the left, and find the largest perfect square which is equal to or smaller than the number in the group.
4. Place this root number above the group and on top of the radical sign.
5. Multiply this root number by itself and place the result below the first group on the left.
6. Subtract, as in division, and bring down the next group to the right of first group.
7. Multiply 20 times the number above the radical sign.

8. Divide this approximate number into the dividend created in step 6. This will give you an approximate quotient.
9. Place this number above the second group on top of the radical sign.
10. Multiply this group by the last digit of the number, and enter it under the number created in step 6.
11. Repeat steps 6 through 10 until the square root has been determined or satisfactorily approximated. Add zeros to form groups of two as necessary to the right of the numbers to the right of the decimal point.

CHAPTER SUMMARY

1. To calculate the *square root* of a number means to determine what number, when multiplied by itself, will equal the original number.
2. *Factors* are numbers which are multiplied together to form another number.
3. An *exponent* is a number written above and to the right of another number and indicates how many times that number is used as a factor.
4. To *square* a number the number is used as a factor two times.
5. To *cube* a number the number is used as a factor three times.
6. When a factor is used more than three times, it is called a *power.*
7. The *radical sign,* $\sqrt[3]{78}$, indicates that a root is to be determined.
8. The *index of the root,* $\sqrt[3]{\ }$, is located inside the radical sign.
9. When no index number appears within the radical sign, $\sqrt{\ }$, square root is understood.
10. The number under the radical sign, $\sqrt{78}$, is called the *radicand.*
11. Numbers which have even square roots are called *perfect squares.*
12. The square of a hypotenuse of a right triangle is equal to the square of the base plus the square of the altitude.
13. The hypotenuse of a right triangle equals the square root of the square of the base plus the square of the altitude.
14. The base of a right triangle is equal to the square root of the square of the hypotenuse minus the square of the altitude.
15. The altitude of a right triangle is equal to the square root of the hypotenuse minus the square of the base.

CHAPTER TEST

T-5-1. Find the square of 46.

T-5-2. $16^2 =$

T-5-3. $(2.4)^2 =$

T-5-4. Find the square root of 10.24.

T-5-5. Find the square of 2.236.

T-5-6. $\sqrt{289} =$

T-5-7. $\sqrt{361} =$

T-5-8. Find the square root of 3844.

T-5-9. $\sqrt{.5184} =$

T-5-10. Find the square root of the fraction $\frac{169}{225}$.

T-5-11. $(23.3)^2 =$

T-5-12. $(.00949)^2 =$

T-5-13. $(7)^3 =$

T-5-14. $\left(\frac{5}{9}\right)^3 =$

T-5-15. $(4)^4 =$

T-5-16. Calculate the square root of 141,376.

T-5-17. $(5.36)^3 =$

T-5-18. Calculate the square root of .22373 to the nearest thousandth.

T-5-19. To the nearest hundredth, calculate the square root of 1609.

T-5-20. Calculate the square root of 26,606 to the nearest thousandth.

T-5-21. Find the diagonal of an 8-by-10 rectangle to the nearest tenth.

T-5-22. Find the hypotenuse of a right triangle if one side is 12.5 units and the other side is 16.4 units (to the nearest tenth).

T-5-23. Find the cube root of 216.

T-5-24. Find the length of the side of a square if the diagonal is 10.6066 (to the nearest tenth).

T-5-25. What is the longest line that could be drawn on a basketball court which is 88 feet by 50 feet?

Section Two

BASIC ALGEBRA

CHAPTER

6

FUNDAMENTAL DEFINITIONS AND OPERATIONS OF ALGEBRA

OBJECTIVES

1. To define fundamental algebraic terms.
2. To perform basic algebraic operations.
3. To determine when algebraic formulas are necessary to the solution of problems.

SELF-TEST

The test which follows will help you determine your need to develop the basic algebraic skills taught in this chapter. Show your work as you proceed so that your difficulties may be easily detected and you may be given the precise help you need.

S-6-1. Subtract: $\begin{array}{r} -11xy \\ \underline{7xy} \end{array}$

S-6-2. Subtract: $\begin{array}{r} +764 \\ \underline{-231} \end{array}$

S-6-3. Add: $\begin{array}{r} +42 \\ \underline{-26} \end{array}$

S-6-4. Add: $\begin{array}{r} -17 \\ \underline{+6} \end{array}$

S-6-5. Divide: $\dfrac{-144}{-3}$

S-6-6. Divide: $\dfrac{126}{-21}$

S-6-7. Multiply: $(-3)(-4)(-5)(-6)$

S-6-8. Multiply: $(-29)(-11)(+5)(-13)$

S-6-9. Simplify: $189 - (69 \div 23) \times 2 + 44$

S-6-10. Simplify: $8 \div 8 \times 8 \div 8 \times 8$

S-6-11. Simplify: $-8 - 4(x - 2y) + 6y$

S-6-12. Simplify: $-x + 3y(y - 3) + 7x$

S-6-13. Simplify: $11 - [6 \div 10(9 - 6 - 3) + 2]$

S-6-14. Simplify:
$35 - [9z - (7z + 2) + 3(5z + 3)]$

S-6-15. Multiply: $y^2y^3y^4$

S-6-16. Multiply: $(3x^2)(2x^3)$

S-6-17. Divide: $\dfrac{25x^3y^4}{5x^2y^3}$

S-6-18. Divide: $\dfrac{3xy^5z^2}{18x^2yz^3}$

S-6-19. Divide: $\dfrac{-36a^3b^2c^1}{-6ab^2c^3}$

S-6-20. Multiply: $(-5a^2)^3$

Express the following as algebraic expressions:

S-6-21. The sum of 3, a, and b^3.

S-6-22. The quotient of seven more than h and 7.

S-6-23. The area of a triangle equals one half the product of the base (b) times the altitude (a).

S-6-24. The side (a) opposite the 30° angle of a 30–60–90 triangle equals one half the hypotenuse (c).

S-6-25. The cost-to-rent money equals the principal (p) times the rate (r) times the time in days (d) divided by 360.

INTRODUCTION

As a child you first learned to crawl, then to walk, then to run, and eventually to ride a bicycle. You found that riding a bicycle to get from one place to another was faster and easier than crawling, walking, or running.

In like manner, when you first became

acquainted with mathematics, you learned numbers, then to count, then to add, subtract, multiply, and divide. After you developed these simple skills, you developed more complicated skills. These new skills helped you solve certain mathematical problems faster and more easily. Algebraic skills will bring solutions to problems which you could not solve with arithmetic. More importantly, you will become skillful in working with formulas which will make your job easier. As you move through the fundamentals of algebra and its definitions and operations, look for ways you can apply these skills to your job.

DEFINITIONS

Algebra may be thought of as a language in which symbols are used to represent numbers and operations with numbers. An *algebraic expression* is an expression which uses signs and symbols to represent numbers or quantities. A *numerical algebraic expression* is made up entirely of numbers. A *literal algebraic expression* combines both letters and numbers, then to count, then to add, subtract, multiply, separated from other terms by a plus or minus sign. Thus in the example $5xy + 3c - z$, the terms are $+5xy$, $+3c$, and $-z$. In the expression $6xy$, the 6, x, and y are *factors* of the expression. The 6 is called the *numerical coefficient* and the x and y are *literal coefficients*. In the example, x^5, x is called the *base* and 5 is the *exponent* or *power*. A *monomial* $(5x)$ is an algebraic expression which has one term, a *binomial* $(4k - j)$ has two terms, a *trinomial* $(l + k - j)$ has three terms, and a *polynomial* $(k + m + n + s)$ has two or more terms. *Like terms* are quantities which have the same literal factors $(4k + 5k - 2k)$. *Unlike terms* are terms which have unlike literal factors or powers.

Algebra has the same operations as arithmetic (addition, subtraction, multiplication, and division). Multiplication is expressed in several different ways in algebra. The following symbols all indicate multiplication: $a \times b = a \cdot b = (a)(b) = a(b) = (a)b$. The *signs of quantity* indicate whether a number is *positive* (+) or *negative* (−). When someone gives you $5.00, the number is positive; when you give someone $5.00, the number is considered negative. The plus sign in front of a number indicates it is positive, the negative or minus sign in front of a number indicates a negative. When neither a plus nor a minus sign appears in front of a number or quantity, it is considered to be positive. Often terms are grouped together. Special signs are used to indicate groups and that the quantities within them are to be treated as one quantity. The grouping signs are as follows:

Parentheses	( )
Brackets	[ ]
Braces	{ }
Vinculum or bar	$\overline{\quad\quad}$

Parentheses, brackets, and braces are placed around groups, and a vinculum or a bar is placed over a group. A vinculum or a bar is generally used in writing complex fractions or with the division sign or radical sign.

CONTENT

When determining the algebraic sum of a quantity, you may obtain a positive or negative value. The algebraic sum of two quantities with like signs is the sum of their values, with the same sign in front of that value. The algebraic sum of two or more quantities with unlike signs is the difference of their values, with the sign of the larger value in front of that value. These concepts can be best seen in the examples which follow.

Example 6-1

PROBLEM: Add +3, +2, +4.

SOLUTION: Note the numbers on the number line below. Plot the numbers given above beginning at zero and moving to the right. The number line proves the result.

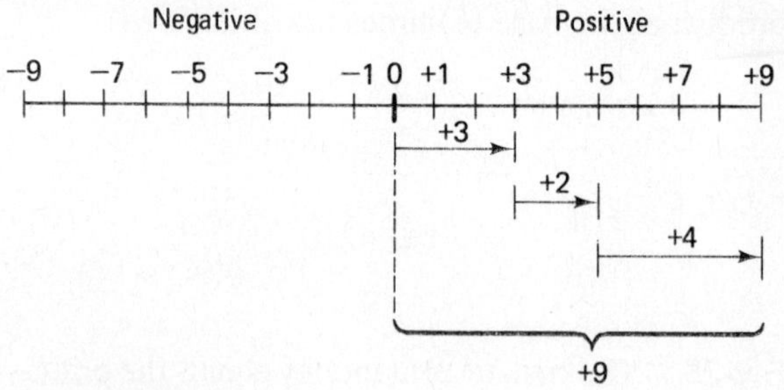

Example 6-2

PROBLEM: Find the value of +5, −4, −3.

SOLUTION: Plot the numbers on the number line in the same order in which they are given, moving to the right from zero for +5 (positive) and to the left for −4 and −3 (negative). When all the numbers have been plotted, note the distance from zero to the right or left. In this case the distance to the left from zero extends to negative 2. Thus the resulting value is −2.

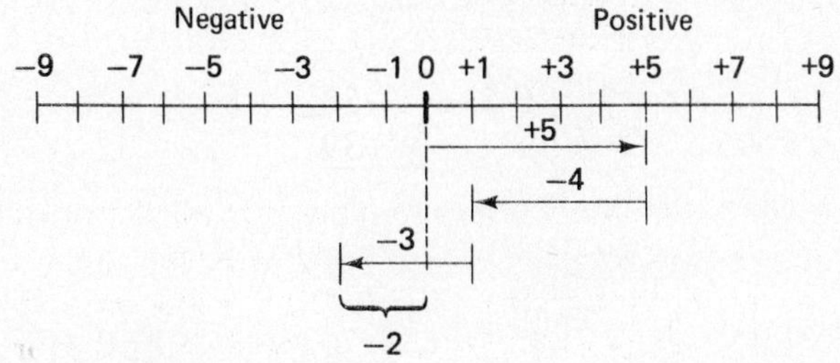

CONTENT

When terms with unlike signs are multiplied or divided, the results are negative. When terms with like signs are multiplied or divided, the results are positive.

Example 6-3

PROBLEM: If you earned \$14 each day for 5 days at a part-time job, how much money have you earned?

SOLUTION: Multiply +5 days × \$14

$$+5 \times 14 = 70$$

Thus, you have earned \$70.

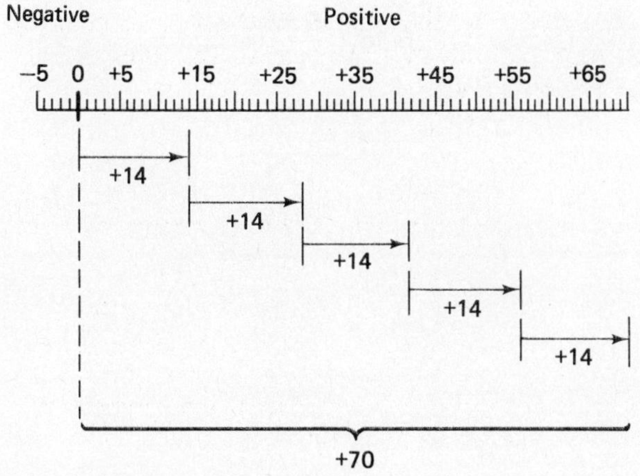

Example 6-4

PROBLEM: If you spent \$3 every day for 7 days, what would be the change in your account?

SOLUTION: Because you are spending, you are subtracting from your account; therefore, you would multiply negative 3 times 7. Think of this multiplication problem as subtracting 3 from zero 7 times.

$$-\$3 \times 7 = -\$21$$

Thus you would have −\$21 in your account. Observe the number chart.

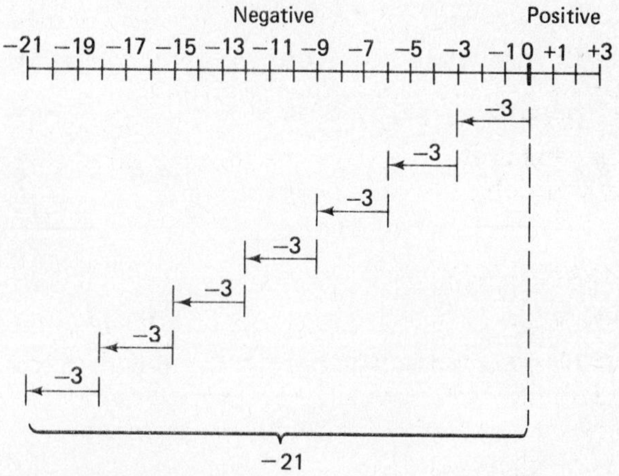

Example 6-5

PROBLEM: Multiply −2 × −3.

SOLUTION: Because subtraction is the reverse of addition, you should think of this multiplication problem as meaning the −3 is subtracted from zero 2 times. Remember that like signs result in a positive sign.

$$(-2) \times (-3) = -(-3) - (-3)$$

$$= +3 + 3$$

$$= 6$$

Observe the number chart.

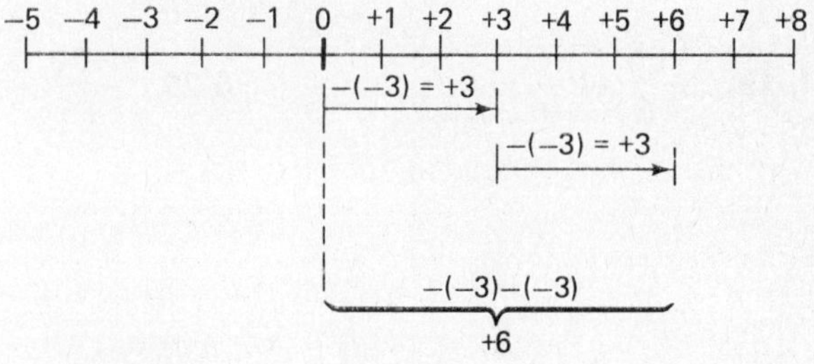

Example 6-6

PROBLEM: Divide −42 by 7.

SOLUTION: Write the problem in a fractional form and simplify.

$$\frac{-42}{7} = \frac{-6}{1} = -6$$

EXERCISE 6-1

As you work through the following problems, carefully note all of the signs in each problem.

6-1. $\begin{array}{r} +5 \\ \underline{+\ +3} \end{array}$

6-2. $\begin{array}{r} +\ 8 \\ \underline{+\ -16} \end{array}$

6-3. $\begin{array}{r} -32 \\ \underline{+\ +39} \end{array}$

6-4. $\begin{array}{r} 16 \\ \underline{+\ \ 1} \end{array}$

6-5. $\begin{array}{r} -\ 9 \\ \underline{-\ -15} \end{array}$

6-6. $\begin{array}{r} -\ 4 \\ \underline{+\ +16} \end{array}$

6-7. $-5 \times +4 =$

6-8. $-6 \times -3 =$

6-9. $+8 \times -3 =$

6-10. $+16 \times +4 =$

6-11. $-4 \times -4 \times -4 =$

6-12. $-5 + 3 - 2 + 4 =$

6-13. $9 - 5 + 6 - 1 =$

6-14. $-11 + 10 - 11 + 10 =$

6-15. $-34 + 72 =$

6-16. $-5 \times -3 \times -2 \times -1 =$

6-17. $\frac{28}{-7} =$

6-18. $\frac{-39}{13} =$

6-19. $\frac{-36}{-6} + \frac{-6}{+6} =$

6-20. $\frac{-20}{+5} + \frac{-25}{-5} =$

CONTENT

The removal of signs of grouping is an essential operation in the evaluation of an algebraic expression. When grouping signs are removed, the sign in front of the grouping determines what operation is to be performed. When there is a negative sign in front of the grouping, change all the signs within the group to the opposite of the intended operation as you remove the grouping sign. When a positive sign is in front of the grouping, do not change any of the signs within the grouping as you remove the grouping sign. When removing several sets of grouping signs, start with the innermost set of groupings and work to the outermost, making positive and negative sign changes as necessary.

After the grouping signs have been removed, perform the arithmetic operations within the group. Perform all multiplication and division operations first, moving from left to right. Then perform all addition and subtraction operations, moving from left to right. Combine only like quantities.

The *signs of relation* show the value of one term with respect to another. The equal sign (=) means "is equal to." For example, $143 = 143$. The equal sign with a slash through it ($\neq$) means "is not equal to." For example, $12 \neq 11$. An arrowhead is used to indicate that a difference exists between two quantities. For example, $5 > 4$ indicates that 5 "is greater than" 4 and $5 < 32$ indicates that 5 "is less than" 32. The arrowhead always points to the smaller quantity.

Example 6-7

PROBLEM: Evaluate $3(16 \div 2) - 6 =$

SOLUTION: First do the operation inside the parentheses.

$$3(16 \div 2) - 6 =$$

$$3\left(\frac{\overset{8}{\cancel{16}}}{\underset{1}{\cancel{2}}}\right) - 6 =$$

$$3(8) - 6 =$$

Second, complete the multiplication indicated and then add the negative 6.

$$3(8) - 6 =$$

$$24 - 6 = 18$$

Example 6-8

PROBLEM: $5 + \{3x - [2 - (5 + 5x)]\} =$

SOLUTION: First, remove the innermost parentheses and change every sign within the parentheses as indicated by the negative sign.

$$5 + \{3x - [2 - (5 + 5x)]\} =$$

$$5 + \{3x - [2 - 5 - 5x]\} =$$

Second, remove the brackets and change every sign within the brackets as indicated by the negative sign.

$$5 + \{3x - [2 - 5 - 5x]\} =$$

$$5 + \{3x - 2 + 5 + 5x\} =$$

Third, remove the braces and do not change any signs because the braces are preceded by a positive sign.

$$5 + \{3x - 2 + 5 + 5x\} =$$

$$5 + 3x - 2 + 5 + 5x =$$

Fourth, combine like quantities.

$$5 + 3x - 2 + 5 + 5x =$$

$$+3$$

$$+8 \qquad + 8x = 8 + 8x$$

Example 6-9

PROBLEM: Evaluate:

$$\{24 \div 6 \div 2 - (3x - 6[3 - 2x]) + 4\} =$$

SOLUTION: First, remove the innermost brackets and multiply the enclosed terms by −6.

$$\{24 \div 6 \div 2 - (3x - 6[3 - 2x]) + 4\} =$$

$$\{24 \div 6 \div 2 - (3x - 18 + 12x) + 4\} =$$

Second, remove the parentheses and change every sign within the parentheses because the parentheses are preceded by a negative sign.

$$\{24 \div 6 \div 2 - (3x - 18 + 12x) + 4\} =$$

$$\{24 \div 6 \div 2 - 3x + 18 - 12x + 4\} =$$

Third, complete the division indicated, moving from left to right.

$$\{24 \div 6 \div 2 - 3x + 18 - 12x + 4\} =$$

$$\left\{\frac{24}{6} \div 2 - 3x + 18 - 12x + 4\right\} =$$

$$\{4 \div 2 - 3x + 18 - 12x + 4\} =$$

$$\left\{\frac{4}{2} - 3x + 18 - 12x + 4\right\} =$$

$$\{2 - 3x + 18 - 12x + 4\} =$$

Fourth, remove the braces but do not change any signs because there is no sign in front of the braces. Thus the sign is understood to be positive.

$$2 - 3x + 18 - 12x + 4 =$$

Fifth, combine like quantities.

$$2 - 3x + 18 - 12x + 4 =$$

$$20$$

$$24 - 15x = 24 - 15x$$

EXERCISE 6-2

Solve the following problems. Note the signs and groupings. Perform the operations indicated as you have learned through the examples.

6-21. $3 \times 4 + 5 \times 6 =$

6-22. $8 \times 7 - 6 \times 6 =$

6-23. $(16 \div 2) - 8 =$

6-24. $60 \div (5 \times 3) =$

6-25. $60 \div 5 \times 3 =$

6-26. $8 \div 8 \times 8 + 8 =$

6-27. $(8 \div 8) \times (8 + 8) =$

6-28. $24 - (90 \div 18) + 5 =$

6-29. $390 \div 13 - 45 + 10 =$

6-30. $x + y - z + z - x + y =$

6-31. $a + b - (3a + 5b) - \{a - b\} =$

6-32. $(6a + 2b) - (4b - 3b - 2b) =$

6-33. $3k + 2 - [2j + 5 - 6j + 3k - 3] =$

6-34. $4r - 8y - (3r - 4y) + 4r - 32y =$

6-35. $\{(r + 3t - 8) - (r - 3t + 8)\} =$

6-36. $\{-3a - 2b - (c - a)\} =$

6-37. $128 \div 2 \div 2 \div 2 \div 2 =$

6-38. $4a - [-5b - (-12b + 3a)] - \{a + 3b\} =$

6-39. $7b - 4\{2b + 6k - [-2b + 6k] - 4b\} =$

6-40. $8x - [4y - (-3x + 3y) - 2x] =$

CONTENT

An *exponent*, you will recall, is a number written above and to the right of a quantity and indicates how many times the *base number* or quantity is used as a factor. Exponents are used to abbreviate algebraic expressions. For example, $k \cdot k \cdot k \cdot k = k^4$. k is the base number or base quantity, and 4 is the exponent. This is one of the laws of exponents. Some of the other laws and their examples are given below.

1. When multiplying quantities with like bases containing exponents, add the exponents.

$$(j^3)(j^2) = j^{3+2} = j^5$$

or

$$(j \cdot j \cdot j)(j \cdot j) = j \cdot j \cdot j \cdot j \cdot j = j^5$$

$$(a^m)(a^n) = a^{m+n}$$

$$(3)^3(3)^2 = 27 \times 9 = (3)^5$$

2. When dividing quantities with like bases containing exponents, subtract the exponent of the denominator from the exponent of the numerator.

$$\frac{y^5}{y^2} = \frac{\cancel{y} \cdot \cancel{y} \cdot y \cdot y \cdot y}{\cancel{y} \cdot \cancel{y}} = y^3$$

or

$$\frac{y^5}{y^2} = y^{5-2} = y^3$$

or

$$\frac{(3)^5}{(3)^2} = \frac{243}{9} = (3)^3$$

3. A quantity with an exponent of zero is equal to 1.

$$1 = \frac{1}{1} = \frac{\overset{1}{\cancel{a}} \cdot \overset{1}{\cancel{a}} \cdot \overset{1}{\cancel{a}} \cdot \overset{1}{\cancel{a}}}{\underset{1}{\cancel{a}} \cdot \underset{1}{\cancel{a}} \cdot \underset{1}{\cancel{a}} \cdot \underset{1}{\cancel{a}}} = \frac{a^4}{a^4} = a^{4-4} = a^0 = 1$$

Study the following examples. They illustrate how exponents are to be handled.

Example 6-10

PROBLEM: Multiply: $x^3 \cdot x^6$.

SOLUTION: $(x^3)(x^6) = x^{3+6} = x^9$

because

$$\overbrace{x \cdot x \cdot x}^{(x^3)} \cdot \overbrace{x \cdot x \cdot x \cdot x \cdot x \cdot x}^{(x^6)} = x^9$$

When you multiply, you add exponents of like terms.

Example 6-11

PROBLEM: Divide: $y^8 \div y^5$.

SOLUTION: $y^8 \div y^5 = y^{8-5} = y^3$

because

$$\frac{y^8}{y^5} = \frac{\cancel{y} \cdot \cancel{y} \cdot \cancel{y} \cdot \cancel{y} \cdot \cancel{y} \cdot y \cdot y \cdot y}{\underset{1}{\cancel{y}} \cdot \underset{1}{\cancel{y}} \cdot \underset{1}{\cancel{y}} \cdot \underset{1}{\cancel{y}} \cdot \underset{1}{\cancel{y}}} = y^3$$

When you divide, subtract exponents (denominator from numerator) of the like terms.

Example 6-12

PROBLEM: Simplify: $(6x^2y^3)^2$.

SOLUTION:

$$\begin{aligned}(6x^2y^3)^2 &= (6x^2y^3)(6x^2y^3)\\ &= (6)(6)x^{2+2}y^{3+3}\\ &= 36x^4y^6\end{aligned}$$

EXERCISE 6-3

As you work through the following problems, keep in mind the skills you have begun to develop in working with exponents.

6-41. $m^3 \cdot m^4 =$

6-42. $x^3 \cdot x^4 \cdot x^2 =$

6-43. $5^4 \cdot 5^1 =$

6-44. $3 \cdot 3 \cdot 3^2 =$

6-45. $a^5 \cdot a^{10} \cdot a^{15} =$

6-46. $xy \cdot xy \cdot xy =$

6-47. $x^3y^3 \cdot x^4y^4 \cdot x^1y^2 =$

6-48. $(xyz)(xyz) =$

6-49. $(xy)^3 =$

6-50. $(-4)^2(-4)^1(-4)^3 =$

6-51. $\frac{x^5}{x^4} =$

6-52. $\frac{x^3y^2}{x^2y^2} =$

6-53. $(x^3)^4 =$

6-54. $(x^3y^2)^3 =$

6-55. $-(x^1y^2z^3)^4 =$

6-56. $\left(\frac{y^2}{y^2}\right)^3 =$

6-57. $\left(\frac{2x^2}{4y^3}\right)^2 =$

6-58. $\left(\frac{3^2x^2y^3}{x^1y^2}\right)^2 =$

6-59. $-\left(\frac{12r^1s^2t^3}{6rst}\right) =$

6-60. $\left(\frac{16x^2r^4s^5}{2^4x^2s^5r^4}\right)^6 =$

CONTENT

One of the more difficult things to do in mathematics is to change a written problem into a formula or an equation. There are several reasons for this, among which are: (1) a lack of understanding of the problem and what is to be done; (2) a lack of ability to translate words and ideas into symbols; (3) a lack of comprehension of the purpose and use of the formula or equation. Mathematical operations are often expressed by words and phrases which mean the same thing. Some of these are as follows:

Addition: sum; combine; total; add.

Subtraction: find the difference; subtracted from; is how much greater; decreased by; the difference between.

Multiplication: find the product; square the quantity; cube the quantity; double the quantities.

Division: divided by; find the quotient.

As you work through the following word problems, keep in mind what the words tell you to do, how the words are translated into symbols, and how the problem is solved.

Example 6-13

PROBLEM: Translate into an equation: The sum of three numbers doubled is 38.

SOLUTION: Define the numbers:

x = the first number

y = the second number

z = the third number

Sum means to add and double means to multiply all the numbers by 2. Therefore:

$$2(x + y + z) = 38$$

Example 6-14

PROBLEM: Translate into an equation: The distance traveled depends on the time and the speed you are traveling.

SOLUTION: Define the quantities involved:

d = the distance

t = the time traveled

r = the rate of travel

Since the distance traveled depends on the hours traveled times the rate of travel, we have:

$$d = rt$$

EXERCISE 6-4

Translate the following word problems into formulas or equations. Review the previous examples if necessary.

6-61. Find the sum of two numbers minus a third number.

6-62. Subtract y from the sum of x^3 and z^4.

6-63. The product of two numbers divided by a third number equals 6.

6-64. To convert liters to quarts, multiply by 1.05.

6-65. To convert feet to inches, multiply by 12.

6-66. The perimeter of a square equals the sum of the lengths of the sides.

6-67. The area of a square equals the length of one side times the length of another side.

6-68. The perimeter of a rectangle equals the sum of the lengths of the sides.

6-69. The circumference of a circle equals the diameter times π.

6-70. Miles per gallon equals the miles driven divided by the number of gallons used.

6-71. Four times k plus 2 times p equals f.

6-72. The surface area of a cube equals 6 times the area of each face.

6-73. The sum of three consecutive numbers is 37.

6-74. Square the sum of a plus b minus c cubed.

6-75. The quotient of j and k times l equals t minus s.

6-76. The number of gallons of water in a container is equal to the product of length, width, height, and 7.5.

6-77. The hypotenuse squared of a right triangle equals the sum of the lengths of the other two sides squared.

6-78. The strength of a board depends on the product of the width and the depth squared, divided by the length.

6-79. The weight in pounds of water in a tank equals the product of the length, width, height, and 62.5.

6-80. The cost of carpeting a room depends on the area in square yards. Determine the cost if the room's dimensions are x yards by y yards.

THINK TIME

One of the symbols frequently used in mathematics is zero, 0. There are several facts about zero which are important when performing algebraic operations.

1. Zero is the point of division between positive and negative numbers. It is the only term which is neither positive nor negative and has no intrinsic value.
2. When determining the powers of ten, each additional zero to the right of the number 10 indicates an additional power of ten. For example:

   $$10 = 10^1, \quad 100 = 10^2, \quad 1000 = 10^3$$

3. The addition or subtraction of zero to a number results in the given number.
4. In division, if the dividend is zero and the divisor is not zero, the quotient is zero.
5. In division, if the dividend is not zero and the divisor is zero, the quotient is undefined.
6. If zero appears as one of the factors in any product, the product is equal to zero.
7. A factor to zero power equals the number, 1 provided the factor does not equal zero.

PROCEDURES TO REMEMBER

1. The algebraic sum of quantities with *like* signs is the sum of their values, with the same sign in front of the combined quantities.
2. The algebraic sum of quantities with *unlike* signs is the difference of their values, with the sign of the greater value in front of the combined quantities.
3. When terms with unlike signs are multiplied or divided, the results are negative.
4. When terms with like signs are multiplied or divided, the results are positive.

5. When removing a negative sign in front of a grouping, change each of the signs within the group to the opposite sign indicated.
6. When removing a positive sign in front of a grouping, do not change any signs within the group.
7. When removing several sets of grouping signs, start with the innermost set of groupings and work to the outermost, making positive and negative sign changes as necessary.
8. After grouping signs have been removed and all necessary sign changes have been made, first perform all multiplication and division operations and then perform all addition and subtraction operations, moving from left to right.
9. When multiplying quantities with like bases containing exponents, add the exponents.
10. When dividing quantities with like bases containing exponents, subtract the exponent of the denominator from the exponent of the numerator.

CHAPTER SUMMARY

1. An *algebraic expression* is an expression which uses signs and symbols to represent numbers or quantities.
2. A *numerical algebraic expression* is made up entirely of numbers.
3. A *term* is part of an expression which is separated from other terms by a plus or minus sign.
4. One or more *factors* may be combined to form a term.
5. A *numerical coefficient* refers to numbers which are part of a term.
6. A *literal coefficient* refers to letters which are part of a term.
7. A *base* is a term or factor apart from indication of power.
8. An *exponent* or *power* indicates the number of times a base, term, or factor is multiplied by itself.
9. A *monomial* is an algebraic expression which has one term.
10. A *binomial* is an algebraic expression which has two terms.
11. A *trinomial* is an algebraic expression which has three terms.
12. A *polynomial* is an algebraic expression which has two or more terms.
13. *Like* terms are quantities which have the same literal factors.
14. *Unlike* terms are quantities which have unlike literal factors.
15. *Signs of quantity* indicate whether a number is positive or negative.
16. *Parentheses, brackets, braces,* and *vincula* or *bars* are used to group terms.
17. *Signs of operation* show the value of one term with respect to another.
18. The equal sign (=) means "is equal to."
19. The equal sign with a slash through it ($\neq$) means "is not equal to."
20. An arrowhead is used to indicate that a difference exists between two quantities. For example $8>4$ indicates that 8 "is greater than" 4 and $8<9$ indicates that 8 "is less than" 9. The arrowhead always points to the smaller quantity.
21. A quantity with a zero exponent is equal to 1.
22. Addition means "combine," "total," "add," and "sum."
23. Subtraction means "find the difference," "subtracted from," "is how much greater than," "decreased by," and "the difference between."
24. Multiplication means "find the product," "square the quantity," "cube the quantity," "double the quantities."
25. Division means "divided by" and "find the quotient."

CHAPTER TEST

T-6-1. Add:
$$\begin{array}{r} -35 \\ \underline{-19} \end{array}$$

T-6-2. Add:
$$\begin{array}{r} -140 \\ 93 \\ 120 \\ \underline{-62} \end{array}$$

T-6-3. Subtract:
$$\begin{array}{r} -23 \\ \underline{+18} \end{array}$$

T-6-4. Subtract: $\begin{array}{r} -98 \\ \underline{-46} \end{array}$

T-6-5. Multiply: $(-4)(+5)(-6) =$

T-6-6. Multiply: $(-30)(-2)(-2) =$

T-6-7. Divide: $\dfrac{-111}{+3}$

T-6-8. Divide: $\dfrac{+27}{-3}$

T-6-9. Subtract: $\begin{array}{r} -66ab \\ \underline{+76ab} \end{array}$

T-6-10. Simplify: $84 + 36 \times 8 \div 4 \times 2 - 12 =$

T-6-11. Simplify: $8 - 0(6 \div 3) - 16 \div 2 =$

T-6-12. Simplify: $3x - 7 - 2(x - 5) + 15 =$

T-6-13. Simplify:
$4 + \{10 - [-6 - (-7 + 4) - 2]\} =$

T-6-14. Simplify:
$(5x - 2) - 2\{3x - [5x + 7(3x - 4) + 3]\} =$

T-6-15. Multiply: $-b^3 \cdot b^5 =$

T-6-16. Multiply: $(12x^3)(13x^2y^3) =$

T-6-17. Divide: $\dfrac{-64x^3y^5z^2}{-16x^3y^4z} =$

T-6-18. Divide: $\dfrac{45j^3k^3l^3}{-5j^3k^2l^1} =$

T-6-19. Multiply: $(-4j^3)^4 =$

Express the following as algebraic expressions:

T-6-20. The difference between x and y divided by r^3.

T-6-21. The sum of a and b minus c multiplied by x^2 divided by y^4 minus 7.

T-6-22. The product of 36 minus z times j.

T-6-23. The area of a trapezoid equals one half the sum of the bases (b_1 and b_2) times the altitude (a).

T-6-24. The perimeter of a 60°–60°–60° triangle equals three times one side (s).

T-6-25. The cost (C) to operate a car equals the cost per gallon (c) times the gallons (g) plus insurance (i) plus oil (o) plus tires (t) plus maintenance (m) plus other expenses (e).

CHAPTER

7

SIMPLE EQUATIONS AND FORMULAS

OBJECTIVES

1. To learn how to use the algebraic axioms or rules of addition, subtraction, multiplication, and division to solve simple equations.
2. To become familiar with the use of formulas to solve problems.

SELF-TEST

This test will determine whether or not you need to study this chapter. The subject matter of this test involves basic and literal equations. Solve for the unknowns and check your work.

S-7-1. $x - 5 = 3$

S-7-2. $4x + 5 = 17$

S-7-3. $x + 3 = 10$

S-7-4. $2x + 3 = x + 4$

S-7-5. $4b = 36$

S-7-6. $17x - 7x = x + 18$

S-7-7. $4x + 5 - 7 = 2x + 6$

S-7-8. $9y - 19 = y - 2y + 11$

S-7-9. $4x - 10 = 2x + 2$

S-7-10. $9x + 9 = -3x + 15$

S-7-11. $300x - 250 = 50x + 750$

S-7-12. $2.5x + .5x = 1.5x + 1.5$

S-7-13. $x + 2x + 3 - 4x = 5x - 9$

S-7-14. $2y + 3y - 4 = 5y + 6y - 16$

S-7-15. $75z - 150 = 80z - 300$

S-7-16. $(4x + 6) - 2x = (x - 6) + 24$

S-7-17. $15y - [3 - (4y + 4) - 57] = -(2 + y)$

S-7-18. $4t - (12t - 24) + 38t - 39 = 0$

S-7-19. $47r - 17 = 235 - 37r$

S-7-20. $ax + b = 7; \quad x =$

S-7-21. $x - 2a = 3x + 4a; \quad x =$

S-7-22. $3(b - y) = 5(b - 2y); \quad y =$

S-7-23. $5(x - 2b) = -3x + 2(x + 2b); \quad x =$

S-7-24. $\frac{3}{4}(3a - 2x) = \frac{2}{5}(x) + 2a; \quad x =$

S-7-25. $5.2(x + 3) + 3.7(2 - x) = 3; \quad x =$

INTRODUCTION

The world of technology uses many formulas in order to solve its problems. As technology advances, many more new formulas will be devised to solve the problems of a new age. It is absolutely necessary that a good technician have complete knowledge of the formulas or methods used to solve the problems he faces in his work. In this chapter you will be given explanations, illustrations, and the opportunity to work with formulas so that you may become skilled in their use. Many people fail to understand the connection between formulas and equations. What the technician or tradesman labels as formulas or methods to solve a problem, the mathematician defines as literal equations.

As we develop the fundamental operations of equations in the following pages, you should be aware of the fact that these are the same procedures you will be using to solve the problems which confront you in your work.

DEFINITIONS

An *equation* is a set of two algebraic expressions which are equal. For example, $ab = 5$ has two algebraic expressions, ab and 5, which are equal. A *formula* is a rule or law used in mathematics, science, or industry to solve specific problems. The formula is generally composed of letters, symbols, and constant terms. *Literal quantities* are letters or symbols which represent numbers or unknown quantities.

CONTENT

Historically, the symbol for justice and equality has been the balance scale, often called the "scales of justice." Perhaps the scales of justice can serve as an aid to understanding equations. The balance point, A, is comparable to the equal sign, =, of an equation. In order to keep the scales balanced, if you add, subtract, multiply, or divide on one side of the scale, you must do the same on the other side of the scale. Failure to do this will result in an imbalance.

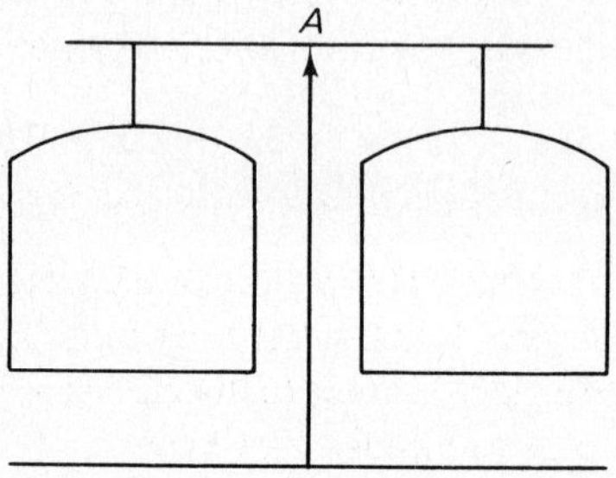

Look at the illustration below. Note that in each equation what is stated on the left side is equal to what is stated on the right side.

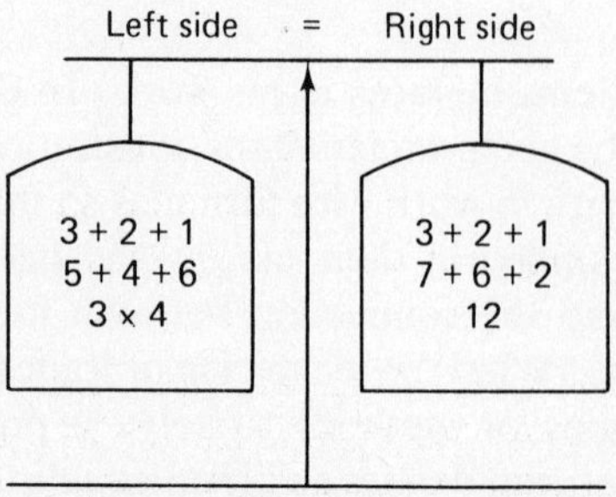

The quantities on the left side of the equation are called the *left member*. Those on the right side are known as the *right member*. The equal sign in the center is the *sign of equality* or the balance point of the equation. Remember, whatever mathematical operation you perform on one side of the equation, you must also perform the same operation on the other side. Failure to practice this principle will result in an "injustice." On the job, the incorrect use of an equation may result in an injustice to yourself or to your customer.

The procedures used in solving equations are the same as those used in solving formulas. However, it is easier to learn how to solve equations than to learn how to solve formulas. Using the skills you will learn in solving equations will be of assistance to you in working with formulas.

The four basic operations used in solving an equation are the same as the fundamental operations of arithmetic: addition, subtraction, multiplication, and division. However, in algebra they are called the *axioms* or *rules* of algebra.

Study each of the four diagrams shown below. Note how the balance scale remains balanced as weights are added, subtracted, multiplied, or divided in equal amounts.

Remember, you should always check your solutions to be certain that you have the correct balance or that both sides of the equation are equal. The solution or the value of the unknown number is called the *root* of the equation. When the root becomes known, the problem has been solved or the

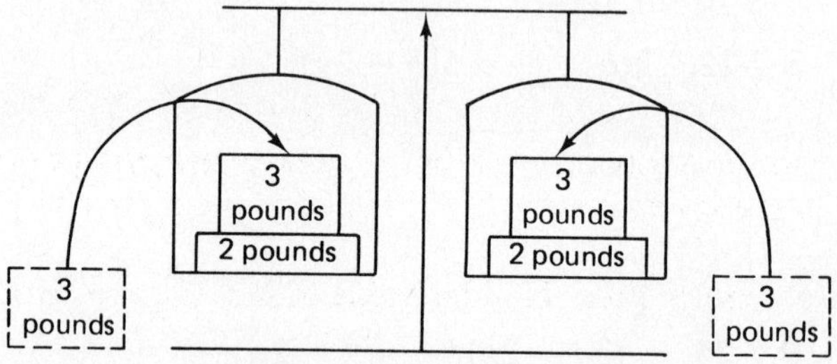

Add equal weight to each side.

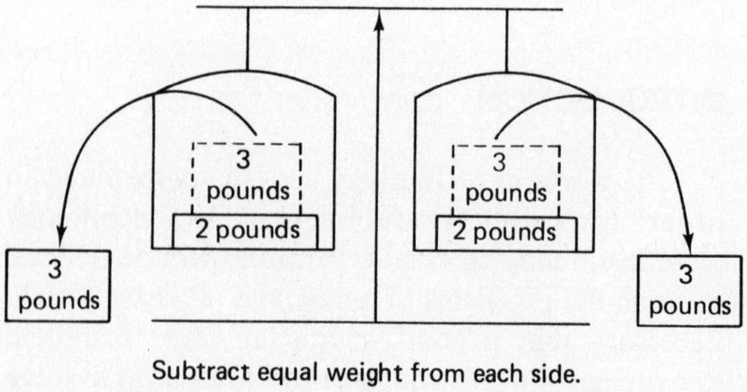

Subtract equal weight from each side.

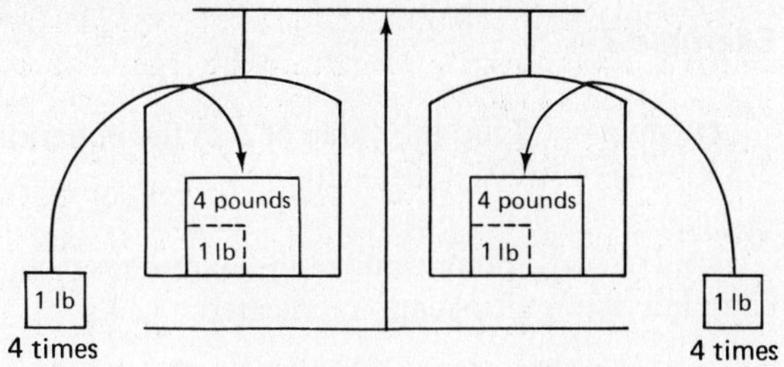

Multiply the weight on each side by an equal quantity.

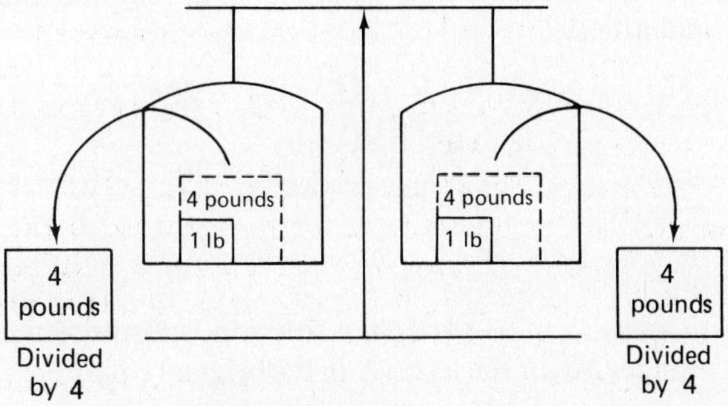

Divide the weight on each side by an equal quantity.

solution set has been found. The wise technician or tradesman always checks his solutions to prevent costly mistakes.

The following examples illustrate the basic axioms of algebra: addition, subtraction, multiplication, and division. Carefully note the use of two different ways to find the root of the equation.

Example 7-1

Addition Axiom: What is added to the left side of the equation must also be added to the right side.

PROBLEM: Find the value of X in the equation $X - 3 = 2$.

SOLUTION A: You can solve this equation by *adding* 3 to each side of the equation.

$$\begin{array}{rcr} X - 3 &=& 2 \\ +\,3 &=& +\,3 \\ \hline X &=& 5 \end{array}$$

CHECK: To verify the solution, substitute the number 5 for the letter X in the original equation.

$$X - 3 = 2$$
$$5 - 3 = 2$$
$$2 = 2$$

SOLUTION B: You could also solve this problem through *transposition*. Through this procedure, a quantity is moved from one side of the equation to the other. In the transposition process, positive quantities become negative quantities, and vice versa. Thus you could solve this equation by moving the -3 to the other side of the equal sign and changing it to $+3$. Observe:

$$X - 3 = 2$$
$$X = 2 + 3$$
$$X = 5$$

Example 7-2

Subtraction Axiom: What is subtracted from the left side of the equation must also be subtracted from the right side.

PROBLEM: Find the value of X in the equation $X + 4 = 6$.

SOLUTION A: You can solve this equation by *subtracting* 4 from each side of the equation.

$$\begin{array}{rcr} X + 4 &=& 6 \\ -\,4 &=& -4 \\ \hline X &=& 2 \end{array}$$

CHECK: To verify the solution, substitute the number 2 for the letter X in the original equation.

$$X + 4 = 6$$
$$2 + 4 = 6$$

SOLUTION B: This problem may also be solved by the process of transposition, by moving the +4 to the opposite side of the equal sign and changing its sign from plus to minus. Observe:

$$X + 4 = 6$$
$$X = 6 - 4$$
$$X = 2$$

Example 7-3

PROBLEM: Find the value of X in the equation $8X - 5 = 7X + 19$.

SOLUTION A: You can solve this equation by first *subtracting* $7X$ from each side of the equation and then *adding* 5 to each side of the equation.

$$\begin{aligned} 8X - 5 &= 7X + 19 \\ -7X \quad &= -7X \\ \hline X - 5 &= +19 \end{aligned}$$

$$\begin{aligned} X - 5 &= 19 \\ +5 &= +5 \\ \hline X \quad &= 24 \end{aligned}$$

CHECK: To verify the solution, substitute the number 24 for the letter X in the original equation.

$$\begin{aligned} 8X - 5 &= 7X + 19 \\ 8(24) - 5 &= 7(24) + 19 \\ 192 - 5 &= 168 + 19 \\ 187 &= 187 \end{aligned}$$

SOLUTION B: This problem may also be solved by the process of transposition. Thus, the $+7X$ is moved to the opposite side of the equal sign and its sign is changed from plus to minus. Then the -5 is moved to the opposite side of the equal sign and its sign is changed from minus to plus. Observe:

$$\begin{aligned} 8X - 5 &= 7X + 19 \\ 8X - 7X - 5 &= +19 \\ \underbrace{8X - 7X} &= \underbrace{19 + 5} \\ X \quad &= \quad 24 \end{aligned}$$

EXERCISE 7-1

Solve the following equations. Check your solutions by substituting your answer for the unknown into the original equation. Use either the axiom method of addition or subtraction or the transposition method. Consult the examples if you have problems. Remember the scales of justice and the need to keep both sides of the equation in balance.

7-1. $X + 3 = 5$

7-2. $X - 6 = 10$

7-3. $3X - 4X = -2X + 10$

7-4. $8y - 12 = 7y - 11$

7-5. $X - 7 = 13$

7-6. $3 + a = 10$

7-7. $k - 12 = 3$

7-8. $r - 4 = -18$

7-9. $X + 14 = 8$

Example 7-4

PROBLEM: Find the value of x in the equation $3(3X - 4) = 4(2X + 6) - 20$.

SOLUTION: First, remove the parentheses and complete the multiplication indicated.

$$\begin{aligned} 3(3X - 4) &= 4(2X + 6) - 20 \\ 9X - 12 &= 8X + 24 - 20 \end{aligned}$$

Second, move the -12 and the $+8X$ to the opposite sides of the equation and combine the quantities.

$$\begin{aligned} 9X - 12 &= 8X + 24 - 20 \\ \underbrace{9X - 8X} &= \underbrace{12 + 24 - 20} \\ X \quad &= \quad 16 \end{aligned}$$

CHECK: To verify the solution, substitute the number 16 for the letter X in the original equation.

$$\begin{aligned} 3(3X - 4) &= 4(2X + 6) - 20 \\ 3(3(16) - 4) &= 4(2(16) + 6) - 20 \\ 3(48 - 4) &= 4(32 + 6) - 20 \\ 3(44) &= 4(38) - 20 \\ 132 &= 152 - 20 \\ 132 &= 132 \end{aligned}$$

7-10. $2(X + 10) = X + 30$

7-11. $3(X - 1) + 2X = 2(3X + 4)$

7-12. $4(3 - 2X) = 7(5 - X)$

7-13. $3(3X + 3) = 4(2X - 4)$

7-14. $3(X - 8) = 2(X - 14)$

CONTENT

As you solve equations, you will generally want to know the value of *one* unit of the unknown. If the unknown is greater than one unit, for example $5X$, the unknown should be divided so that it becomes one unit. However, you must remember to divide both sides of the equation by the same number.

Example 7-5

Division Axiom: Each side of the equation must be divided by the same number or value.

PROBLEM: Find the value of X in the equation $5X = 15$.

SOLUTION: Divide both sides of the equation by 5.

$$5X = 15$$

$$\frac{\overset{1}{\cancel{5}}X}{\underset{1}{\cancel{5}}} = \frac{\overset{3}{\cancel{15}}}{\underset{1}{\cancel{5}}}$$

$$X = 3$$

CHECK: Verify by substituting $X = 3$ into the original equation.

$$5X = 15$$

$$5(3) = 15$$

$$15 = 15$$

CONTENT

If the unit value of the unknown is less than one, or a fractional value, multiply *both* sides of the equation by the reciprocal of that fraction. A reciprocal is a number which, when multiplied by another number, will equal one or unity.

Example 7-6

Multiplication Axiom: Each side of the equation must be multiplied by the same number or value.

PROBLEM: Find the value of X in the equation $\frac{2}{3}X = 12$.

SOLUTION: Multiply both sides of the equation by $\frac{3}{2}$, which is the reciprocal of $\frac{2}{3}$.

$$\frac{2}{3}X = 12$$

$$\left(\frac{\overset{1}{\cancel{3}}}{\underset{1}{\cancel{2}}}\right)\frac{\overset{1}{\cancel{2}}}{\underset{1}{\cancel{3}}}X = \overset{6}{\cancel{12}}\left(\frac{3}{\underset{1}{\cancel{2}}}\right)$$

$$X = 18$$

CHECK: Verify by substituting $X = 18$ into the original equation.

$$\frac{2}{3}X = 12$$

$$\frac{2}{\underset{1}{\cancel{3}}}(\overset{6}{\cancel{18}}) = 12$$

$$12 = 12$$

Study the following examples which use the algebra axioms.

Example 7-7

PROBLEM: Find the value of X in the equation

$$8X - 5(4X + 3) = (-3) - 4(2X - 7)$$

SOLUTION: First, remove parentheses.

$$8X - 20X - 15 = (-3) - 8X + 28$$

Second, combine like quantities as indicated by the grouping symbols.

$$8X - 20X - 15 = -3 - 8X + 28$$

$$-12X - 15 = -8X + 25$$

Third, transpose quantities as necessary to opposite sides of the equal sign.

$$-12X - 15 = -8X + 25$$

$$-25 - 15 = -8X + 12X$$

Fourth, combine like quantities as indicated by the grouping symbols.

$$-25 - 15 = -8X + 12X$$

$$-40 = 4X$$

Fifth, divide both sides by 4.

$$-40 = 4X$$

$$\frac{\overset{-10}{-\cancel{40}}}{\underset{1}{\cancel{4}}} = \frac{\overset{1}{\cancel{4}X}}{\underset{1}{\cancel{4}}}$$

$$-10 = X$$

CHECK: Verify by substituting $X = -10$ into the original equation.

$$8X - 5(4X + 3) = (-3) - 4(2X - 7)$$

$$8(-10) - 5(4(-10) + 3) = (-3) - 4(2(-10) - 7)$$

$$-80 - 5(-40 + 3) = (-3) - 4(-20 - 7)$$

$$-80 - 5(-37) = (-3) - 4(-27)$$

$$-80 + 185 = -3 + 108$$

$$105 = 105$$

Example 7-8

PROBLEM: Find the value of k in the equation

$$4k + 3[2k - 4(k - 2)] = 72 - 6k$$

SOLUTION: First, remove inside parentheses.

$$4k + 3[2k - 4k + 8] = 72 - 6k$$

Second, combine like quantities.

$$4k + 3[-2k + 8] = 72 - 6k$$

Third, remove outside parentheses.

$$4k - 6k + 24 = 72 - 6k$$

Fourth, combine like quantities.

$$-2k + 24 = 72 - 6k$$

Fifth, transpose quantities as necessary to opposite sides of the equation.

$$-2k + 6k = 72 - 24$$

Sixth, combine like quantities.

$$4k = 48$$

Seventh, divide both sides by +4.

$$\frac{\overset{1}{\cancel{4}k}}{\underset{1}{\cancel{4}}} = \frac{\overset{12}{\cancel{48}}}{\underset{1}{\cancel{4}}}$$

$$k = 12$$

CHECK: Verify by substituting $k = 12$ in the original equation.

$$4k + 3[2k - 4(k - 2)] = 72 - 6k$$

$$4(12) + 3[2(12) - 4(12-2)] = 72 - 6(12)$$

$$48 + 3[24 - 4(10)] = 72 - 72$$

$$48 + 3[24 - 40] = 0$$

$$48 + 3[-16] = 0$$

$$48 - 48 = 0$$

$$0 = 0$$

Example 7-9

PROBLEM: Find the value of y in the equation

$$16.5 - 1.5(2y - .5) - 15.6 + 2.1(y + .3) = .03$$

SOLUTION: First, remove parentheses.

$$16.5 - 3.0y + .75 - 15.6 + 2.1y + .63 = .03$$

Second, combine like quantities.

$$-.9y + 2.28 = .03$$

Third, transpose y quantities to one side of the equal sign.

$$-.9y = .03 - 2.28$$

Fourth, combine like quantities.

$$-.9y = -2.25$$

Fifth, divide both sides by −.9.

$$\frac{\overset{1}{\cancel{-.9}}y}{\underset{1}{\cancel{-.9}}} = \frac{\overset{2.5}{\cancel{-2.25}}}{\underset{1}{\cancel{-.9}}}$$

$$y = 2.5$$

CHECK: Verify by substituting $y = 2.5$ into the original equation

$$16.5 - 1.5(2y - .5) - 15.6 + 2.1(y + .3) = .3$$

$$16.5 - 1.5(2(2.5) - .5) - 15.6 + 2.1(2.5 + .3) = .03$$

$$16.5 - 1.5(5 - .5) - 15.6 + 2.1(2.8) = .03$$

$$16.5 - 1.5(4.5) - 15.6 + 5.88 = .03$$

$$16.5 - 6.75 - 15.6 + 5.88 = .03$$

$$.03 = .03$$

Some of the steps may be combined when you become more proficient at solving equations. However, while you are learning, it may be a good idea to write down each operation you perform.

EXERCISE 7-2

Solve for the unknown quantity and check your results in the original equation. The following problems involve the algebra axioms of addition, subtraction, multiplication, and division. Remember the scales of justice.

7-15. $3a - 9 = 9$

7-16. $5L + 3 = 2 + 4L$

7-17. $\frac{2}{3}X - 9 = \frac{1}{3}X$

7-18. $k - 10 = 5 + 4k$

7-19. $2t + 8 = t + 13$

7-20. $4d + 15 = 6d + 5$

7-21. $12R - 18 + 42 = 10(R + 1)$

7-22. $4X + 8 = 2(4X + 2)$

7-23. $11X + 3 - 4X = 16 - 2X + 2$

7-24. $18 - 5(3 - 2z) = z - 4(z + 9)$

7-25. $4 + 3(J - 7) = 16 + 2(J + 1) + 3J$

7-26. $4(a - 5) - 3(a - 2) = 2(a - 1)$

7-27. $(X + 1)^2 = X^2 + 9$

7-28. $5X - (3X - 2) = 10$

7-29. $(a + 1)(a - 2) = a^2 + 5$

7-30. $\frac{1}{5}n + 7 = 13$

7-31. $\frac{1}{3}X + \frac{1}{4}X = \frac{7}{2}$

7-32. $10(y - 2) - 10(2 - y) = 4y - 40$

7-33. $(b + 4)(5 - 2b) - b(10 - 2b) = -6$

7-34. $(a + 5)(a - 4) + 4a^2$
$= (5a + 3)(a - 4) + 2(a - 4) + 64$

Are you sure you checked all your solutions? If not, you know what to do. CHECK THEM!

DEFINITIONS

A *literal equation* is an equation in which some or all of the quantities are represented by letters instead of numbers. Letters from the beginning of the alphabet usually represent known quantities, and those toward the end of the alphabet represent unknown quantities. When you are solving an equation for x, y, or z, you are generally attempting to solve a literal equation.

CONTENT

The same procedures are used in solving literal equations that are used in solving other equations. Furthermore, the same procedures are used to solve formulas that are used to solve basic equations and literal equations, namely, the algebra axioms of addition, subtraction, multiplication, and division. Letters in a literal equation are handled just as if they were numbers in a regular equation. Perform the same operations with literal equations that you performed with the basic equations.

After you have learned how to work with literal equations, you will be better able to work with the formulas of the technician and tradesman. This section will help you gain the knowledge you need to work with the formulas.

You will avoid difficulty with this section if you simply remember that the early letters in the alphabet represent numbers and should be treated as if they were numbers. Study the following examples with this in mind.

Example 7-10

PROBLEM: Find the value of x in the equation $ax = b$.

SOLUTION: Divide both sides of the equation by a.

$$\frac{\overset{1}{\cancel{a}}x}{\underset{1}{\cancel{a}}} = \frac{b}{a}$$

$$x = \frac{b}{a}$$

CHECK: Verify by substituting $x = \frac{b}{a}$ into the original equation.

$$ax = b$$

$$\overset{1}{\cancel{a}}\left(\frac{b}{\underset{1}{\cancel{a}}}\right) = b$$

$$b = b$$

Example 7-11

PROBLEM: Find the value of x in the equation $b(x + 1) = c$.

SOLUTION: First, remove parentheses.

$$bx + b = c$$

Second, transpose the $+b$ so that the unknown x will remain on one side of the equal sign.

$$bx + b = c$$

$$bx = c - b$$

Third, divide both sides by b.

$$bx = c - b$$

$$\frac{\overset{1}{\cancel{b}}x}{\underset{1}{\cancel{b}}} = \frac{c - b}{b}$$

$$x = \frac{c - b}{b}$$

CHECK: Verify your solution by substituting $x = \frac{c - b}{b}$ into the original equation.

$$b(x + 1) = c$$

$$b\left(\frac{c - b}{b} + 1\right) = c$$

Perform operations inside the parentheses first. Note that the number 1 is changed to another form of unity, $\frac{b}{b}$, which also equals 1.

$$b\left(\frac{c - b}{b} + \frac{(1)b}{b}\right) = c$$

$$b\left(\frac{c - b + b}{b}\right) = c$$

$$\overset{1}{\cancel{b}}\left(\frac{c}{\underset{1}{\cancel{b}}}\right) = c$$

$$c = c$$

Example 7-12

PROBLEM: Find the value of y in the following equation:

$$9y - 12k = 3y + 9k - y$$

SOLUTION: First, combine like quantities.

$$9y - 12k = 2y + 9k$$

Second, transpose the unknown quantity to one side of the equation.

$$9y - 2y = 9k + 12k$$

Third, combine like quantities.

$$7y = 21k$$

Fourth, divide both sides by 7.

$$7y = 21k$$

$$\frac{\overset{1}{\cancel{7}}y}{\underset{1}{\cancel{7}}} = \frac{\overset{3}{\cancel{21}}k}{\underset{1}{\cancel{7}}}$$

$$y = 3k$$

CHECK: Verify by substituting $y = 3k$ into the original equation.

$$9y - 12k = 3y + 9k - y$$

$$9(3k) - 12k = 3(3k) + 9k - 3k$$

$$27k - 12k = 9k + 9k - 3k$$

$$15k = 15k$$

EXERCISE 7-3

Solve and check the following exercises for the unknowns. Remember to treat the beginning letters of the alphabet as if they were known numbers and the latter letters of the alphabet as unknowns.

7-35. $6abx = 9abc$

7-36. $k - nx = j$

7-37. $3az = 5az + 2$

7-38. $7x + 8a - 22a = 0$

7-39. $5a^3x = 30$

7-40. $4w = 4a \times 8b$

7-41. $5(p - x) = 3(p - x)$

7-42. $20a = 14a - (x - 2a)$

7-43. $\frac{2w}{3} - 4 = 20$

7-44. $7x + 8a - 3x = 3a + 2x$

7-45. $\frac{N}{7} + 2 = 8$

7-46. $ax + b - c = 0$

7-47. $8y + (d - 2y) = 4(d - y)$

7-48. $\frac{5}{6}(3a - 2x) - \frac{3}{4}x = a$

7-49. $(2x - b)(2x + b) + b(b - 8) = 4x(x - 5) - 48b$

THINK TIME

Make a test of formulas used in your vocation. Practice using some of the simple formulas to solve current or potential problems. Can you think of some areas of your vocation for which formulas might be of value but for which there are no formulas presently? If you have an opportunity, discuss some of these areas that could use formulas with a teacher or friend. You may find that there are some formulas which could be used to solve problems in these areas.

PROCEDURES TO REMEMBER

1. Solutions of equations by addition and subtraction.
   (a) Remove parentheses where necessary, starting with the innermost set of parentheses.
   (b) Combine like quantities on each side of the equation.
   (c) Using either the axiom method of addition and subtraction or the transposition method, move the unknown quantities to one side of the equation and the known quantities to the other side of the equation.
   (d) Combine like quantities.
   (e) Solve for the unknown.
   (f) Check by substituting your answer for the unknown into the original equation.
2. Solutions of equations by multiplication and division.
   (a) Remove parentheses, starting with the innermost set of parentheses.
   (b) Combine like quantities; starting from the left, perform multiplication, division, subtraction, and addition in that order as necessary.
   (c) Transpose quantities as necessary to opposite sides of the equation.
   (d) Combine like quantities.
   (e) Solve for the unknown by either the multiplication or division axiom, when necessary.
   (f) Check your solution by substituting the value of the unknown into the original equation.
3. Solving literal or simple equations.
   (a) Handle all literal factors as if they were numbers.
   (b) Remove parentheses, starting from the innermost set of parentheses.
   (c) Combine like quantities; starting from the left, perform multiplication, division, subtraction, and addition in that order.
   (d) Move the unknown quantity or quantities to one side of the equal sign. This is called transposition.
   (e) Combine like quantities.
   (f) Solve for the unknown quantity using either the multiplication or division axiom as necessary.
   (g) Make certain the unknown is a positive value. This may be done by multiplying each side of the equation by -1 or making an interchange of the left and right members.
   (h) Check your solution by substituting what you have discovered to be the value of the unknown into the original equation.

CHAPTER SUMMARY

1. An *equation* is defined as two algebraic expressions which are equal.
2. A *formula* is a rule, or law, generally referring to some scientific relationship expressed as an equation by letters, symbols, and constant terms.
3. *Literal quantity* means letters or symbols representing numbers.
4. The *left side* of an equation is that portion of the equation to the left of the equal sign, and the *right side* is the portion to the right of the equal sign.
5. *Addition axiom*: What is added to the left side of the equation must also be added to the right side.
6. *Subtraction axiom*: What is subtracted from the left side of the equation must also be subtracted from the right side.
7. *Multiplication axiom*: Each side of the equation must be multiplied by the *same* number or value.
8. *Division axiom*: Each side of the equation must be divided by the *same* number or value.
9. The solution or the value of the unknown that satisfies an equation is defined as the *root* of the equation; in modern mathematics, it is known as the *solution set.*
10. A *literal equation* is an equation in which some or all of the known quantities are represented by letters instead of numbers; frequently both letters and numbers are included.

11. *Transposition* is a process through which a quantity is moved from one side of the equation to the other. In the transposition process, positive quantities become negative quantities and negative quantities become positive quantities.

CHAPTER TEST

Solve for the unknowns and check.

T-7-1. $x - 7 = 17$

T-7-2. $2x - 3 = 13$

T-7-3. $x + 1 = -8$

T-7-4. $3k + 15 = 5k + 11$

T-7-5. $\frac{J}{5} = -20$

T-7-6. $8x = -72 + 168$

T-7-7. $4x - 6 = 5x - 10$

T-7-8. $8y = 35 + y$

T-7-9. $4z - 2(3 + z) = z - 4$

T-7-10. $6n - 8 = 2(2n - 1)$

T-7-11. $y + 3(y - 2) = 2(y + 4)$

T-7-12. $3(x - 4) = 2(x - 2)$

T-7-13. $3x - 2 = 5x + 7$

T-7-14. $\frac{1}{2}x + 8 = 15$

T-7-15. $9 + \frac{x}{8} = 11$

T-7-16. $\frac{2}{3}y + \frac{1}{6} = \frac{1}{2}(y - 8 + 5)$

T-7-17. $z + b = 5$

T-7-18. $4.3y + 2.5 = -.7y + 22.5$

T-7-19. $y - 2a = 3y + 4a$

T-7-20. $2[(y - 3) - 3(5 - 3y)] = 4(3 + y)$

T-7-21. $3(m - 10) - 2(45 - m) = 0$

T-7-22. $\frac{1}{4x} = 16$

T-7-23. $5(a - x) = 3(a - 3x)$

T-7-24. $8(x - 3) - 5(3x + 1) - 6 = 0$

T-7-25. $3(2y - 4) + 2(y - 3) = 5(y + 3)$

CHAPTER

8

FORMULA EVALUATION

OBJECTIVE

To evaluate formulas.

SELF-TEST

This test will determine whether or not you need to study this chapter. The subject matter of this test is the evaluation of formulas. Evaluate the following formulas. Use the given values to solve for the value of the unknown. Be sure to show your work.

S-8-1. $C = fgh;\ f = 8,\ g = 9,\ h = 10$

S-8-2. $y = f^2 + 3g;\ f = 8,\ g = 9,\ h = 10$

S-8-3. $m = \frac{h}{3} + \sqrt{g};\ h = 10,\ g = 9$

S-8-4. $p = \frac{(fg)^2}{h};\ f = 8,\ g = 9,\ h = 10$

S-8-5. $K = f^2 + g^2 + h^2 - fg - gh;$
$f = 8,\ g = 9,\ h = 10$

S-8-6. $V = \frac{abh}{2};\ a = 2,\ b = 7.5,\ h = 15\frac{1}{4}$

S-8-7. $V = \pi r^2 h$; $\pi = 3.14$, $r = 16$, $h = 5$

S-8-8. $A = \dfrac{(B + b)a}{2}$; $B = 10$, $b = 32$, $a = 28$

S-8-9. $c = \sqrt{a^2 + b^2}$; $a = 10$, $b = 24$

S-8-10. $a = \sqrt{c^2 - b^2}$; $c = 52$, $a = 48$

S-8-11. $H = \left(\dfrac{F + 1}{1.5}\right)$; $F = 14$

S-8-12. $D = \sqrt{\dfrac{HP}{6}}$; $HP = 54$

S-8-13. $B = \sqrt{(h + a)(h - a)}$; $h = 30$, $a = 18$

S-8-14. $C = \dfrac{5(F - 32)}{9}$; $F = 95$

S-8-15. $P = .44d^2Kn$; $d = \dfrac{1}{2}$, $K = 100$, $n = 3$

S-8-16. $L = \dfrac{EFD}{2(D - d)}$;

$E = .60$, $F = 130$, $D = 6$, $d = 4$

S-8-17. $I = prt$;
$p = \$8000$, $r = 8\%$, $t = 5\frac{1}{2}$ years

S-8-18. $A = .7854d^2$; $d = .875$

S-8-19. $V = \dfrac{4}{3}\pi r^3$; $\pi = \dfrac{22}{7}$, $r = 7$

S-8-20. $A = \sqrt{s(s - a)(s - b)(s - c)}$;

$s = \dfrac{1}{2}(a + b + c)$, $a = 13$, $b = 11$, $c = 16$

INTRODUCTION

Perhaps like other students you wonder, "How will I ever use algebra once I finish school?" Strange as it may seem, algebra is a frequent tool of men and women in industry. In fact the algebraic formulas they use can often be traced back to some of their early experiences with arithmetic.

Consider, for example, the problem Joel faced. As an employee of the electric company, he was given the task of building a fence around an electric power substation. To determine how much fencing he would need, he used an algebraic formula he had learned as a child. The formula was $P = 2l + 2w$. In other words, the distance around the substation, or the perimeter (P), is equal to 2 times the length ($2l$) plus 2 times the width ($2w$). Knowing this formula saved Joel time, for he did not need to make all of the measurements. By measuring one length and one width, he could let the formula do the rest of the work for him.

Formulas save time and money. They are essential to the world of work. Formulas are algebra.

DEFINITIONS

In Chapter 7, one definition of a formula was given. Here is another way it may be defined. A *formula* uses mathematical language to express the relationship between two or more variables. For example, the formula for determining the number of miles a car will travel on 1 gallon of gasoline is:

$$\text{Miles per gallon} = \frac{\text{miles traveled}}{\text{gallons of gasoline used}}$$

$$\text{Miles per gallon} = \frac{m}{g}$$

CONTENT

Evaluating a formula is not a difficult task. It is important that you have the formula in the proper form and follow the given procedures.

Most students who have difficulty with mathematics in general and formula evaluation in particular have this difficulty because they fail to read the problem correctly. Consequently, they are not able to perform the designated operations properly. This is evidenced by the skipping of essential steps, failure to show the work process, and a lack of understanding of what is to be accomplished.

As you proceed through the following pages, carefully study the formulas given in the examples, noting particularly the steps involved in reaching a solution. As you are given an opportunity to work with formulas, follow the same steps, show your work process, and check back when necessary to the earlier examples.

The following are examples of formulas which use the operations of addition and subtraction.

Example 8-1

Formula: The formula for the perimeter (distance around the outside edge) of a triangle is $P = a + b + c$. This means that the perimeter of a triangle is equal to the length of side a plus the length of side b plus the length of side c.

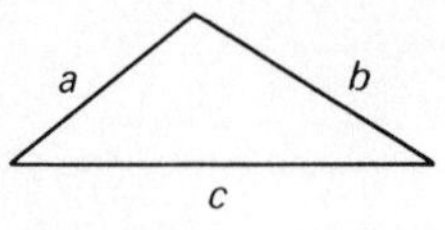

PROBLEM: What is the perimeter of a triangle when $a = 5$, $b = 6$, and $c = 9$?

SOLUTION:
$$\begin{aligned} P &= a + b + c \\ &= 5 + 6 + 9 \\ &= 20 \end{aligned}$$

We call this process the *evaluation* of a formula.

Example 8-2

PROBLEM: The formula for a certain kind of friction is $K = x + y - z$. Evaluate the formula if $x = 3$, $y = 7$, and $z = 4$.

SOLUTION:
$$\begin{aligned} K &= x + y - z \\ &= 3 + 7 - 4 \\ &= 6 \end{aligned}$$

EXERCISE 8-1

Evaluate the following formulas using addition and subtraction as necessary.

8-1. Find the perimeter of the triangle.

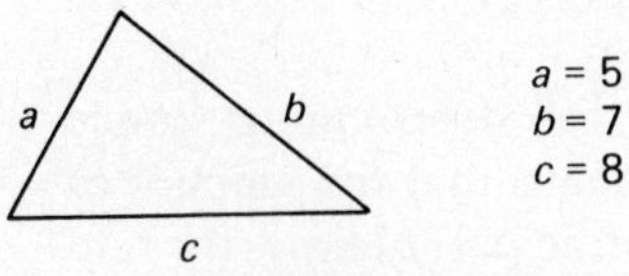

$P = a + b + c$

8-2. Find the perimeter of the trapezoid.

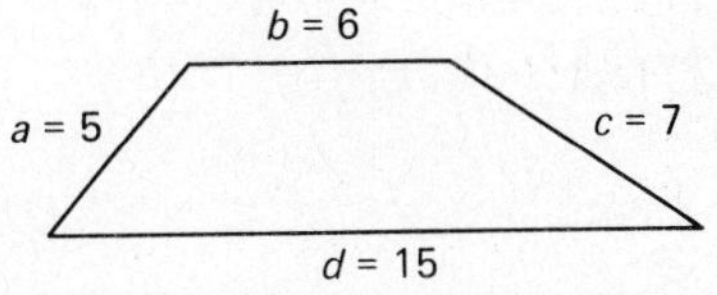

$P = a + b + c + d$

8-3. Find the perimeter of a 6-inch square.

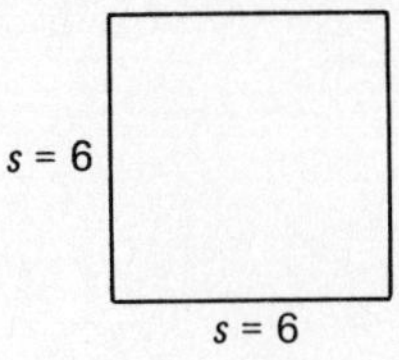

$P = 4s$

8-4. Find the missing dimension x.

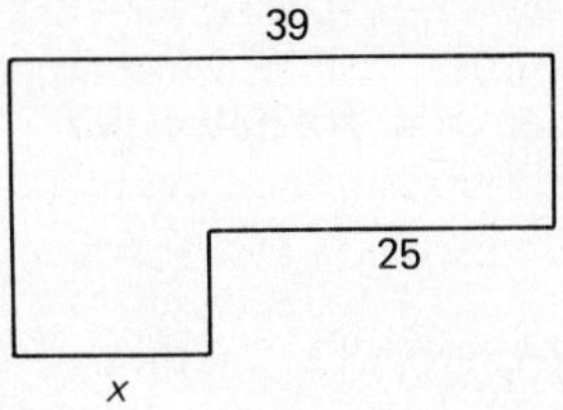

8-5. Find the missing dimension y.

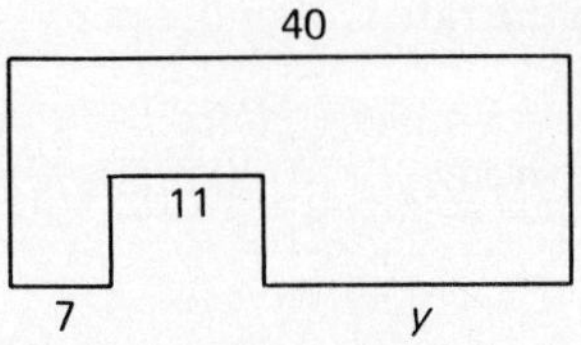

8-6. Using the formula for a certain kind of friction, $K = x + y - z$, find the value of y. $K = 20$, $x = 5$, and $z = 4$.

CONTENT

The formulas you used in the previous section made use of addition and subtraction. The formulas in this section will make use of multiplication.

Example 8-3

Formula: The formula for determining the area of a rectangle is $A = lw$. This means that the area (A) of a rectangle is equal to the length (l) of the rectangle times the width (s) of the rectangle.

PROBLEM: Evaluate when the length (l) is 10 feet and the width (w) is 8 feet.

SOLUTION: $A = lw$

$= 10 \text{ feet} \times 8 \text{ feet}$

$= 80$ square feet or 80 feet^2

Example 8-4

Formula: The formula for determining the area of a square is $A = s^2$. This means that the area (A) of a square is equal to the length (s) of one side squared (^{2}).

PROBLEM: Evaluate when $s = 4$ feet.

SOLUTION: $A = s^2$

$= (4 \text{ feet})(4 \text{ feet})$

$= 16$ square feet

Example 8-5

Formula: The formula for determining the circumference of (distance around) a circle is $C = d\pi$. This means that the circumference (C) is equal to the diameter (d) times $\frac{22}{7}(\pi)$.

PROBLEM: Evaluate when $d = 7$ feet.

SOLUTION: $C = d\pi$

$= (7 \text{ feet})\left(\frac{22}{7}\right)$

$= 22$ feet

Example 8-6

Formula: The formula for determining distance is $d = rt$. This means that distance (d) is equal to the rate of movement (r) times the amount of time involved (t).

PROBLEM: Evaluate when $r = 70$ miles per hour and $t = 8$ hours.

SOLUTION: $d = rt$

$= (70 \text{ miles per hour})(8 \text{ hours})$

$= 560$ miles

Example 8-7

Formula: The formula for determining simple interest is $I = prt$. This means that the amount of interest (I) is equal to the principal (p) times the rate of interest (r) times the amount of time involved (t).

PROBLEM: Evaluate when $p = \$5000$, $r = .08$, $t = 3$ years.

SOLUTION: $I = prt$

$= (\$5000)(.08)(3)$

$= \$1200.00$

EXERCISE 8-2

Solve the following problems, making certain to perform the operations as the signs indicate.

8-7. Find the area of a rectangle when the dimensions are 15 inches and 11 inches.

$A = lw$; $l = 15$ inches, $w = 11$ inches

8-8. Find the distance when the rate is 40 miles per hour and the time is 20 hours.

$d = rt$; $r = 40$ miles per hour, $t = 20$ hours

8-9. Find the simple interest for 25 years when the principal is \$1000 and the rate is 7%.

$I = prt$; $p = \$1000$, $r = .07$, $t = 25$ years

8-10. Find the volume of a rectangular solid which is 30 feet by 12 feet by 8 inches.

$V = lwh$; $l = 30$ feet, $w = 12$ feet, $h = 8$ inches

8-11. Find the circumference of a circle 8 yards in diameter.

$C = d\pi$; $d = 8$ yards, $\pi = 3.1416$

8-12. Convert 27 feet to inches.

1 foot = 12 inches

8-13. Find the circumference of a circle when the radius is 34 inches.

$C = 2\pi r;\quad r = 34$ inches, $\quad \pi = 3.14$

8-14. Find the volume of a rectangular solid with a square base containing 36 square feet and a height of 24 inches.

$V = Bh;\quad B = 36$ square feet, $\quad h = 24$ inches

8-15. Find the perimeter of a rectangle 7 feet long and 3 feet wide.

$p = 2l + 2w;\quad l = 7,\quad w = 3$

8-16. Find the area of a 41-foot square.

$A = s^2,\quad s = 41$ feet

8-17. Find the circumference in *feet* of a circle when the radius is 28.5 inches.

$C = 2\pi r;\quad r = 28.5$ inches; $\quad C = \dfrac{2\pi r}{12}$ feet

8-18. Find the surface area of a 3-foot cube.

$S = 6e^2;\quad e = 3$ feet

8-19. Find the surface area of a rectangular solid 6 feet by 5 feet by 3 feet.

$S = 2lw + 2hl + 2hw;\quad l = 6$ feet,

$w = 5$ feet, $\quad h = 3$ feet

8-20. Find the simple interest for $298 at 5.5% for 3 years.

$I = prt;\quad p = \$298,\quad r = 5.5\%,\quad t = 3$ years

CONTENT

You have worked with formulas requiring you to use addition, subtraction, and multiplication. You should now be ready to work with more difficult formulas which will include the additional arithmetic operations of division and roots. As you work through the following examples, take careful note of the procedures involved.

Example 8-8

Formula: The formula for determining the area of a triangle is $A = \dfrac{bh}{2}$. This means that the area (A) of the triangle is equal to the base (b) times the height (h) divided by 2.

PROBLEM: Find the area of a triangle when $b = 4$ feet and $h = 12$ feet.

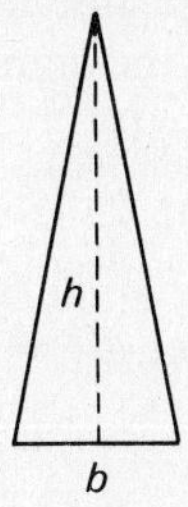

SOLUTION: $A = \frac{bh}{2}$

$= \frac{4 \text{ feet} \times 12 \text{ feet}}{2}$

$= 24$ square feet

Example 8-9

Formula: The formula for determining the area of a trapezoid is $A = \frac{1}{2}(b_1 + b_2)h$. This means that the area (A) of the trapezoid is equal to one half of the base 1 (b_1) plus one half of the base 2 (b_2) times the height (h).

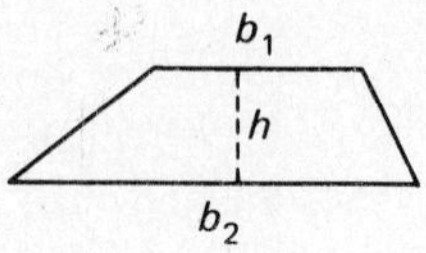

PROBLEM: Find the area of the trapezoid when $b_1 = 10$ inches, $b_2 = 20$ inches, and $h = 5$ inches.

SOLUTION:

$A = \frac{1}{2}(b_1 + b_2)h$

$= \frac{1}{2}(10 \text{ inches} + 20 \text{ inches})\ 5 \text{ inches}$

$= \frac{1}{2}(30 \text{ inches})(5 \text{ inches})$

$= 75$ square inches

Example 8-10

Formula: The formula for determining the depth of a certain type of thread is

$$D = d - \frac{1.732}{N}$$

This means that the depth of the thread (D) is equal to the depth of one thread (d) minus 1.732 divided by the number of threads (N).

PROBLEM: Find the depth of the thread when $d = .75$ and $N = 5$.

SOLUTION: $D = .75 - \frac{1.732}{5}$

$= .75 - .3464$

$= .4036$

EXAMPLE 8-11

Formula: The formula for determining the height of an isosceles triangle given the equal sides and base is

$$h = \sqrt{a^2 - \frac{b^2}{4}}$$

This means that the height (h) of the isosceles triangle is equal to the square root ($\sqrt{\ }$) of one of the equal sides squared (a^2) minus the base squared divided by 4 $\left(\frac{b^2}{4}\right)$.

SOLUTION: $h = \sqrt{a^2 - \frac{b^2}{4}}$

$= \sqrt{(26)^2 - \frac{(20)^2}{4}}$

$= \sqrt{676 - \frac{400}{4}}$

$= \sqrt{676 - 100}$

$= \sqrt{576}$

$= 24$ feet

EXERCISE 8-3

Solve the following problems, making certain to perform the operations the signs indicate.

8-21. $A = \frac{1}{2}(b_1 + b_2)h$

Find A when $b_1 = 2$ feet, $b_2 = 36$ inches, and $h = 24$ inches.

8-22. $E = \frac{360}{n}$

Find E when $n = 4$.

8-23. $A = \frac{.7854nd^2}{360°}$

Find A when $n = 60°$ and $d = 12$ inches.

8-24. $K = \frac{L(D - d)}{l}$

Find K when $L = 3.5$, $D = .5$, $d = .25$, and $l = .25$.

8-25. $M = (D - 1.732P) + 3W$

Find M when $D = .75$, $P = 10$, and $W = .06$.

8-26. $A = .7854d^2$

Find A when $d = 11$ feet.

8-27. $S = 2\pi r(r + h)$

Find S when $\pi = \frac{22}{7}$, $r = 2$ feet, and $h = 1\frac{1}{2}$ feet.

8-28. $C = \frac{5}{9}(F - 32°)$

Find C when $F = 131°$.

8-29. $V = \frac{4\pi r^3}{3}$

Find V when $r = 6$ yards and $\pi = 3.14$.

8-30. $F = \frac{9}{5}C + 32°$

Find F when $C = 115°$.

8-31. $W = \frac{Kbd^2}{L}$

Find W when $K = 6000$, $b = 6$, $d = 8$, and $L = 24$.

8-32. $S = \frac{gt^2}{2}$

Find S when $g = 32$ and $t = 4$.

8-33. $S = \frac{n}{2}(a + l)$

Find S when $a = 10$, $l = 6$, and $n = 50$.

8-34. $H = \frac{DV}{375}$

Find H when $D = 187.5$ and $V = 200$ miles per hour.

8-35. $K = 2a - 5(n - 1)$

Find K when $a = 8$ and $n = 3$.

8-36. $R = \dfrac{1}{\dfrac{1}{r_1} + \dfrac{1}{r_2}}$

Find R when $r_1 = 3$ and $r_2 = 4$.

8-37. $A = \dfrac{D}{.5(D - d)}$

Find A when $D = 10$ and $d = 4$.

8-38. $A = 3.1416\,(R + r)(R - r)$

Find A when $R = \frac{3}{2}$ and $r = \frac{3}{8}$.

8-39. $E = \dfrac{(N)P}{NP + R}$

Find E when $N = 2$, $P = 4$, and $R = 5.3$.

The exercises you have completed have dealt with formula evaluation. In the next chapter you will learn how to manipulate or transform formulas into the desired forms you need.

THINK TIME

Write four of the most important formulas used in your vocation on a sheet of paper. Using each of these formulas, together with the knowledge and skills learned in this chapter, solve for one of the unknowns. Think of some problem situations in which you could use each of these formulas. Practice solving these problems with the formulas.

PROCEDURES TO REMEMBER

To evaluate formulas:

1. Read the formula to determine its purpose.
2. Reread the formula, making special note of the given quantities and what you are to solve.
3. Substitute the given quantities into the formula, making certain you have used the correct dimensions.
4. Estimate your answer so that you will have a reasonable idea of the possible solution.
5. Evaluate; show all necessary mathematical operations in order to prevent costly errors.
6. Be sure that your solution is written in the correct dimensions.

CHAPTER SUMMARY

A formula uses mathematical language to express the relationship between two or more variables. Thus a formula is a mathematical abbreviation for an accepted operational procedure. Formula evaluation is the process of obtaining the value of a formula from the given variables.

CHAPTER TEST

Evaluate the following using the quantities stated in the problem.

T-8-1. When $x = 5$, $y = 3$, and $z = 2$, $\frac{x + y}{z} =$

T-8-2. When $a = 15$, $b = 14$, and $c = 12$, $a^2 + b^2 - c^2 =$

T-8-3. When $x = 1$, $y = 3$, and $z = 2$, $x^2 + y^2 + z^3 =$

T-8-4. When $w = -2$ and $k = 3$, $\frac{2w + 5k}{7} =$

T-8-5. When $x = 1$, $y = 3$, $z = 2$, and $w = -2$, $xy + yz + zw =$

T-8-6. Evaluate $\frac{1}{x} + \frac{3}{y} =$

when $x = 1$ and $y = -1$.

T-8-7. Evaluate $(x + z)(w + k) =$ when $x = 1$, $z = 2$, $w = -2$, and $k = 3$.

T-8-8. Evaluate $(ab)^2(c - b)^2 =$ when $a = 4$, $b = 0.5$, and $c = 2.5$.

T-8-9. Evaluate $(a + b)^2(a - b)^2 =$ when $a = 6$ and $b = 4$.

T-8-10. Evaluate $(x + y - z)^3 =$ when $x = 5$, $y = 12$, and $z = -7$.

T-8-11. The formula for the area of an ellipse is $A = \pi ab$. Evaluate A when $\pi = 3.14$, $a = 20$, and $b = 12$.

T-8-12. The formula for the perimeter of an ellipse is $P = \pi(a + b)$. Find P when $\pi = 3.14$, $a = 24$, and $b = 21$.

T-8-13. If $l = 2\sqrt{h(2r - h)}$, find l when $h = 4$ and $r = 10$. This is the formula for the length of a segment when you know the radius of the circle and the height of the segment.

T-8-14. The formula for the area of a ring is $A = \pi(R + r)(R - r)$. Find A when $R = \frac{7}{8}$, $r = \frac{3}{4}$, and $\pi = 3.14$.

T-8-15. The formula for the radius of a circle, given the area, is $r = \sqrt{\frac{A}{\pi}}$. Find r if $\pi = 3.14$ and $A = 502.4$.

T-8-16. The formula for the volume of a sphere is $V = \frac{4\pi r^3}{3}$. Find the volume of a sphere (in cubic yards) when the radius is 12 feet.

T-8-17. The formula for the volume of a cylinder is $V = \pi r^2 h$. Find the volume of a cylinder when $\pi = 3.14$, $r = 5$, and $h = 12$.

T-8-18. The formula for the area of a sector of a circle is $A = \frac{\theta}{360}\pi r^2$. Find A when $\theta = 75$, $\pi = 3.14$, and $r = 8$.

T-8-19. The formula for the diameter of a circle given the area is: $d = \sqrt{A \div \frac{\pi}{4}}$. Find d when $A = 12.56$ and $\pi = 3.14$.

T-8-20. The formula for the area of a triangle, given all the sides, is

$$A = \sqrt{s(s - a)(s - b)(s - c)}, \quad s = \tfrac{1}{2}(a + b + c).$$

Find A when $a = 28$, $b = 25$, and $c = 17$.

CHAPTER

9

FORMULA TRANSPOSITION

OBJECTIVE

To change formulas into more usable forms.

SELF-TEST

This test will determine whether or not you need to study this chapter. If you are able to compute your answers accurately, you are ready to move to the next chapter and the next self-test. Be sure to show your work. This will enable you or someone else to check over your work to determine where you made your mistakes and what changes you need to make so you will not make the same mistakes again.

Solve these equations for the unknown quantities.

S-9-1. Solve $P = a + b + c$ for c.

S-9-2. Solve $V = \frac{1}{3}Bh$ for B.

S-9-3. Solve $L = 2\pi rh$ for h.

S-9-4. Solve $P = \frac{w}{t}$ for t.

S-9-5. Solve $HP = \frac{fl}{t}$ for t.

S-9-6. Solve $H = \frac{D^2N}{2.5}$ for D.

S-9-7. Solve $V = \frac{1}{3}\pi r^2 h$ for h.

S-9-8. Solve $A = \frac{h}{2}(b_1 + b_2)$ for b_2.

S-9-9. Solve $S = 2\pi rh + 2\pi r^2$ for h.

S-9-10. Solve $V = \pi r^2 h$ for r.

S-9-11. Solve $V = \frac{4}{3}\pi r^3$ for r.

S-9-12. Solve $DS = ds$ for d.

S-9-13. Solve $S = \frac{CD}{12d}$ for D.

S-9-14. Solve $W = \frac{5bd^2K}{2L}$ for K.

S-9-15. Find a when $T = 384$ and $T = 6a^2$.

S-9-16. Find r when $A = \frac{5\pi}{16}$, $R = \frac{3}{4}$, and $A = \pi R^2 - \pi r^2$.

S-9-17. Find b when $P = 51\pi$, $a = 32$, and $P = \pi(a + b)$.

S-9-18. Find r when $V = \pi R^2 h - \pi r^2 h$, $R = 8$, $h = 10$, $V = 77.5\pi$, and $\pi = 3.14$.

S-9-19. Solve $S = 4\pi r^2$ for r and then find r when $S = 484$.

S-9-20. Find d if $V = \frac{32\pi}{3}$ and $V = \frac{\pi d^3}{6}$.

S-9-21. Solve $A = \frac{\theta \pi r^2}{360}$ for r.

S-9-22. Solve $r = \frac{\left(\frac{w}{2}\right)^2 + h^2}{2h}$ for w.

S-9-23. Solve $V = \frac{\pi h}{3}(a^2 + ab + b^2)$ for h, and then find h when $a = 24$, $b = 18$, and $V = 3552$.

S-9-24. Solve $P = \pi\sqrt{2(a^2 + b^2)}$ for b.

S-9-25. Find r_3 when $R = \frac{12}{13}$, $r_1 = 2$, $r_2 = 3$, and $\frac{1}{R} = \frac{1}{r_1} + \frac{1}{r_2} + \frac{1}{r_3}$.

INTRODUCTION

Unfortunately, formulas which you are given or which you may find in a handbook are not always in their most useful forms. It is therefore important that you know how to change formulas so that they may work for you. This is not an impossible task if you move through the process step-by-step. Many students find the changing or transposition of formulas very difficult because they fail to read the formulas correctly. For example, the formula $A = \frac{1}{2}(b_1 + b_2)h$ could be misread as $A = \frac{1}{2}b_1 + b_2 \times h$. Failure to first complete work inside the parentheses could cause this problem. Careful reading of the formula can help to prevent confusion and error.

CONTENT

Reading the original formula correctly becomes very important when you wish to change the formula in order to find the solution. Once the formula is read correctly, you will be able to make the necessary changes. These changes are often made through inverse operations. This means, for example, that division may be used to solve an equation containing multiplication. When this is done, it is important to perform the same operations on both sides of the equation.

Study the following examples and note how the inverse operations are used to solve the equations.

EXAMPLE 9-1

Formula: The area of a rectangle equals the length times the width, or $A = lw$.

PROBLEM: Solve the formula for w.

SOLUTION: Remember that division is the inverse operation of multiplication. In this formula, divide both sides of the equation by l in order to find the value of w.

$$A = lw$$

$$\frac{A}{l} = \frac{\not{l}w}{\not{l}}$$

$$\frac{A}{l} = w$$

This means the width (w) equals the area (A) divided by the length (l).

$$w = \frac{A}{l}$$

PROBLEM: Find w when $A = 44$ square units and $l = 11$ units.

$$w = \frac{A}{l}$$

$$w = \frac{\overset{4}{\not{44}} \text{ square units}}{\underset{1}{\not{11}} \text{ units}}$$

$$w = 4 \text{ units}$$

Example 9-2

This example involves the operations of both multiplication and addition.

Formula: The perimeter of a rectangle equals 2 times the width plus 2 times the length, or

$$P = 2l + 2w$$

PROBLEM: Solve the formula for w.

SOLUTION: First, isolate the quantity you are looking for (w) on one side of the equation by subtracting $2l$ from each side of the equation.

$$P = 2l + 2w$$

$$P - 2l = 2l + 2w - 2l$$

$$P - 2l = 2w$$

Second, as $2w$ is isolated on one side of the equation, solve the equation for w, by dividing each side of the equation by 2.

$$\frac{\not{2}w}{\not{2}} = \frac{P - 2l}{2}$$

$$w = \frac{P - 2l}{2}$$

PROBLEM: Find w when $P = 38$ and $l = 12$.

SOLUTION:

$$w = \frac{P - 2l}{2}$$

$$= \frac{38 - 2(12)}{2}$$

$$= \frac{38 - 24}{2}$$

$$= \frac{\overset{7}{\not{14}}}{\underset{1}{\not{2}}}$$

Example 9-3

This example involves the operations of both multiplication and division.

Formula: The area of a triangle equals one half the base times the height, or $A = \frac{1}{2}bh$.

PROBLEM: Solve the formula for h.

SOLUTION: When a formula involves a fraction, it should be changed to a whole number through the process of multiplication. In order to accomplish this task, first determine the lowest common denominator, which in this formula is 2. Then multiply each term on both sides of the equation by the lowest common denominator (2).

$$A = \frac{1}{2}bh \quad \text{or} \quad A = \frac{bh}{2}$$

$$2A = (\overset{1}{\not{2}})\frac{1}{\underset{1}{\not{2}}}bh$$

The fraction has now been eliminated. In order to isolate h, each side of the equation must be divided by b.

$$2A = bh$$

$$\frac{2A}{b} = \frac{\overset{1}{\not{b}}h}{\underset{1}{\not{b}}}$$

$$h = \frac{2A}{b}$$

PROBLEM: Find b when $A = 360$ square feet and $b = 40$ feet.

SOLUTION:

$$h = \frac{2A}{b}$$

$$= \frac{2(230 \text{ square feet})}{40 \text{ feet}}$$

$$= \frac{\overset{18 \text{ feet}}{\cancel{720}} \text{ square feet}}{\underset{1}{\cancel{40}} \text{ feet}} = 18 \text{ ft}$$

EXERCISE 9-1

Solve the following problems, making sure you read them correctly and solve for the correct letter. Show all necessary work.

9-1. Solve $Q = \frac{WL}{T}$ for T. Find T when $Q = 1250$, $W = 125$, and $L = 20$.

9-2. Solve $I = prt$ for t. Find t when $p = \$500$, $r = .08$, and $I = \$120$.

9-3. Solve $C = 2\pi r$ for r. Find r when $C = 154$ and $\pi = \frac{22}{7}$.

9-4. Solve $p = a + b + c$ for c. Find c when $p = 100$, $a = 21$, and $b = 17$.

9-5. Solve $V = Bh$ for B. Find B when $V = 1000$ cubic feet and $h = 8$ feet.

9-6. Solve $Q = .000477\ EIT$ for T. Find T when $Q = 954$, $E = 1000$, and $I = 20$.

9-7. Solve $C = \frac{5}{9}(F - 32°)$ for F. Find F if $C = 100°$.

9-8. Solve $x - y = xy$ for y. Find y when $x = 8$.

9-9. Solve $S = \pi rs$ for s. Find s when $S = 4000$ and $r = 462$.

9-10. Solve $P = 2a + b$ for a. Find a when $P = 28$ and $b = 12$.

9-11. Solve $I = prt$ for r. Find r when $I = \$80$, $P = \$500$, and $t = 2$ years.

9-12. Solve $C = \pi d$ for d. Find d when $C = 120$ feet and $\pi = 3.14$.

9-13. Solve $L = \frac{\pi r A}{180}$ for A. Find A when $L = 314$, $r = 60$, and $\pi = 3.14$.

9-14. Solve $T = \frac{1}{2}ps + B$ for B. Find B when $T = 580$, $p = 58$, and $s = 12$.

9-15. Solve $T = 2B + pH$ for H. Find H when $T = 860$, $B = 45$, and $p = 11$.

CONTENT

The process of transposition becomes more difficult when the extraction of roots becomes necessary. However, if you remember what you have learned about the extraction of roots in previous chapters and what you have learned about transposition in this chapter, you will have little difficulty. Carefully study the following examples.

Example 9-4

Formula: The surface area of a cube equals 6 times the length of one edge squared, or $A = 6e^2$.

PROBLEM: Solve the formula for e.

SOLUTION: First, divide both sides by 6 in order to isolate e^2 on one side of the equation.

$$A = 6e^2$$

$$\frac{A}{6} = \frac{\overset{1}{\cancel{6}}e^2}{\underset{1}{\cancel{6}}}$$

$$\frac{A}{6} = e^2$$

Second, solve the equation for e. This means that it is necessary to determine the square root of each side of the equation.

$$e^2 = \frac{A}{6}$$

$$\sqrt{e^2} = \sqrt{\frac{A}{6}}$$

$$e = \sqrt{\frac{A}{6}}$$

PROBLEM: Find e if $A = 216$ square feet.

SOLUTION:

$$e = \sqrt{\frac{A}{6}}$$

$$= \sqrt{\frac{\overset{36}{\cancel{216}} \text{ square feet}}{\underset{1}{\cancel{6}}}}$$

$$= \sqrt{36 \text{ square feet}}$$

$$= 6 \text{ feet}$$

Example 9-5

Formula: The area of a circle equals π (approximately 3.14) times the radius of the circle squared, or $A = \pi r^2$.

PROBLEM: Solve the formula for r.

SOLUTION: First, divide both sides by π in order to isolate r^2 on one side of the equation.

$$A = \pi r^2$$

$$\frac{A}{\pi} = \frac{\cancel{\pi} r^2}{\cancel{\pi}}$$

$$\frac{A}{\pi} = r^2$$

Second, solve the formula for r. In order to do this, it will be necessary to determine the square root of each side of the equation.

$$r^2 = \frac{A}{\pi}$$

$$\sqrt{r^2} = \sqrt{\frac{A}{\pi}}$$

$$r = \sqrt{\frac{A}{\pi}}$$

PROBLEM: Find r if the area of a circle is 50.24 square units and $\pi = 3.14$.

SOLUTION:

$$r = \sqrt{\frac{A}{\pi}}$$

$$= \sqrt{\frac{50.24 \text{ square units}}{3.14}}$$

$$= \sqrt{16 \text{ square units}}$$

$$= 4 \text{ units}$$

Example 9-6

Formula: The volume of a sphere equals 4 times π times the radius of the sphere cubed divided by 3, or $V = \dfrac{4\pi r^3}{3}$.

PROBLEM: Solve the formula for r.

SOLUTION: Note that r is cubed. Consequently, in the third step of the solution we will need to find the cube root of r^3. The first two steps will lead up to that point. First, clear the fraction, $\dfrac{4\pi r^3}{3}$, by multiplying both sides of the equation by 3.

$$V = \frac{4\pi r^3}{3}$$

$$(3)V = (\overset{1}{\cancel{3}})\frac{4\pi r^3}{\underset{1}{\cancel{3}}}$$

$$3V = 4\pi r^3$$

Second, isolate r^3 by dividing both sides of the equation by 4π.

$$3V = 4\pi r^3$$

$$\frac{3V}{4\pi} = \frac{\overset{1}{\cancel{4}}\cancel{\pi} r^3}{\underset{1}{\cancel{4}\cancel{\pi}}}$$

$$\frac{3V}{4\pi} = r^3$$

Third, solve the formula for r. In order to do this, it will be necessary to determine the cube root of each side of the equation.

$$r^3 = \frac{3V}{4\pi}$$

$$\sqrt[3]{r^3} = \sqrt[3]{\frac{3V}{4\pi}}$$

$$r = \sqrt[3]{\frac{3V}{4\pi}}$$

PROBLEM: Find r if $V = 523.333$ cubic units and $\pi = 3.14$.

SOLUTION: $r = \sqrt[3]{\frac{3v}{4\pi}}$

$$= \sqrt[3]{\frac{(3)(523.333) \text{ cubic units}}{4(3.14)}}$$

$$= \sqrt[3]{\frac{1570 \text{ cubic units}}{12.56}}$$

$$= \sqrt[3]{125 \text{ cubic units}}$$

$$= 5 \text{ units}$$

Example 9-7

Formula: The height of an isosceles triangle is equal to the square root of one of the equal sides squared minus the base squared divided by 4, or

$$h = \sqrt{a^2 - \frac{b^2}{4}}$$

PROBLEM: Solve the formula for a.

SOLUTION: Even though it would appear simple to find the square root of a^2, it must be remembered that a^2 is still a part of the *whole* formula and cannot be dealt with separately at this point. First, the process of finding the value of a calls for the squaring of both sides of the equation.

$$(h)^2 = \left(\sqrt{a^2 - \frac{b^2}{4}}\right)^2$$

$$h^2 = a^2 - \frac{b^2}{4}$$

Second, add $\frac{b^2}{4}$ to both sides of the equation.

$$\frac{b^2}{4} + h^2 = a^2 - \frac{b^2}{4} + \frac{b^2}{4}$$

$$\frac{b^2}{4} + h^2 = a^2$$

Third, take the square root of each side.

$$\sqrt{a^2} = \sqrt{\frac{b^2}{4} + h^2}$$

$$a = \sqrt{\frac{b^2}{4} + h^2}$$

PROBLEM: Find the value of a when $h = 24$ and $b = 20$.

SOLUTION: $a = \sqrt{\frac{b^2}{4} + h^2}$

$$= \sqrt{\frac{(20)^2}{4} + (24)^2}$$

$$= \sqrt{\frac{400}{4} + 576}$$

$$= \sqrt{100 + 576}$$

$$= \sqrt{676}$$

$$= 26 \text{ units}$$

EXERCISE 9-2

Solve the following problems. Study each problem carefully to make sure you understand exactly how the problem is to be solved. Should you have difficulty, check the example problems.

9-16. Solve $A = \frac{\pi r^2}{2}$ for r. Find r when $A =$ 154 square units and $\pi = \frac{22}{7}$.

9-17. Solve $x = ky^2$ for y. Find y when $x = 128$ and $k = 8$.

9-18. Solve $F = \frac{kM}{d^2}$ for d. Find d when $k = 28$, $M = 8$ and $F = 14$.

9-19. Solve $S = \frac{at^2}{2}$ for t. Find t when $a = 4$ and $S = 128$.

9-20. Solve $F = \frac{kmM}{d^2}$ for M. Find M when $d = 3$, $F = 42$, $k = 14$, and $m = \frac{1}{3}$.

9-21. Solve the mass-energy equation $e = mc^2$ for c. Find c if $e = 6348$ and $m = 12$.

9-22. Solve $V = e^3$ for e. This is the formula for the volume of a cube. Find e when $V = 729$ cubic yards.

9-23. Solve $S = 4\pi r^2$, which is the formula for the surface area of a sphere, for r. Find r when $S = 784$ square feet.

9-24. Solve the volume of a sphere formula, $V = \frac{\pi d^3}{6}$, for d. Find d when $V = \frac{171.5\pi}{3}$.

9-25. Solve $r = \sqrt{\frac{A}{\pi}}$ for A. This is the formula for the radius of a circle when you know the area. Find A when $r = 7$ and $\pi = \frac{22}{7}$.

9-26. Solve the algebra equation $(y - a)^2 - b = 2ab$ for $(y - a)$. Find y when $a = 6$ and $b = 4$.

9-27. Solve $A = \frac{\theta \pi r^2}{360}$ for r. This is the formula for the area of a sector of a circle. Find r when $A = 20$ and $\theta = 72$.

CONTENT

Many formulas are used in business and industry today. However, each development in research and technology brings new problems with new formulas to solve these problems. Consequently, no matter how many formulas you may learn over the years, either in school or on the job, there will always be new formulas to be learned. It is important for you to learn how to work with formulas in order to solve the problems that confront you. You will find that formulas can become extremely complex, but once you learn the techniques of working with formulas, many of your problems will diminish in their degree of difficulty.

The next group of formulas are more complex. It is important for you to make use of skills you have already learned regarding formulas and literal equations.

Example 9-8

Formula: The amount of current flowing through the armature of a generator can be determined by the formula $I = \frac{E - e}{R}$.

PROBLEM: Solve this formula for E.

SOLUTION: First, remove the fraction from the equation by multiplying each side of the equation by R. Note that the numerator of the fraction represents more than one unknown quantity. In order to protect the value of each quantity, the entire numerator is placed within parentheses. When these parentheses are no longer needed, they can be removed as shown in the second step.

$$I = \frac{(E - e)}{R}$$

$$(R)I = \frac{(E - e)}{\cancel{R}_{1}}(\cancel{R}^{1})$$

Second, add e to each side of the equation.

$$RI = E - e$$

$$e + RI = E - e + e$$

$$E = e + RI$$

Example 9-9

Formula: The drop in voltage can be determined by the formula $\frac{E}{e} = \frac{R + r}{r}$.

PROBLEM: Solve this formula for r.

SOLUTION: First, remove the fraction from the equation by multiplying each side of the equation by the lowest common denominator, er.

$$\frac{E}{e} = \frac{R + r}{r}$$

$$(\cancel{e}r)^{1}\frac{E}{\cancel{e}_{1}} = \frac{(R + r)}{\cancel{r}_{1}}(e\cancel{r})^{1}$$

$$(r)E = (R + r)(e)$$

$$rE = Re + re$$

Second, in order to get all of the terms which include r on one side of the equation, subtract re from each side of the equation.

$$rE = Re + re$$

$$rE - re = Re + re - re$$

$$rE - re = Re$$

Note that the quantities which contain r can be shown in another way.

$$rE - re = Re$$

$$r(E - e) = Re$$

The formula can now be divided by the quantity $(E - e)$ in order to find r.

$$\frac{r\cancel{(E - e)}^{1}}{\cancel{(E - e)}_{1}} = \frac{Re}{(E - e)}$$

$$r = \frac{Re}{E - e}$$

Example 9-10

Formula: Geometric progression can be determined by the formula $S = \frac{r - a}{r - 1}$.

PROBLEM: Solve this formula for r.

SOLUTION: First, remove the fraction from the equation by multiplying each side of the equation by the lowest common denominator, $r - 1$.

$$S = \frac{r - a}{r - 1}$$

$$(r - 1)S = \frac{(r - a)}{\cancel{(r - 1)}_{1}}\cancel{(r - 1)}^{1}$$

$$(r - 1)S = r - a$$

$$rS - S = r - a$$

Second, in order to get all the terms which include r on one side of the equation, add $(S - r)$ to each side of the equation.

$$rS - S + (S - r) = r - a + (S - r)$$

$$rS - S + S - r = r - a + S - r$$

$$rS - r = S - a$$

Third, factor r from all terms on the left side of the equation.

$$r(S - 1) = S - a$$

Fourth, the formula can now be divided by the quantity $(S - 1)$ in order to find r.

$$r(S - 1) = S - a$$

$$\frac{r(S - 1)}{(S - 1)} = \frac{S - a}{(S - 1)}$$

$$r = \frac{S - a}{S - 1}$$

Example 9-11

Formula: The amount of parallel resistance can be determined by the formula $\frac{1}{R} = \frac{1}{r_1} + \frac{1}{r_2}$.

PROBLEM: Solve this equation for r_1.

SOLUTION: First, remove the fractions from the equation by multiplying each side of the equation by the lowest common denominator, Rr_1r_2.

$$\frac{1}{R} = \frac{1}{r_1} + \frac{1}{r_2}$$

$$(Rr_1r_2)\frac{1}{R} = (Rr_1r_2)\frac{1}{r_1} + (Rr_1r_2)\frac{1}{r_2}$$

$$r_1r_2 = Rr_2 + Rr_1$$

Second, in order to get all of the terms which include r_1 on one side of the equation, subtract Rr_1 from each side of the equation.

$$r_1r_2 = Rr_2 + Rr_1$$

$$r_1r_2 - Rr_1 = Rr_2 + Rr_1 - Rr_1$$

$$r_1r_2 - Rr_1 = Rr_2$$

Third, in order to further isolate r_1, factor each term on the left side of the equation for r_1.

$$r_1r_2 - Rr_1 = Rr_2$$

$$r_1(r_2 - R) = Rr_2$$

Fourth, divide each side of the equation by $(r_2 - R)$.

$$r_1(r_2 - R) = Rr_2$$

$$\frac{r_1(r_2 - R)}{(r_2 - R)} = \frac{Rr_2}{(r_2 - R)}$$

$$r_1 = \frac{Rr_2}{(r_2 - R)}$$

Rework this example and carefully note each step in the process. These methods will be of value to you in other formula solutions.

EXERCISE 9-3

Solve the following problems. Review the methods used in the examples if you have any difficulty. Show all necessary work.

9-28. The formula for determining latent heat vaporization is $Q = \frac{WL}{T}$. Solve the formula for T when $Q = 100$, $W = 5$, and $L = 800$.

9-29. The formula for determining the average speed of a uniformly accelerating body is $V = \frac{V_t + V_0}{2}$. Solve the formula for V_0 when $V = 312$ and $V_t = 112$.

9-30. The formula for determining temperature conversion is $T = \frac{1}{a} + t$. Solve the formula for a when $T = 96$ and $t = 46$.

9-31. Using the electrical equivalent heat formula, $Q = .000477EIT$, solve for T when $E = 1000$, $I = 200$, and $Q = 95.4$.

9-32. Solve the thickness-of-pipe formula, $A = \frac{m}{t}(p + t)$, for t when $A = 1$, $m = \frac{1}{4}$, and $p = 12$.

9-33. The formula for determining magnetic intensity if $H = \frac{.4\pi NI}{L}$. Solve the formula for I when $N = 4$, $L = 1.256$, $H = 2$, and $\pi = 3.14$.

9-34. Using the formula for determining the theoretical amount of air required to burn solid fuel, $M = 10.5C + 35.2\left(w - \frac{C}{8}\right)$, solve for C when $M = 78.6$ and $w = 5$.

9-35. Solve the photographic enlargement formula, $\frac{1}{x} + \frac{1}{nx} = \frac{1}{f}$, for x when $f = 12$ and $n = 3$.

9-36. The formula for determining the tap-size drill for a U.S. Standard thread is $S = T - \frac{1.299}{N}$. Solve the formula for N when $T = 5.433$ and $S = 5$.

9-37. Solve the differential pulley formula, $W = \frac{2PR}{R - r}$, for R when $P = 12$, $r = 75$, and $W = 48$.

9-38. Solve the algebraic equation, $x - y = xy$, for y when $x = \frac{1}{8}$.

9-39. Using the prismoidal formula, $V = \frac{h}{6}(B + 4M + b)$, solve for M. In this problem $B = 25$, $b = 5$, $h = 36$, and $V = 420$.

9-40. Solve the formula for expansion of gases, $V_1 = V_0(1 + .00365t)$, for t when $V_1 = 112.5$ and $V_0 = 3$.

9-41. The formula for finding the area of an ellipse is $A = \pi ab$. Solve the formula for b when $A = 113.04$ square units and $a = 18$.

9-42. Using the formula for determining the perimeter of an ellipse, $P = \pi\sqrt{2(a^2 + b^2)}$, solve for b when $P = 16\pi$ and $a = 9.59$.

9-43. Solve the formula for finding the width of a segment of a circle, $W = 2\sqrt{h(2r - h)}$, for r. In this problem $h = 8$ and $W = 8\sqrt{8}$.

9-44. The formula for determining the volume of a torus is $V = 2\pi^2Rr^2$. Solve the formula for r when $R = 49$, $\pi = \frac{22}{7}$, and $V = 968$ square units.

9-45. Using the formula for finding the total surface area of a right circular cylinder is $T = 2\pi r(r + H)$. Solve for H when $T = 320\pi$ and $r = 8$.

9-46. Solve the formula for determining the volume of a sphere, $V = \frac{\pi d^3}{6}$, for d when $V = 288\pi$.

9-47. The formula for finding the distance an object travels when acted upon by gravity is $S = 16t^2 + vt$. Solve the formula for v when $S = 1000$ feet and $t = 3$ seconds.

9-48. Using the formula for determining the distance an object travels when acted upon by gravity and an initial velocity, $S = 16t^2 + Vt$, solve for V when $S = 1100$ and $t = 4$.

9-49. The formula for determining the exposed surface area of a cylinder is

$$S = \left[\left(\frac{\pi d^2}{2}\right) + \left(\frac{\pi dl}{r}\right)\right]\left(\frac{4rc}{d^2l}\right)$$

Solve the formula for S when $d = 4$, $l = 12$, $c = 2$, and $r = 2$.

9-50. Using the formula for finding the volume of a cylinder, $V = \pi h(R + r)(R - r)$, solve for V when $R = 8$, $h = 20$, and $r = 4$.

THINK TIME

Make a list of five well-known formulas used in your vocation. Study each formula carefully to determine your ability to work with it efficiently. Think of several problems you might face on the job in which you might use the formula. Use the formulas to solve these problems. Ask your teacher or friend to check your work.

Because formula transposition skills are vital to many occupations, the following practice problems are provided for your in-depth development.

1. Solve $F = \dfrac{Nmv^2}{3}$ for v.

2. Solve $T = ph + 2A$ for h.

3. Solve $S = \dfrac{E - IR}{0.220}$ for I.

4. Solve $\dfrac{1}{f} = \dfrac{1}{p} + \dfrac{1}{q}$ for q.

5. Solve $I = \dfrac{En}{R + nr}$ for r.

6. Solve $E = RI + \dfrac{rI}{n}$ for I.

7. Solve $A = \frac{2}{3}hw$ for h.

8. Solve $V = 2\pi^2 Rr^2$ for R.

9. Solve $V = \pi R^2 h - \pi r^2 h$ for r^2.

10. Solve $T = \dfrac{(P + p)s}{2}$ for p.

11. Solve $V = \dfrac{4\pi r^3}{3}$ for r.

12. From problem 11, find r when $V = 1333.333\pi$.

13. Given $F = \dfrac{16mx}{T^2}$, find x.

14. From problem 13, find x when $m = 6$, $T = 4$, and $F = 90$.

15. Solve $\frac{1}{R} = \frac{1}{r_1} + \frac{1}{r_2} + \frac{1}{r_3}$ for r_3.

16. Solve $w = \frac{2PR}{R - r}$ for R.

17. Solve $P = \pi\sqrt{2(a^2 + b^2)}$ for a.

18. From problem 17, find b when $P = 12\pi$ and $a = 4.796$.

19. Solve $S = 16t^2 + Vt$ for V.

20. Solve $V = \frac{\pi d^3}{6}$ for d.

PROCEDURES TO REMEMBER

To accomplish formula transposition:

1. Read the formula in order to gain an understanding of the use of the formula.
2. Reread the formula, carefully noting which unknown quantity is tc be sought.
3. Determine what mathematical procedures you will need to use to solve the formula for the unknown quantity.
4. Remember the scales of justice as you select the mathematical procedure to solve the formula. When you add, subtract, multiply, or divide on one side of the equation, you must perform the same action on the other side of the equation.
5. Show all necessary steps and work as you perform the mathematical operations. This will help you to prevent careless errors.
6. Check through your work to make certain you have reached a reasonable solution and that your answer is stated in the correct form, i.e., square feet, cubic inches, etc.

CHAPTER SUMMARY

Formula transposition is the process of solving a formula for the desired quantity by mathematically changing its original form.

CHAPTER TEST

Solve the following problems using the most efficient procedure. Be sure to show all necessary work in order to prevent costly errors. Remember to label your results correctly.

T-9-1. Solve $P = 2l + 2w$ for w.

T-9-2. Solve $A = \frac{bh}{2}$ for b.

T-9-3. Solve $C = 2\pi r$ for r.

T-9-4. Solve $Q = .7ELT$ for L.

T-9-5. Find s when $S = 4000$ and $r = 462$ in the formula $S = \pi rs$ ($\pi = 3.14$).

T-9-6. Solve $P = 2a + b$ for a.

T-9-7. Solve $a - b = ab$ for b.

T-9-8. Find r when $I = \$800$, $p = \$5000$, and $t = 2$ in the formula $I = prt$.

T-9-9. Solve $T = 2B + pH$ for H.

T-9-10. Solve $S = \frac{\pi rs}{2}$ for r.

T-9-11. Solve $C = \frac{5(F - 32)}{9}$ for F.

T-9-12. Find F if $C = 100$ in the formula of problem 11.

T-9-13. Solve $L = \frac{\pi rA}{180}$ for A.

T-9-14. Find A when $L = 314$ and $r = 60$ in problem T-9-13.

T-9-15. Solve $A = 6e^2$ for e.

T-9-16. Solve $A = \pi r^2$ for r.

T-9-17. Solve $S = 4\pi r^2$ for r.

T-9-18. From problem T-9-17, find r when $S = 900\pi$.

T-9-19. Find y when $a = 6$ and $b = 4$ in the formula $(y - a) - b^2 = 2ab$. (First, solve the formula.)

T-9-20. Find e when $V = 343$ cubic inches in the formula $V = e^3$.

T-9-21. Solve $V = \dfrac{\pi r^2 h}{3}$ for h.

T-9-22. Solve $M = \dfrac{YI}{r}$ for r.

T-9-23. Solve $\dfrac{p_1 V_1}{T_1} = \dfrac{p_2 V_2}{T_2}$ for T_1.

T-9-24. Solve $A = \dfrac{h(B + b)}{2}$ for b.

T-9-25. Solve $F = \dfrac{4^2 n^2 Wx}{g}$ for n.

CHAPTER

10

RATIO AND PROPORTION

OBJECTIVES

1. To define ratios and proportions.
2. To manipulate ratios and proportions.
3. To apply ratios and proportions to the needs of technology, industry, and everyday life.

SELF-TEST

The successful completion of this self-test indicates that you do not need to further develop the skills taught in this chapter. Read each problem carefully and carry out the operations as directed. Be sure to show your work so that problem areas may be more easily identified.

Simplify the ratios in fractional form:

S-10-1. 39 to 3

S-10-2. $18y$ to $3y$

S-10-3. $1\frac{3}{4} \div 1\frac{1}{2}$

S-10-4. 5 hours to 50 minutes

Solve for the unknown in the following problems.

S-10-5. $\frac{4}{10} = \frac{N}{40}$

S-10-6. $\frac{4}{12} = \frac{Y}{9}$

S-10-7. $\frac{2.5}{K} = \frac{10}{6}$

S-10-8. $4:X = X:16$

S-10-9. $27:r = 18:12$

S-10-10. $y:15 = 75:15$

S-10-11. What number is to 14 as 2.5 is to 6?

S-10-12. Nylon tricot sells at the rate of \$4.98 for 3 yards. How much will $\frac{1}{2}$ yard cost?

S-10-13. How much will 3 pounds of coffee cost if 2 pounds of coffee costs \$3.28?

S-10-14. A company needs 5 workers to produce 31 air conditioners per day. How many workers would be needed to produce 217 per day?

S-10-15. A motorist finds he averages 450 miles per day at a speed of 45 miles per hour. If he increases his speed to 55 miles per hour, how many miles would he average per day?

S-10-16. A stock pays a dividend of \$5.50 per share every 180 days. What would the dividends amount to in 1080 days?

S-10-17. A restaurant serves 175 people on the average each evening. Of those served, 82 order steak. The restaurant owner remodeled so that he now serves 255 people on the average each evening. If the ratio remains the same, how many steaks will he serve?

S-10-18. If you travel 594 miles in 11 hours, how long would it take to travel 486 miles?

S-10-19. Referring to the previous problem, how many miles per hour would you be traveling?

S-10-20. How far is it across the pond from point A to point B?

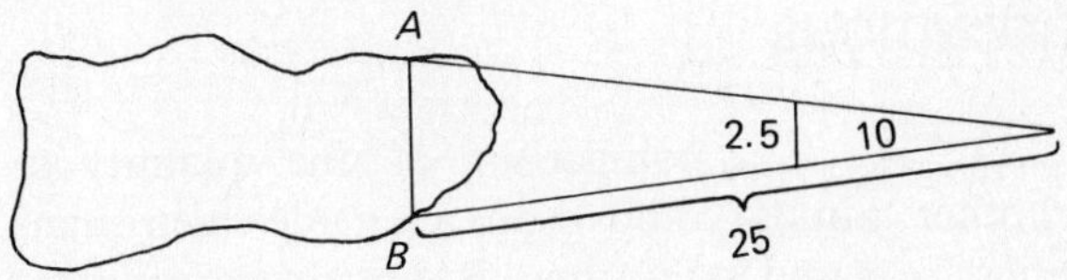

S-10-21. A blueprint lists the scale: $\frac{1}{4}$ inch to 6 inches. Determine the height of a doorway which measures 4 inches on the blueprint.

S-10-22. A copper wire is 360 feet long and has a resistance of 3.82 ohms. What is the resistance of 1000 feet of this wire?

S-10-23. The larger of two gears in mesh makes 35 revolutions per minute. The smaller gear has a diameter of 38 centimeters and makes 180 revolutions per minute. What is the diameter of the larger gear?

S-10-24. Craig shoots a score of 195 on a rifle range, using 25 rounds. What would his score be if he fired 35 rounds?

S-10-25. Two speed boats start at the same time from the same place and travel in opposite directions. The ratio of their rates is 1:1.5. In 3 hours, they are 60 miles apart. What is the rate of speed of the faster boat?

INTRODUCTION

We often compare one thing to another. For example, we may compare the gasoline consumption of a small car to that of a large car. If a small car uses 1 gallon of gasoline to travel 20 miles and a large car uses 1 gallon of gasoline to travel 10 miles, we could say the small car travels twice as far as the large car on 1 gallon of gasoline. The ratio of mileage is 2 to 1. We could also say that the ratios of miles traveled is 20 to 1 for the small car and 10 to 1 for the large car.

We also use ratios to compare energy efficiency, gears, cost effectiveness, inputs to outputs, and many other things in technology and industry.

DEFINITIONS

A *ratio* is a comparison of one quantity to another. Similar comparisons are made when quantities are stated in fractions. Ratios always compare like items. As the old saying goes, apples must be compared to apples and oranges to oranges—not apples to oranges.

Examples of ratios and how they may be stated are:

$$\frac{15}{60} = 15:60 \quad \text{or} \quad 15 \text{ is to } 60 = \frac{1}{4}$$

$$\frac{144}{12} = 144:12 \quad \text{or} \quad 144 \text{ is to } 12 = \frac{12}{1}$$

$$\frac{3x}{6y} = 3x:6y \quad \text{or} \quad 3x \text{ is to } 6y = \frac{x}{2y}$$

When two quantities are compared, they must be given in the same units so that, when they are simplified through division or cancellation, both numbers and units may be divided. As a result of the simplification process, a *true ratio* will be determined. When a ratio is inverted, an *inverse ratio* results.

CONTENT

When ratios are simplified, it is important for them to be in the same units. If they are not in the same units, convert them to like units. Generally this will require conversion to smaller units. Simplification or reduction should be only to whole numbers and not to decimals or fractions.

Example 10-1

PROBLEM: What is the ratio of 12 to 48?

SOLUTION: The ratio of 12 to 48 should be written in a fractional form and then reduced.

$$\frac{12}{48} = \frac{1}{4} = 1:4$$

Example 10-2

PROBLEM: Write the ratio of 50¢ to $3.00.

SOLUTION: First, convert $3.00 to cents by multiplying $3.00 by 100, because there are 100 cents in 1 dollar.

$$\begin{array}{rl} \$3.00 & \\ \times\ \ 100 & \text{cents per dollar} \\ \hline 000 & \\ 000\ & \\ 300\ \ & \\ \hline 300.00 & \text{cents} \end{array}$$

Second, reduce to lowest terms.

$$\frac{50}{300} = \frac{1}{6} = 1:6$$

Example 10-3

PROBLEM: What is the ratio of $5\frac{1}{2}$ to $\frac{1}{2}$?

SOLUTION: First, write the ratio as a fraction.

$$5\tfrac{1}{2} \text{ to } \frac{1}{2} = \frac{5\frac{1}{2}}{\frac{1}{2}} = 5\tfrac{1}{2} \div \frac{1}{2}$$

Second, carry out the indicated division.

$$5\tfrac{1}{2} \div \frac{1}{2} = \frac{11}{\cancel{2}_1} \times \frac{\cancel{2}^1}{1} = \frac{11}{1} = 11:1$$

Example 10-4

PROBLEM: What is the ratio of 6 hours to 45 minutes?

SOLUTION: First, convert 6 hours to minutes.

$$\begin{array}{rl} 6 & \text{hours} \\ \times\ 60 & \text{minutes per hour} \\ \hline 360 & \text{minutes} \end{array}$$

Second, set up the ratio and reduce to lowest terms.

$$\frac{\cancel{360}^{40}}{\cancel{45}_{5}} = \frac{\cancel{40}^{8}}{\cancel{5}_{1}} = \frac{8}{1} = 8:1$$

EXERCISE 10-1

Simplify the following ratios, reducing them to lowest terms

10-1. $\frac{4}{8}$

10-2. $\frac{8}{24}$

10-3. $\frac{32}{24}$

10-4. $\frac{48}{12}$

10-5. $\frac{2.4}{.3}$

10-6. $\frac{6\frac{1}{2}}{\frac{1}{2}}$

10-7. 7 to 42

10-8. 96 to 12

10-9. $\frac{5}{12}$ to $\frac{8}{15}$

10-10. 6.5 to .5

10-11. $\dfrac{3 \text{ minutes}}{1 \text{ hour}}$

10-12. 8 centimeters to 40 centimeters

10-13. 18 games to 2 games

10-14. 5 nickels to 4 dimes

10-15. 6 feet to 144 inches

10-16. 5 days to 6 weeks

10-17. \$10 to 25¢

10-18. 300 turns to 37.5 turns

10-19. a^2b^2 to ab

10-20. 6 feet 6 inches to 3 yards

DEFINITIONS

When one ratio is said to be equal to another ratio, a proportion is formed. A *proportion* is a statement of equality between two ratios.

There are several ways a proportion may be stated. Two of the more common are:

proportional form $\quad 2:3 = 8:12$

fraction form $\quad \dfrac{2}{3} = \dfrac{8}{12}$

A proportion has four terms. Using the above example, 2 is the *first term*, 3 is the *second term*, 8 is the *third term*, and 12 is the *fourth term*. The product of the first and fourth terms is called the product of the *extremes* (outside terms), and the product of the second and third terms is called the product of the *means* (inside terms). The product of the means equals the product of the extremes.

CONTENT

There are times when only three of the four terms of a proportion are known. In order to determine the fourth term, obtain the product of the extremes and set it equal to the product of the means. This process is often referred to as setting the cross products equal to each other. Once the products have been determined, the unknown can be determined.

Example 10-5

PROBLEM: Solve the proportion $X:8 = 6:24$ for the unknown.

SOLUTION: First, determine the product of the means and the product of the extremes.

$X:8 = 6:24 \qquad (X)(24) = (8)(6)$

(8 and 6: means; X and 24: extremes)

Second, divide both sides by 24.

$$\frac{(X)(\overset{1}{\cancel{24}})}{\underset{1}{\cancel{24}}} = \frac{(\overset{1}{\cancel{8}})(\overset{2}{\cancel{6}})}{\underset{\underset{1}{\cancel{3}}}{24}}$$

$$X = 2$$

Third, enter the now-known first term, and check for correctness.

$$2:8 \;=\; 6:24$$

means (8 and 6); extremes (2 and 24)

$$(2)(24) = (8)(6)$$
$$48 = 48$$

Example 10-6

PROBLEM: Solve the proportion for the unknown.

$$\frac{2}{3} = \frac{K}{12}$$

SOLUTION: First, determine the cross products of the fractional proportion.

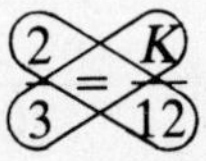

$$(2)(12) = (3)(K)$$

Second, divide both sides by 3.

$$\frac{(2)(\overset{4}{\cancel{12}})}{\underset{1}{\cancel{3}}} = \frac{(\overset{1}{\cancel{3}})(K)}{\underset{1}{\cancel{3}}}$$

$$8 = K$$

Third, enter the now-known term, and check for correctness.

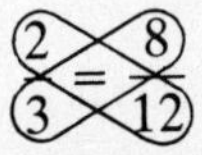

$$(2)(12) = (8)(3)$$
$$24 = 24$$

Example 10-7

PROBLEM: Solve the proportion for the unknown.

$$\frac{30}{5} = \frac{10}{X}$$

SOLUTION: First, determine the cross products of the fractional proportion.

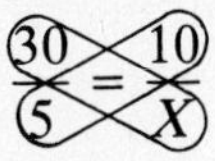

$$(30)(X) = (10)(5)$$

Second, divide both sides by 30.

$$\frac{(\cancel{30})(X)}{\cancel{30}} = \frac{(\cancel{10})(5)}{\underset{3}{\cancel{30}}}$$

$$X = \frac{5}{3} \text{ or } 1.66$$

Third, enter the now-known term, and check for correctness.

$$(30)(X) = (10)(5)$$

$$\frac{(\overset{10}{\cancel{30}})(5)}{\underset{1}{\cancel{3}}} = (10)(5)$$

$$50 = 50$$

EXERCISE 10-2

Solve the following propertions for the unknown and check for correctness.

10-21. $\frac{5}{8} = \frac{j}{24}$

10-22. $\frac{3}{7} = \frac{X}{49}$

10-23. $\frac{80}{r} = \frac{60}{5}$

10-24. $\frac{60}{360} = \frac{k}{45}$

10-25. $\frac{10}{s} = \frac{.4}{36}$

10-26. $\frac{y}{15} = \frac{40}{3}$

10-27. $\frac{6}{x} = \frac{27}{36}$

10-28. $\frac{.3}{24} = \frac{.2}{a}$

10-29. $4:2 = 13:x$

10-30. $16:y = 52:8$

10-31. $x:40 = \frac{1}{4}:2$

10-32. $y:12 = 9:27$

10-33. $112:128 = 42:x$

10-34. $36:9 = 28:x$

10-35. $\frac{(x-4)}{20} = \frac{14}{70}$

10-36. $\frac{9}{x} = \frac{x}{1}$

10-37. $\frac{36}{4x} = \frac{9x}{49}$

10-38. $\frac{x}{4.3} = \frac{19.7}{21.1775}$

10-39. $\frac{45}{5y} = \frac{9y}{64}$

10-40. $\frac{.400}{.014} = \frac{.08}{p}$

CONTENT

Many occupations within industry require the skills necessary to solve problems involving ratio and proportion. The following examples relate to some of the daily needs of industry.

Example 10-8

PROBLEM: If a copying machine produces 48 copies in 30 seconds, how many seconds will it take to produce 240 copies?

SOLUTION: First, set up the proportion in fractions and let X represent the unknown.

$$\frac{48}{30} = \frac{240}{X}$$

$$(48)(X) = (30)(240)$$

Second, solve the proportion, remembering that cross products are equal.

$$\frac{48}{30} = \frac{240}{X}$$

Third, divide each side of the equation by 48.

$$\frac{\overset{1}{\cancel{(48)}}(X)}{\underset{1}{\cancel{48}}} = \frac{(30)\overset{5}{\cancel{(240)}}}{\underset{1}{\cancel{48}}}$$

Fourth, multiply through.

$$\frac{(1)(X)}{1} = \frac{(30)(5)}{1}$$

$$X = 150$$

Thus, it will take 150 seconds to produce 240 copies.
Fifth, check for correctness.

$$\frac{48}{30} = \frac{240}{150}$$

$$(48)(150) = (30)(240)$$

$$7200 = 7200$$

Example 10-9

PROBLEM: A man wishes to lift a balled tree out of a shallow hole. He estimates that the balled tree weighs 420 pounds. If a 12-foot plank is used, where will the fulcrum need to be placed if the man weighs 180 pounds?

SOLUTION: First, recall the principle of the lever.

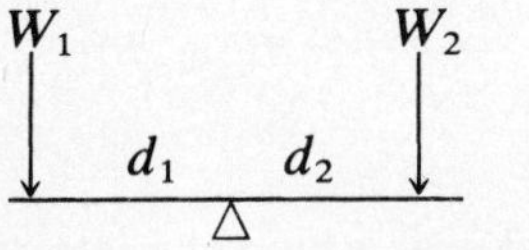

$$(W_1)(d_1) = (d_2)(W_2)$$

or

$$\frac{\text{Distance to the left}}{\text{Distance to the right}} = \frac{\text{Weight to the right}}{\text{Weight to the left}}$$

Second, draw a sketch of the problem.

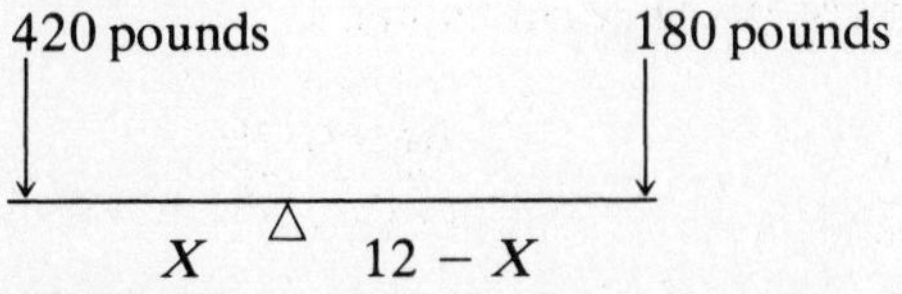

Third, recalling the principle of the lever, write the problem as a proportion.

$$(420)(X) = (12 - X)(180)$$

Fourth, solve for the unknown.

$$420X = 12(180) - 180X$$

$$420X + 180X = (12)(180)$$

$$600X = (12)(180)$$

$$\frac{\cancel{600X}}{\cancel{600}} = \frac{\overset{1}{\cancel{(12)}}\overset{18}{\cancel{(180)}}}{\underset{5}{\underset{\cancel{50}}{\cancel{600}}}}$$

$$X = \frac{18}{5}$$

$$X = 3.6$$

Fifth, check for correctness.

$$420(3.6) = (12 - 3.6)(180)$$

$$1512 = (8.4)(180)$$

$$1512 = 1512$$

Example 10-10

PROBLEM: A brick wall measures 2.8 millimeters on a drawing. How long is the wall if the scale of the drawing is 1 millimeter = 25 meters.

SOLUTION: First, recall metric measurement.

1000 millimeters = 1 meter

Second, set up the proportion

$$\frac{\text{Actual size}}{\text{Scale size}} = \frac{\text{Actual length}}{\text{Map length}}$$

$$\frac{25}{.001} = \frac{X}{2.8}$$

Third, solve for the unknown.

$$(25)(2.8) = (X)(.001)$$

$$\frac{(25)(2.8)}{.001} = \frac{(X)(.001)}{(.001)}$$

70,000 millimeters = X

or

70 meters = X

Fourth, check for correctness.

$$\frac{25}{.001} = \frac{70{,}000}{2.8}$$

$$(.001)(70{,}000) = (25)(2.8)$$

$$70 = 70$$

Example 10-11

PROBLEM: Two pulleys are connected by a belt. The diameter of the larger pulley is 60 centimeters and the diameter of the smaller pulley is 45 centimeters. If the larger pulley rotates at 1500 revolutions per minute, how fast does the smaller pulley rotate?

SOLUTION: First, sketch the pulleys and set up a proportion. Remember that pulleys and gears are to be set up in an inverse ratio.

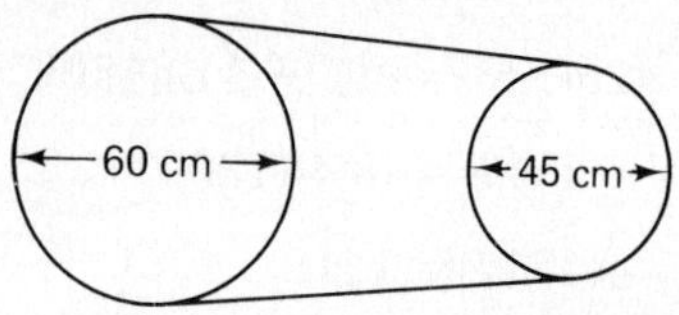

$$\frac{\text{Diameter of larger}}{\text{Diameter of smaller}} = \frac{\text{Rev. per minute of smaller}}{\text{Rev. per minute of larger}}$$

X = the revolutions per minute of smaller pulley

$$\frac{60}{45} = \frac{X \text{ revolutions per minute}}{1500}$$

Second, solve the ratio for X.

$$\frac{60}{45} = \frac{X \text{ revolutions per minute}}{1500}$$

$$(60)(1500) = 45(X \text{ revolutions per minute})$$

$$\frac{\overset{20}{\cancel{(60)}}\overset{100}{\cancel{(1500)}}}{\underset{\underset{1}{\cancel{3}}}{\cancel{45}}} = \frac{(\cancel{45})(X \text{ revolutions per minute})}{\cancel{45}}$$

$$2000 = X \text{ revolutions per minute}$$

Third, check for correctness.

$$\frac{60}{45} = \frac{2000}{1500}$$

$$(60)(1500) = (2000)(45)$$

$$90{,}000 = 90{,}000$$

EXERCISE 10-3

Work through the following problems which are similar to those which might arise in industry and daily living.

10-41. A baker sells 3 doughnuts for 49¢. How much will $1\frac{1}{2}$ dozen doughnuts cost?

10-42. A laborer is paid at a rate of $147 for every 3 days that he works. How much will he earn in 21 days?

10-43. A drawer is $1\frac{1}{2}$ feet wide. In a drawing the drawer is 1 inch wide. Determine the ratio of the actual width of the drawer to the width of the drawer in the drawing.

10-44. If 12 men can complete a job in 75 days, how many men would be required to complete the job in 10 days?

10-45. The scale on a map is 1 inch = 12 miles. How many miles is it from one town to another if the distance measures $7\frac{1}{4}$ inches on the map?

10-46. In Arkansas, the ratio of harvested land to the total area of the state is 1:4.8. If the total area of the state is 53,225 square miles, how many square miles is harvested land?

10-47. A machine can produce 60,000 nails in 12 hours. At this rate, how long would it take to produce 160,000 nails?

10-48. The population of a certain country is 50,000,000. The area of the country is 215,000 square miles. Determine the density of population (number of people per square mile).

10-49. A train travels 84 miles in $1\frac{2}{5}$ hours. How long will it take to travel 480 miles?

10-50. If 6 men take $10\frac{1}{2}$ days to complete a job, how long will it take 10 men? (Remember the inverse ratio concept.)

10-51. A cubic foot of water weighs 62.5 pounds and a cubic foot of ice weighs 57.5 pounds. What is the ratio of the weight of ice to water?

10-52. Find BC when $AD = 8$, $AB = 18$, and $DE = 6$.

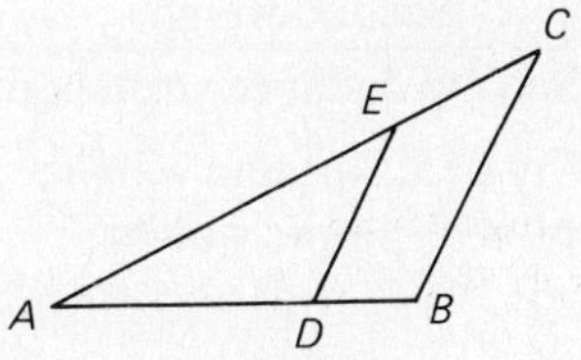

10-53. A picture is 24 inches long and 18 inches wide. If a reduced copy of this picture is 5.5 inches long, determine its width.

10-54. The volume of a given amount of gas is inversely proportional to the pressure on the gas. If a gas occupies 175 cubic inches when under a pressure of 20 pounds per square inch, what will be its volume if the pressure is decreased to 11 pounds per square inch?

10-55. The weight of dried apples to the weight of fresh apples is in the ratio of 2:5. How many pounds of fresh apples are needed to produce 490 pounds of dried apples?

10-56. A 40-acre field yields 880 bushels of wheat. When fertilizer is applied, the yield increases by $1\frac{1}{2}$ times. How much yield could be expected from an 84-acre field when fertilizer is applied?

10-57. A large gear with 30 teeth is turning at 400 revolutions per minute and turns a smaller gear with 12 teeth. If the revolutions per minute of the larger gear are tripled, determine the speed of the smaller gear.

10-58. If 16 pounds of cement are used to make 80 pounds of concrete, how many pounds of concrete can be made with 120 pounds of cement.

10-59. An important scientific principle of machinery concerning the stretching of springs according to weight is known as *Hooke's law.* It is stated as a proportion as follows:

$$\frac{\text{First weight}}{\text{First distance stretched}} = \frac{\text{Second weight}}{\text{Second distance stretched}}$$

A spring is stretched .5 inch by a 15.5-pound weight. How far will it be stretched by a 90-pound weight?

10-60. A small pulley, 4.5 inches in diameter, is belted to a 16-inch-diameter pulley. The smaller pulley is turning at 3500 revolutions per minute and is decreased by $\frac{1}{7}$; how fast is the larger pulley turning after the decrease?

10-61. A pry bar uses the same principles as the lever. If a bar is 60 inches long and a fulcrum is placed 6 inches from the end of a 1500-pound weight, how much weight must be applied to lift the weight?

10-62. Find the distance from the fulcrum to the base of a 2500-pound milling machine if the pry bar is 80 inches long and 220 pounds of force is applied.

10-63. A regular claw hammer, when used to pull a nail, also uses the principle of the lever (see sketch). If 80 pounds force (W_2) is applied on the handle which is 12 inches long, and if the distance from the claws to the nail is $2\frac{1}{4}$ inches, how much force can be applied to pull the nail?

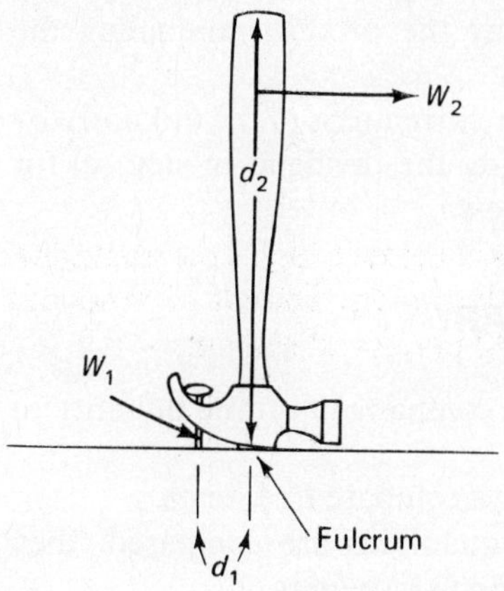

10-64. Referring to the sketch in problem 10-63. How much force is needed if the nail's downward force is 224 pounds?

10-65. The breaking force of the hammer handle (problem 10-63) is 125 pounds. What is the maximum pulling force that could be obtained from the hammer?

10-66. A chef prepares a banquet for 60 people. He uses 18 pounds of potatoes. How many pounds of potatoes would he need for a banquet of 210 people?

10-67. A painter uses 1 quart of paint to cover 225 square feet of wall. How much paint would he need to cover 1215 square feet?

10-68. Thunder is heard 48 seconds after lightning flashes 10 miles away. When would a person 6 miles away hear the thunder?

10-69. What will it cost to serve 99 people ground beef that costs $1.39 per pound, if 5 pounds will serve 22 people.

10-70. The maximum height reached by a bullet fired from a certain gun is proportional to the square of the time it takes to reach that height. If a bullet reaches a height of 5 feet .01 seconds after it is fired, how high would a bullet reach in half a second?

THINK TIME

Ratios and proportions may be used to solve a wide variety of problems in industry. For example, they may be used to determine screw thread depth and number, horsepower, mechanical efficiency of engines, taper and diameter, speed of pulleys and gears, strength of materials, velocity, density, and pressure. Think of ways in which you can use the principles of ratio and proportion to solve problems in your occupation.

One of the common uses of proportion is to determine the portion of one's home which can be used as a business expense for tax purposes. Follow through the simple example given below.

PROBLEM: Brian owns his home and uses one room exclusively as an office for his service station business. In this room he keeps his accounts and makes out required reports; at night he uses it to receive calls for emergency road service. He determines that the size of the room is 108 square feet. The total square footage of his home is 960. If his electric, fuel, and water bills total $755 for the year, how much of this can be allowed for the office as business expenses.

SOLUTION: First, set up a proportion.

$$\frac{108 \text{ square feet}}{960 \text{ square feet}} = \frac{X}{\$755}$$

Second, determine the product of the means and the product of the extremes and divide to determine the amount allowed.

$$(108)(\$755) = (x)(960)$$

$$\$81{,}540 = 960x$$

$$x = \$84.94$$

PROCEDURES TO REMEMBER

1. To simplify a ratio:
   (a) Convert to like terms.
   (b) Set up a fraction.
   (c) Reduce to lowest terms.
2. To solve a proportion for an unknown:
   (a) Set up the proportion, converting to like terms if necessary.
   (b) Determine the product of the means and the product of the extremes.
   (c) Divide the product containing the knowns by the product containing the unknown.
   (d) Check for correctness of the unknown by substituting the dividend in step (c) for the unknown.

CHAPTER SUMMARY

1. A *ratio* is a comparison of one quantity to another.
2. Ratios always compare like items.
3. When two quantities are compared, they must be given in like units.
4. A *true ratio* is determined through the simplification process.
5. When a ratio is inverted, an *inverse ratio* results.
6. Simplification or reduction should only be to whole numbers and not to decimals or fractions.
7. A *proportion* is a statement of equality between two ratios.
8. There are two common ways of expressing a proportion:
   (a) Proportional form ($3:4 = 9:12$).
   (b) Fractional form ($\frac{3}{4} = \frac{9}{12}$).
9. A proportion has four terms. In the example, $6:8 = 18:24$, 6 is the *first term*, 8 is the *second term*, 18 is the *third term*, and 24 is the *fourth term*.
10. The product of the first and fourth terms is called the product of the *extremes* (outside terms).
11. The product of the second and third terms is called the product of the *means* (inside terms).
12. The product of the means equals the product of the extremes.

CHAPTER TEST

Simplify the ratios and fractional forms in problems T-10-1 through T-10-4.

T-10-1. 4 inches to 12 inches

T-10-2. 7 to 63

T-10-3. 5 feet 8 inches to 6 feet 2 inches

T-10-4. $\dfrac{6\frac{1}{2}}{\frac{7}{8}} =$

T-10-5. $\dfrac{x}{7} = \dfrac{1.5}{1}$

T-10-6. $\dfrac{1.5}{2} = \dfrac{3.5}{p}$

T-10-7. $\dfrac{E}{10.5} = \dfrac{9}{31.5}$

T-10-8. $\dfrac{84}{20} = \dfrac{24}{N}$

T-10-9. $j:8 = 18:j$

T-10-10. $18:21 = M:63$

T-10-11. What number is to 28 as 5 is to 12?

T-10-12. A trucker finds he used 63.5 gallons of fuel to drive 762 miles. How many gallons will he need to drive 4575 miles?

T-10-13. Burlap material sells at a rate of $5.88 for 3 yards. How much will $6\frac{1}{2}$ yards cost?

T-10-14. Dividends of $.375 per share are paid every 90 days. How many days would it take to accumulate a total of $375.00 on 200 shares?

T-10-15. What is the ratio of two gears if one gear has 105 teeth and the other has 15 teeth?

T-10-16. Joel shoots a score of 164 on a rifle range using 30 rounds. What would his score be if he fires 45 rounds?

T-10-17. A jet plane travels 3972 miles in 6 hours. At this rate, how far can the plane travel in $4\frac{1}{2}$ hours?

T-10-18. Referring to the previous problem, what is the average rate of speed of the plane in miles per hour?

T-10-19. If there are 160 pounds of nickel in one ton of an alloy, how much nickel would be required to make 4500 pounds of the alloy?

T-10-20. A copying machine produces 35 copies in 20 seconds. How many copies will it produce in 5 minutes?

T-10-21. Find the missing dimension of the A-frame cottage.

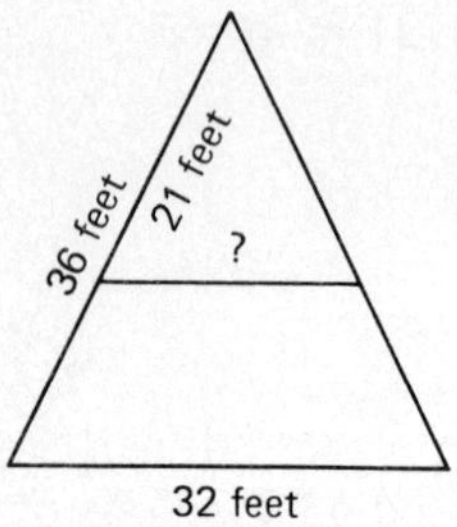

T-10-22. The scale on a drawing is $\frac{1}{8}$ inch = 3 feet. Determine the length in feet of a rafter if it measures $1\frac{3}{8}$ inches on the drawing.

T-10-23. A 20-inch pulley makes 1800 revolutions per minute and drives a larger pulley at 300 revolutions per minute. What is the diameter of the larger pulley?

T-10-24. A photograph has a length of 16 inches and a width of 11 inches. An enlargement is made having a length of 45 inches. Determine the width of the enlargement.

T-10-25. A chef uses 22 pounds of flour to bake the cakes he serves during the day. He generally serves cake to about 160 people. He decides to cut the price he charges for a piece of cake. Now 276 people order cake each day. How much flour does he use?

Section Three

APPLIED GEOMETRY

CHAPTER

11

Basic Foundations of Geometry

OBJECTIVES

1. To review the fundamental foundations of geometry.
2. To develop the fundamental skills required to solve geometric problems.

SELF-TEST

The test which follows will measure your understanding and ability to work with basic geometric symbols and forms. If you are able to complete the questions and problems successfully, you are ready to move on to more difficult geometric concepts. It is most important that you have a firm grasp of these geometric symbols and forms before you undertake more vigorous geometric problems.

S-11-1. The symbol $\perp$ indicates ___________?

S-11-2. The symbol $\overleftrightarrow{AJ}$ means ___________?

S-11-3. Write the symbol for line *NL*.

S-11-4. Write the symbol for angle *AOK*.

S-11-5. How long is a line?

S-11-6. If you were sailing due north and made a 270° turn to the right, in what direction would you be going?

S-11-7. How many degrees would the needle of the compass go through if you turned from north to south?

S-11-8. Three fourths of a complete revolution gives an angle of ____ degrees.

S-11-9. Minutes used to measure angles are subdivided into ___________?

S-11-10. The symbol for a straight line is ___________?

S-11-11. A triangle with all sides of equal length is called __________?

S-11-12. What angle do the hands of a clock make at 7 o'clock?

S-11-13. Measure angle *AOQ* with a protractor.

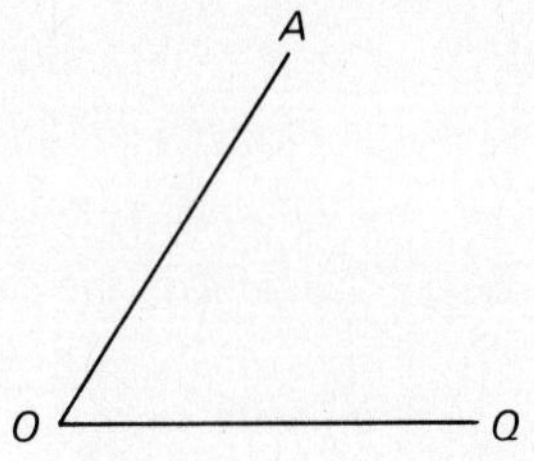

S-11-14. Measure angle *j* with a protractor.

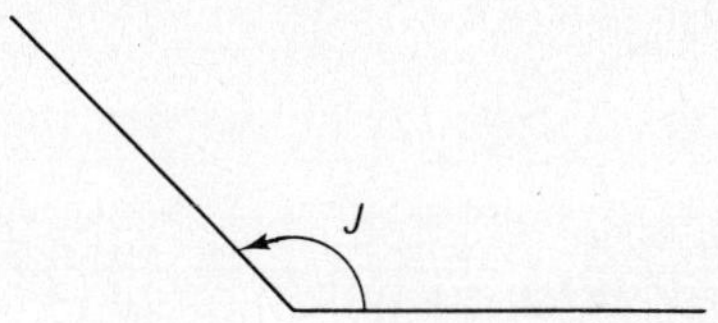

S-11-15. Construct a 153° angle with a protractor.

S-11-16. Construct a 73° angle with a protractor.

S-11-17. Find the complement of a 36° angle.

S-11-18. Find the supplement of a 46° angle.

S-11-19. How many degrees are there in one fourth of a right angle?

S-11-20. Measure ∠*WOZ*.

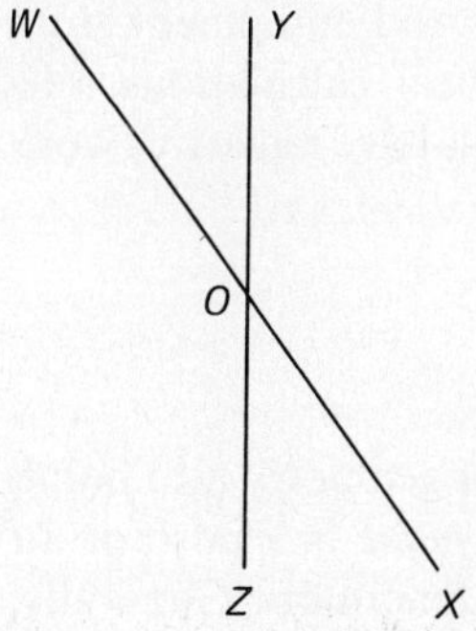

S-11-21. Measure ∠*WOY*.

S-11-22. Angles *WOZ* and *WOY* are called __________ angles.

S-11-23. Measure ∠*PON*.

S-11-24. Measure ∠*POM*.

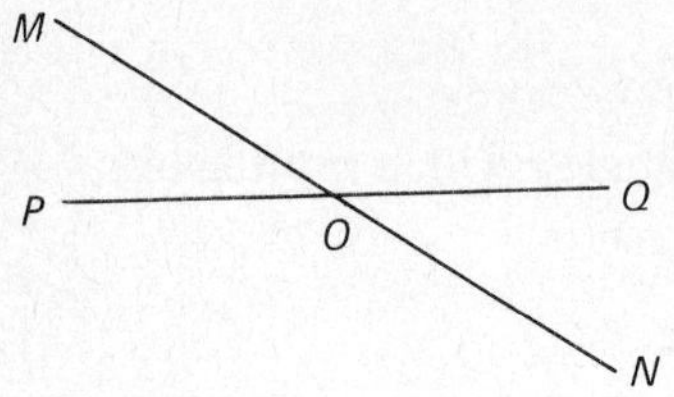

S-11-25. Measure ∠*POQ*.

INTRODUCTION

Geometry is an ancient science. Its oldest known roots can be traced back five or six thousand years to the ancient Egyptian and Babylonian civilizations. The precise construction of the pyramids is a monument to the geometric knowledge and skill of the Egyptians. These same skills were also used to reestablish boundary lines that were washed away each spring by the flooding of the Nile river.

The first great geometer was the Greek, Euclid, who lived about 300 B.C. Euclid gathered together the geometric principles from the past, combined them with his own concepts of geometry, and wrote a geometry textbook. This book contains geometric principles which have changed very little over the centuries. In fact, Euclid's textbook could be used today with very few changes. It is a rarity that any book of science or mathematics could be useful after such a long period of time. But Euclid did write an almost perfect geometry textbook even though he had but the simplest of tools and equipment from which to make his geometrical calculations. His geometric knowledge and skills have made our work and life easier.

DEFINITIONS

The four basic concepts of geometry are: point, line, straight, and plane. A *point* is a position or location which has no size or dimension. Generally, a point is represented by a dot with a capital letter beside it (.*P*). A *line* may be defined as a series or set of points which, when connected, extend indefinitely in opposite directions or may be arranged to form a straight, broken, or curved line. A line does not have width or depth. It has only length and is generally identified by a lowercase letter beside it:

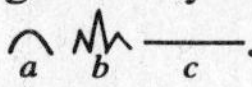

. A *straight* line is the shortest distance between two points. Straight does not mean curved, broken, or nonexistent. In general, when the word "line" is used, it means a straight line. The symbol for a straight line is $\underset{b}{\leftrightarrow}$. This symbol means that straight line *b* extends indefinitely in both directions. Generally it is assumed a line extends indefinitely, and consequently the arrows are not added. Two lines which never meet, no matter how far they are extended, are called *parallel* lines and are identified by the symbol ||. A *plane* is a flat surface which has two dimensions, length and width. The sides of a box are plane surfaces.

There are two basic types of geometry: plane and solid. *Plane geometry* deals with two dimensions: length and width. *Solid geometry* has three dimensions: length, width, and height. Plane geometry may be likened to a two-dimensional art form such as painting, while solid geometry may be likened to a three-dimensional art form such as a piece of sculpture.

CONTENT

The basic definitions given above will be of distinct value when you consider the more complex geometric concepts and forms given in this and the following two chapters.

As indicated previously, a plane is a flat surface that extends indefinitely in all directions. A *plane figure* is any figure that can be drawn on a flat surface. A *plane closed figure* is any plane figure that completely encloses an area or surface. Any plane figure can be described by lines, line segments, points, and angles. Observe the plane closed figures illustrated below. Note that they may be drawn with straight or curved lines.

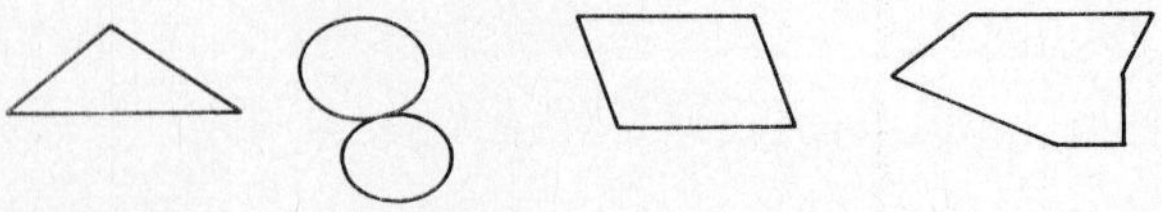

There are special names for certain types of plane closed figures. A *polygon* is a plane closed figure described by straight line segments and angles. A *triangle* is any polygon described by three straight line segments and angles. A *right triangle* is a triangle in which one of the angles is a right angle. An *equilateral triangle* is a triangle in which all three sides are of equal length. An *isosceles triangle* is a triangle with two sides of the same length. Observe the illustrations of the triangles below.

Equilateral triangle   Isosceles triangle   Right triangle

Four-sided polygons which are formed with parallel lines have special names. A *parallelogram* is a four-sided plane closed figure whose opposite sides are parallel. A *rectangle* is a parallelogram in which all angles are right angles. A *rhombus* is a parallelogram in which all sides are of equal length and no angles are right angles. A *square* is a parallelogram in which all sides are of equal length and all angles are right angles. Observe the illustrations of parallelograms.

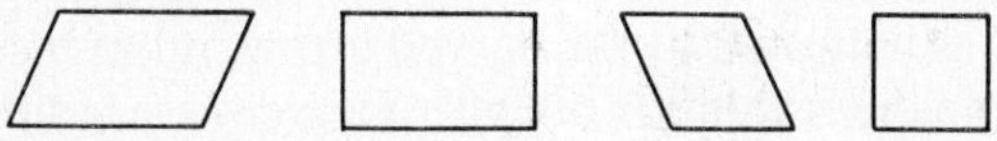

Parallelogram   Rectangle   Rhombus   Square

Observe the drawing of a wooden block. Because it has length, width, and height, it is called a *rectangular solid.* A solid is a figure which has depth as well as length and width. Letters have been placed at the corners (points) of the wooden block.

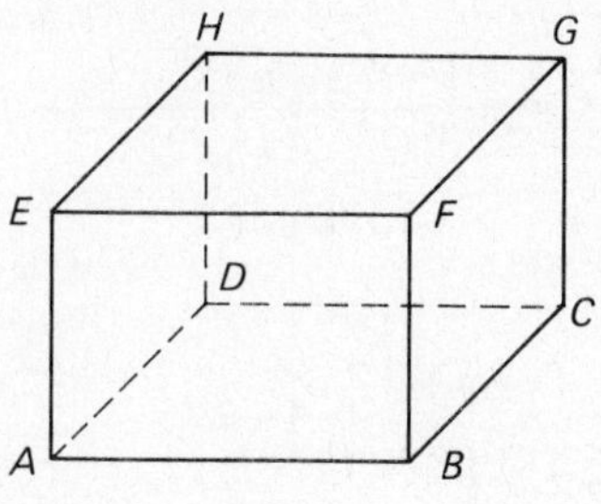

The line from point A to point B is called a *line segment* because it has *end points* A and B. The symbol for line segment AB is $\overline{AB}$. This means that line segment AB is only a portion of a line which extends indefinitely in both directions from the line segment AB. Line segment AB may be referred to as the edge of a rectangular solid.

Lines AB, BF, FE, and EA form a plane or a *face* of the rectangular solid. Generally, the plane would be called $ABFE$. Point F, where the lines EF, BF, and GF meet and where planes $ABFE$, $BCGF$, and $EFGH$ meet is called the *vertex* of the rectangular solid.

Example 11-1

PROBLEM: Sketch a rectangular solid, letter it, and determine its planes, line segments, and vertices.

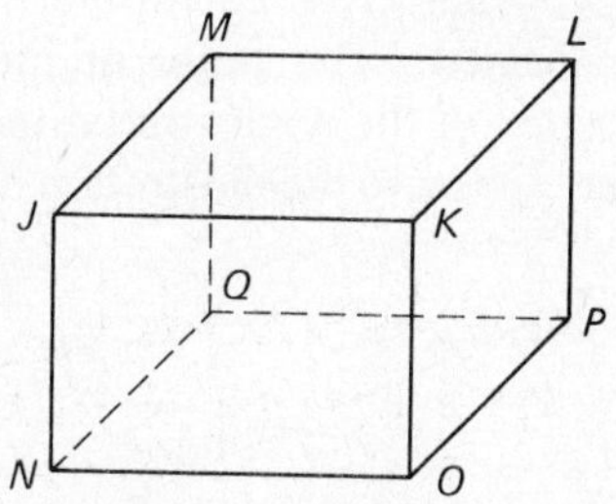

SOLUTION: First, draw the rectangular solid.
Second, letter the rectangular solid.
Third, determine the planes. The planes are $JKLM$, $JKON$, $JMQN$, $KLPO$, $LMQP$, and $NOPQ$.
Fourth, determine the line segments. The line segments are JK, JN, JM, KL, KO, LM, LP, MQ, NO, NQ, OP, and PQ.
Fifth, determine the eight vertices. The vertices are:

J, formed by planes $JKLM$, $JKON$, $JMQN$

K, formed by planes $JKLM$, $JKON$, $KIPO$

L, formed by planes $JKLM$, $LMQP$, $KLPO$

M, formed by planes $JKNO$, $JMQN$, $LMQP$

N, formed by planes $JKNO$, $JMQN$, $NOPQ$

O, formed by plames $JKNO$, $KLPO$, $NOPQ$

P, formed by planes $LMQP$, $KLPO$, $NOPQ$

Q, formed by planes $JMNQ$, $LMQP$, $NOPQ$

EXERCISE 11-1

Complete the following exercises which give you an opportunity to test your understanding of geometric definitions.

11-1. List three types of parallelograms.

11-2. List three types of triangles.

11-3. Draw and label a line segment.

11-4. Draw and label a rectangular solid.

11-5. What are the three dimensions of a solid geometry?

11-6. A triangle with three sides of equal length is called ___________?

11-7. Draw the symbol for parallel lines.

11-8. Draw the symbol for straight line *f*.

11-9. A plane closed figure described by straight line segments and angles is called ___________.

11-10. A triangle with a right angle is called ___________?

CONTENT

When two lines cross each other or intersect, angles are formed. The point of intersection is called the *vertex* of the angles and is indicated by a capital letter. Observe the illustration below. Lines

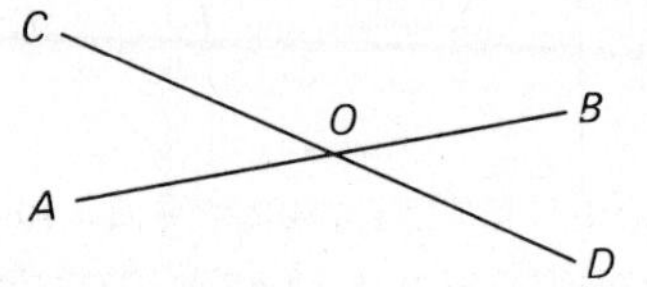

AB and *CD* intersect at point *O*. Point *O* is the vertex of the angles. Each angle is indicated by the angle symbol (∠) and the three letters which make up the angle. The angles for the illustration are ∠*COB*, ∠*COA*, ∠*AOD*, and ∠*BOD*. Note that the middle letter of each angle is always the vertex of the angle. Angles may also be indicated by a lowercase letter inside an angle (∠*a*). Angles may also be formed by rotating one line from another, provided they have a common end point. These lines are called *rays*. By definition, a *ray* is a part of a line that begins at a given point and goes in one direction without ending. The symbol for a ray is $\overrightarrow{JK}$. Observe the illustration below.

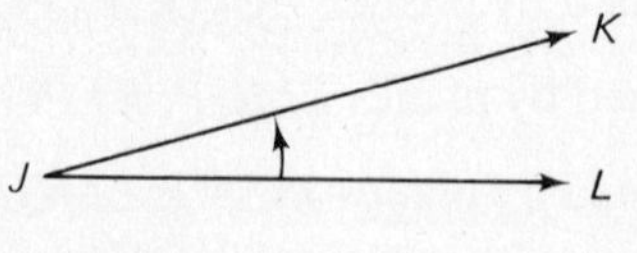

Angle *KJL* is formed by rotating ray *JK* in a counterclockwise direction from ray *JL*. Unless otherwise indicated, angles are always generated in a counterclockwise direction as indicated by the curved arrow between the rays *JK* and *JL*. If a ray or line segment were rotated completely around another ray or line segment from a common vertex, it would make a complete revolution. A complete revolution is divided into 360 parts called *degrees.* Note the illustration.

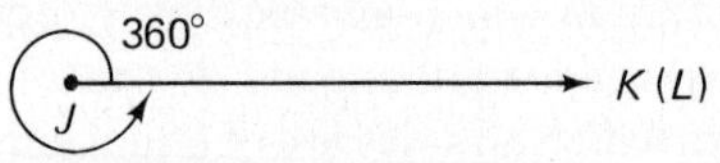

When ray *JK* is rotated completely around from ray *JL* as indicated by the curved arrow, it has made a complete revolution of 360°. Degrees are indicated by the symbol °. Thus 35 degrees is written as 35°. For more accurate measurement, each degree is divided into sixty parts called *minutes* and is indicated by the symbol ′. Each minute is divided into sixty parts, called *seconds,* and is indicated by the symbol ″. When a line or ray rotates 180° or half of a complete revolution, it forms a *straight angle.* Note the illustration below. When a line or ray rotates

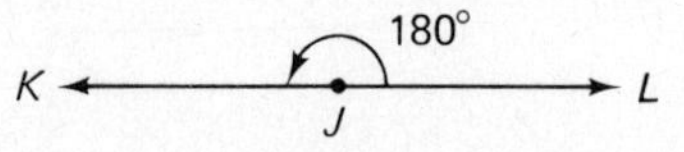

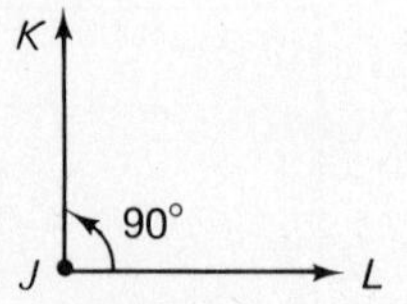

90° or a quarter of a complete revolution, it forms a *right angle.* The symbol for a right angle is ∟. Lines or rays that form a right angle are said to be *perpendicular* to each other. The symbol for perpendicular is ⊥.

An angle greater than 90° and less than 180° is called an *obtuse* angle. An angle less than 90° is called an *acute* angle. Two angles whose sum is 90° are called *complementary* angles. Two angles whose sum is 180° are called *supplementary* angles.

A simple illustration of angles is the hands on the face of a clock. Each hour, the minute hand of a clock moves 360°. If the point where the hands of the clock are attached is considered to be the vertex of an angle and the hands are considered as sides or rays of the angle, an infinite number of angles would be formed. Since a clock has twelve numbers, the degrees between numbers, $\frac{360°}{12}$, is 30°. If a clock has indications for minutes, the degrees between each minute would be $\frac{360°}{60}$, or 6°.

Example 11-2

PROBLEM: What central angle is formed if the hands of a clock indicate that the time is 4 o'clock?

SOLUTION: If necessary, draw a sketch of a clock indicating a time of 4 o'clock.

First, count the number of hours between 12 o'clock and 4 o'clock. The number is 4.

Second, multiply the number of hours (4) times the degrees between numbers on the face of the clock (30°) to arrive at the angle indicated by the hands of the clock (120°).

$$4 \times 30° = 120°$$

EXERCISE 11-2

Work through the following exercises which deal with geometric definitions.

11-11. Sketch and label two intersecting line segments and their point of intersection.

11-12. Sketch three lines and have them intersect at three different points. Label the points of intersection *A*, *B*, and *C*.

11-13. How many degrees are there in a half revolution?

11-14. How many degrees are there in the complement of a 68° angle?

11-15. How many degrees are there in a quarter revolution?

11-16. How many degrees are there in the complement of a 36° angle?

11-17. How many degrees are there in the supplement of a 58° angle?

11-18. How many degrees are there in an acute angle?

11-19. How many degrees are there in a straight angle?

11-20. How many degrees are there in a right angle?

11-21. How many degrees are there in a fourth of a right angle?

11-22. The number of degrees in a circle is equal to ____ straight angles and ____ right angles.

11-23. How many degrees are there in an obtuse angle?

11-24. Name this angle.

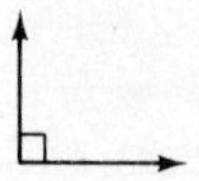

11-25. What type of angle is this?

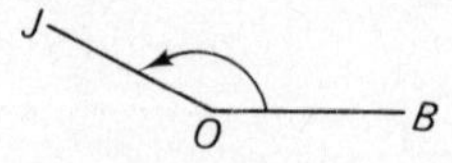

11-26. What type of angle is this?

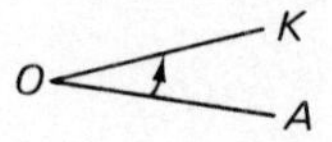

11-27. What central angle will the hands of a clock form at 2 o'clock?

11-28. What central angle will the hands of a clock form at 5 o'clock?

11-29. What obtuse central angle will the hands of a clock form at 8 o'clock?

11-30. If you were driving due north and made a 90° turn to the right, what would be your new direction?

CONTENT

Angles are often constructed and measured through the use of a *protractor*. Study the illustration of a protractor below. The protractor is divided into 10-degree and 1-degree increments.

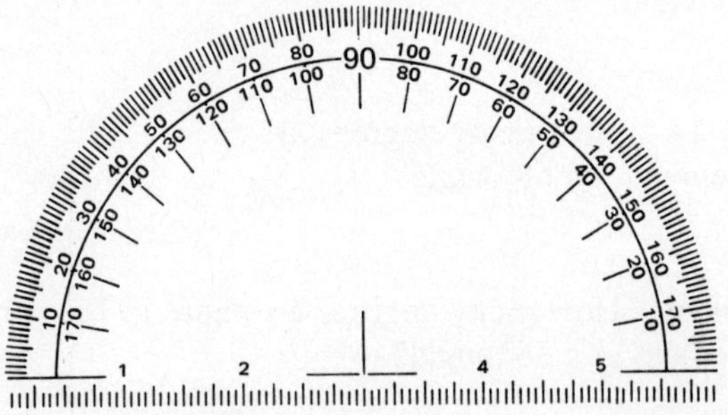

Example 11-3

PROBLEM: Measure the given acute angle using a protractor.

SOLUTION: First, place the protractor on the acute angle as illustrated. The flat edge of the protractor should be placed on ray *JL*, with the midpoint of the flat edge directly on point *J*.

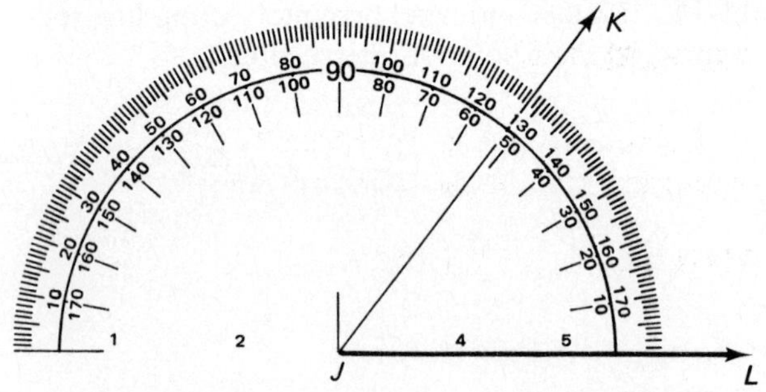

Second, read the inside degrees from right to left. Thus, *KJL* is 52°.

Example 11-4

PROBLEM: Measure the given obtuse angle using a protractor.

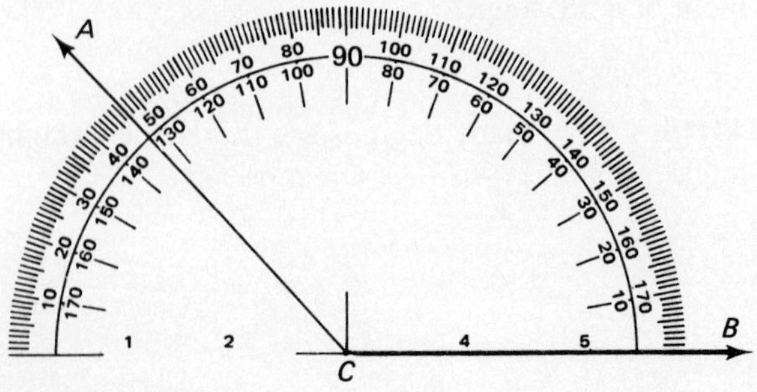

SOLUTION: First, place the protractor on the obtuse angle as illustrated. The flat edge of the protractor should be placed on ray *CB*, with the midpoint of the flat edge directly on point *C*.

Second read the inside degrees from right to left. Thus, *ACB* is 133°.

Example 11-5

PROBLEM: Construct an angle of 37°.

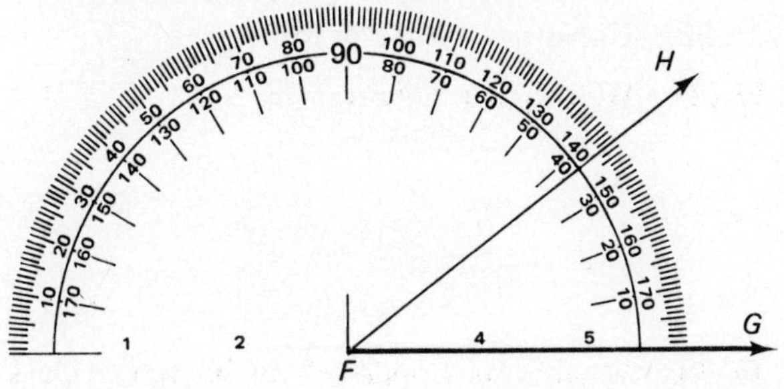

SOLUTION: First, draw a "working" ray, *FG*, and assign the vertex of the angle to point *F* on "working" ray *FG*.

Second, place the protractor on ray *FG*, with the midpoint of the flat edge directly at point *F*.

Third, read the inside degrees from right to left and extend a line from 37° to form ray *FH*.

Fourth, draw in ray *FH* to form ∠*HFG*.

Example 11-6

PROBLEM: Construct an angle of 162°.

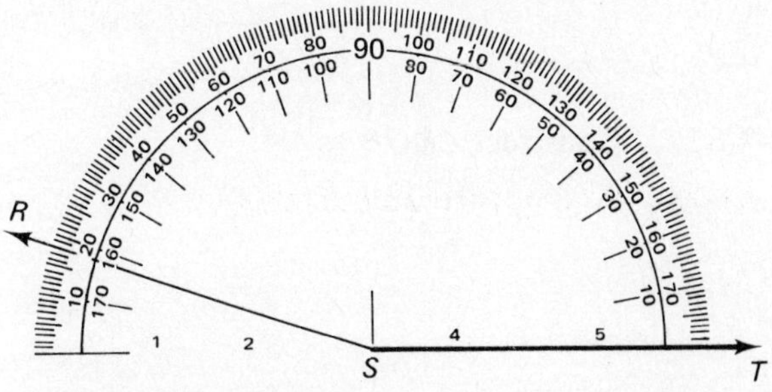

SOLUTION: First, draw a "working" ray, *ST*, and assign the vertex of the angle to point *S* on "working" ray *ST*.

Second, place the protractor on ray *ST*, with the midpoint of the flat edge directly at point *S*.

Third, read the inside degrees from right to left and extend a line from 162° to form ray *SR*.

Fourth, draw in ray *SR* to form ∠*RST*.

EXERCISE 11-3

Using a protractor, determine the degrees of the angles and draw angles as indicated.

11-31. In Figure A, ∠*AOB* = __________°.

11-32. In Figure A, ∠*BOD* = __________°.

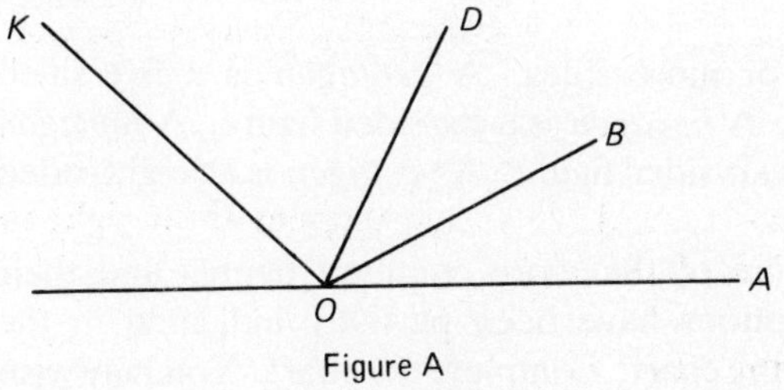

Figure A

11-33. In Figure A, ∠*AOK* = __________°.

11-34. In Figure B, ∠*RST* = __________°.

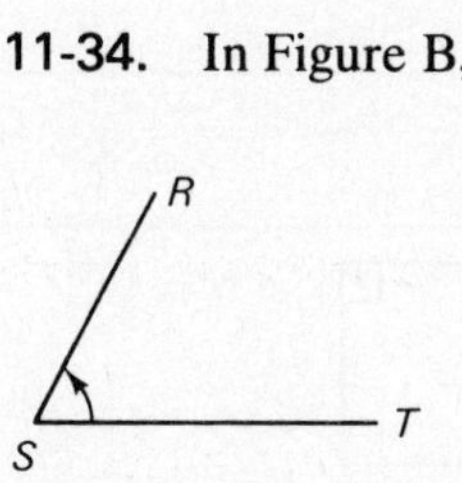

Figure B

11-35. In Figure C, $\angle PQR$ = __________°.

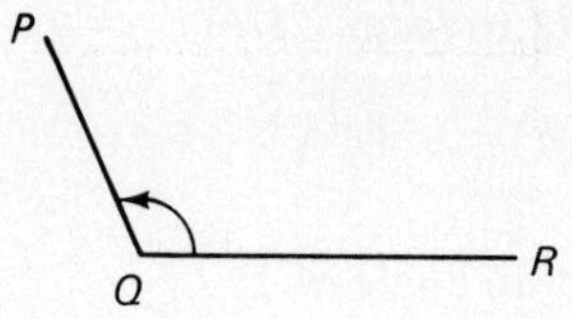

Figure C

11-36. In Figure D, $\angle XYZ \doteq$ __________°.

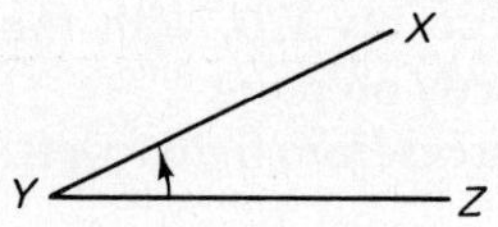

Figure D

11-37. Construct $\angle KOB$ as 48°.

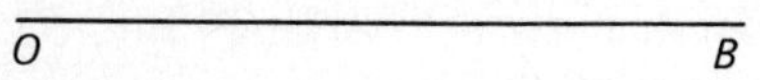

11-38. Construct $\angle ROT$ as 163°.

11-39. Construct the supplement of a 51° angle.

11-40. In Figure E, $\angle AOC$ = __________°.

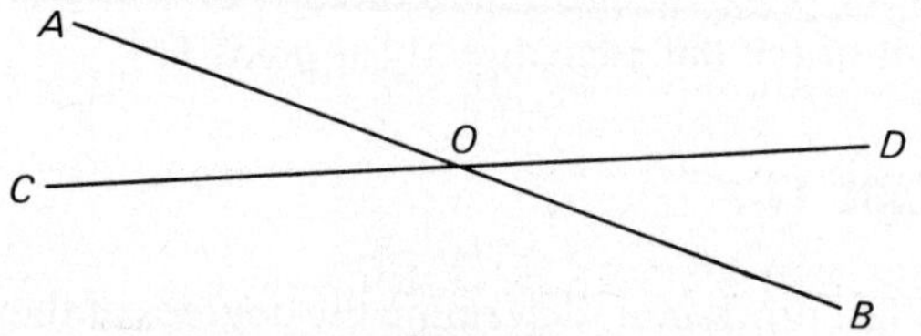

Figure E

THINK TIME

The number of possible geometric figures is infinite. The names for geometric figures often indicate the unique use or arrangement of sides or angles. Sometimes the figure is described by the number of sides used to construct a plane closed figure. A *polygon* is a geometric figure which has three or more sides. A *pentagon* is a five-sided figure. A *hexagon* is a six-sided figure. A *septagon* is a seven-sided figure. An *octagon* is an eight-sided figure.

Some of the more common figures and their descriptions have been partially indicated in the following chart. Complete the chart. You may wish to use it as a reference in the future.

Name	*Figure*	*Number of Sides*	*Special Properties*
1.	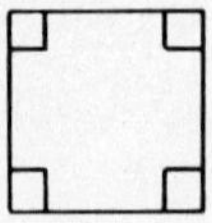		Equal-length sides Four 90° angles
2. Rectangle			

Name	*Figure*	*Number of Sides*	*Special Properties*
3.			
4.	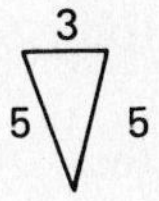		
5. Parallelogram			Both pairs of sides are parallel
6. Rhombus			Equal-length sides Opposite angles equal
7.	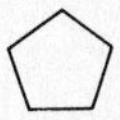	5	
8.	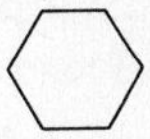		
9.	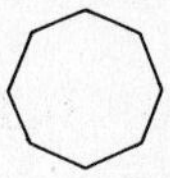		
10. Circle	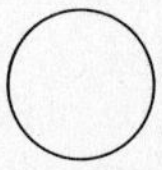		Curved line in which all points are equally distant from center

PROCEDURES TO REMEMBER

1. To measure the degrees of an angle, place the flat edge of a protractor on one of the rays or sides of the polygon so that the midpoint of the flat edge of the protractor is directly on the vertex. Read the inside degrees of the protractor if reading from right to left. Read the outside degrees if reading from left to right.
2. To construct an angle of a given number of degrees, draw a working ray and vertex, place the flat edge of the protractor on the working ray or side with the midpoint of the flat edge of the protractor on the vertex. Read the inside degrees from right to left and make a mark at the given number of degrees. Draw a line from the vertex through the mark to form the angle.

CHAPTER SUMMARY

1. The following geometric terms and their definitions are significant to the use of geometry.

(a) *Point* is a position or location which has no size or dimension.

(b) A *line* is a series or set of points which, when connected, extend indefinitely in opposite directions or may be arranged to form a straight, broken, or curved line.

(c) A *straight line* is the shortest distance between two points.

(d) A *plane* is a flat surface which has length and width but no depth or height.

(e) *Plane geometry* deals with two dimensions: length and width.

(f) *Solid geometry* deals with three dimensions: length, width, and height.

(g) A *plane figure* is any figure that can be drawn on a flat surface.

(h) A *plane closed figure* is any plane figure that completely encloses an area.

(i) A *polygon* is a plane closed figure described by straight line segments and angles.

(j) A *triangle* is any polygon described by three straight line segments and angles.

(k) A *right triangle* is a triangle in which one of the angles is a right angle.

(l) An *equilateral triangle* is a triangle in which all three sides are of equal length.
(m) An *isosceles triangle* is a triangle with two sides of the same length.
(n) A *parallelogram* is a four-sided plane closed figure whose opposite sides are parallel.
(o) A *rectangle* is a parallelogram in which all angles are right angles.
(p) A *square* is a parallelogram in which all sides are of equal length and all angles are right angles.
(q) A *rhombus* is a parallelogram in which all sides are of equal length and no angles are right angles.
(r) A *rectangular solid* is a geometric figure which has length, width, and height and whose six face planes are rectangular.
(s) A *line segment* is a portion of a line indicated by end points.
(t) A *vertex* of a solid is the point of intersection of three planes.
(u) A *ray* is a part of a line that begins at a given point and goes in one direction without ending.
(v) A *degree* is one of three hundred and sixty equal parts that indicate a complete revolution.
(w) A *straight angle* contains 180°.
(x) A *right angle* contains 90°.
(y) An *obtuse angle* is an angle greater than 90° and less than 180°.
(z) *Complementary angles* are two angles whose sum is 90°.
(aa) An *acute angle* is an angle less than 90°.
(bb) *Supplementary angles* are two angles whose sum is 180°.
(cc) A *protractor* is an instrument used to measure and construct angles.

2. The following geometric symbols are important for use in working problems in geometry.
   (a) Curved line a: 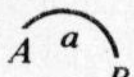

   (b) Broken line b: 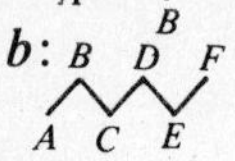

   (c) Straight line c: $\underset{c}{\leftrightarrow}$ or $\underset{c}{\overline{\quad}}$
   (d) Point D: $.D$
   (e) Parallel lines: $\parallel$
   (f) Line segment AB: $\overline{AB}$
   (g) Angle ABC: $\angle ABC$ or $\angle B$
   (h) Ray JK: $\overrightarrow{JK}$
   (i) 276 degrees: 276°
   (j) 53 minutes: 53′
   (k) 16 seconds: 16″
   (l) A right angle: ∟
   (m) Perpendicular lines: $\perp$

CHAPTER TEST

T-11-1. The symbol ∟ means __________ angle.

T-11-2. The symbol $\overrightarrow{AB}$ means __________.

T-11-3. Write the symbol for line segment LK.

T-11-4. Write the symbol for line n.

T-11-5. How long is a ray?

T-11-6. If you were flying south and made a 90° turn to the right, in what direction would you be going?

T-11-7. The symbol for perpendicular lines is __________?

T-11-8. An instrument used to measure degrees is called a __________?

T-11-9. Degrees used to measure angles are subdivided into __________?

T-11-10. Angles are labeled by three capital letters. The middle letter indicates the __________ of the angle.

T-11-11. A triangle with two equal length sides is a (an) __________ triangle.

T-11-12. What central angle, in degrees, would the hands of a clock make at 4 o'clock?

T-11-13. Measure angle r with a protractor.

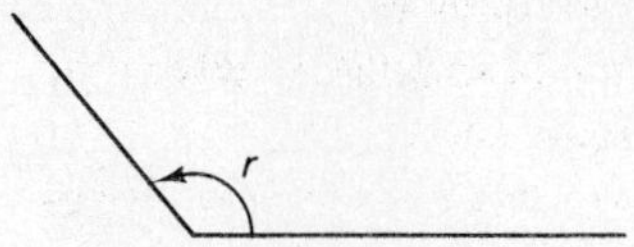

T-11-14. Measure angle PQR with a protractor.

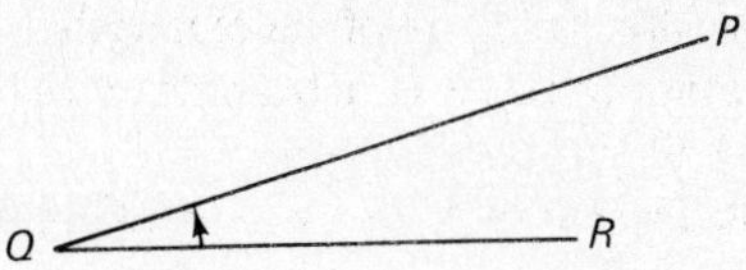

T-11-15. Construct a 39° angle with a protractor.

T-11-16. Construct a 121° angle with a protractor.

T-11-17. What is the complement of a 26° angle?

T-11-18. What is the supplement of a 46° angle?

T-11-19. How many degrees are there in one third of a straight angle?

T-11-20. A position or location which has no size or dimension is called __________?

T-11-21. The point of intersection of three planes or angles is called __________?

T-11-22. A triangle with equal-length sides is called __________?

T-11-23. Measure $\angle JOL$.

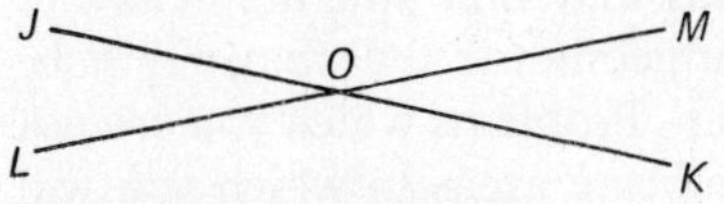

T-11-24. Measure $\angle JOM$.

T-11-25. Angles JOL and JOM are called __________ angles.

CHAPTER
12

PERIMETERS AND AREAS OF PLANE GEOMETRIC FIGURES

OBJECTIVES

1. To develop the skills necessary to determine the perimeter, or circumference, and the area of plane geometric figures.
2. To solve practical problems of industry and daily living which involve the determination of the perimeter, circumference, or area of a geometric figure.

SELF-TEST

The following test will measure your ability to determine the perimeter, circumference, or area of a plane geometric figure. Your successful completion of these problems will indicate that you have the knowledge and skills necessary to determine perimeters and areas and that you are ready to further develop your geometric skills beyond those taught in this chapter. Problems which you are not able to solve will indicate areas in which you will need further skill development.

S-12-1. What kind of geometric figure is Figure A?

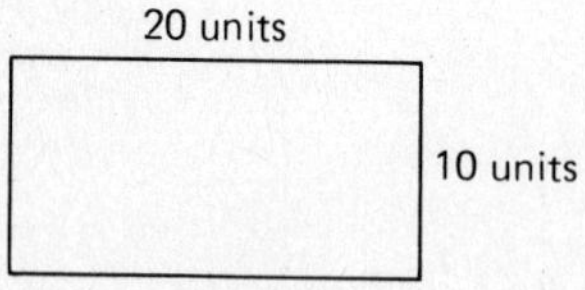

Figure A

S-12-2. Determine the perimeter of Figure A.

S-12-3. Find the area of Figure A.

S-12-4. What kind of a geometric figure is Figure B?

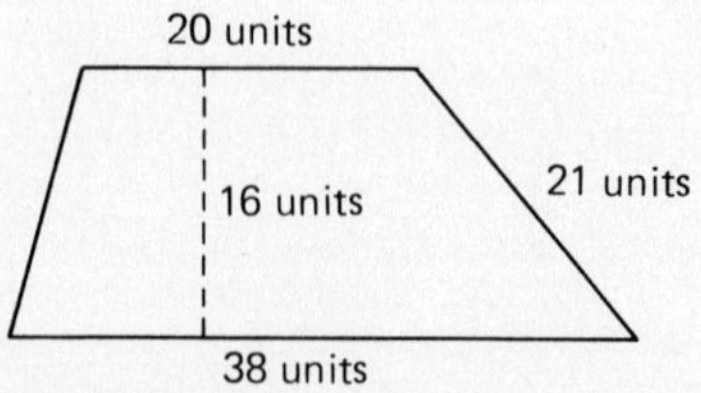

Figure B

S-12-5. Determine the perimeter of Figure B.

S-12-6. Find the area of Figure B.

S-12-7. What kind of a geometric figure is Figure C?

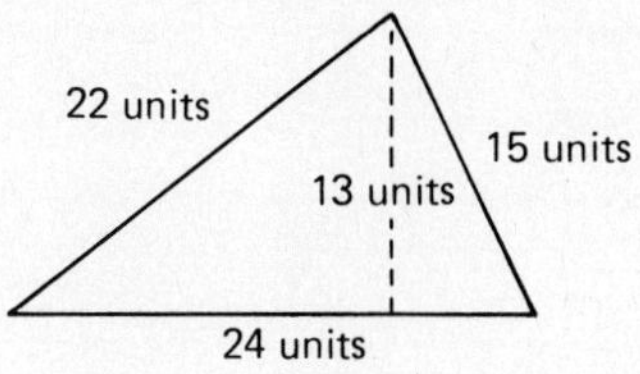

Figure C

S-12-8. Determine the perimeter of Figure C.

S-12-9. What is the altitude of Figure C?

S-12-10. What is the area of Figure C?

S-12-11. What is the diameter of Figure D?

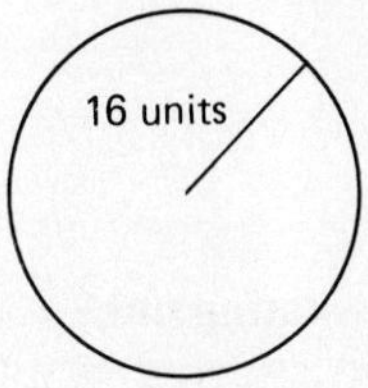

Figure D

S-12-12. Find the area of Figure D.

S-12-13. Determine the circumference of Figure D.

S-12-14. Determine the amount of fencing needed to fence in the garden area represented by Figure E.

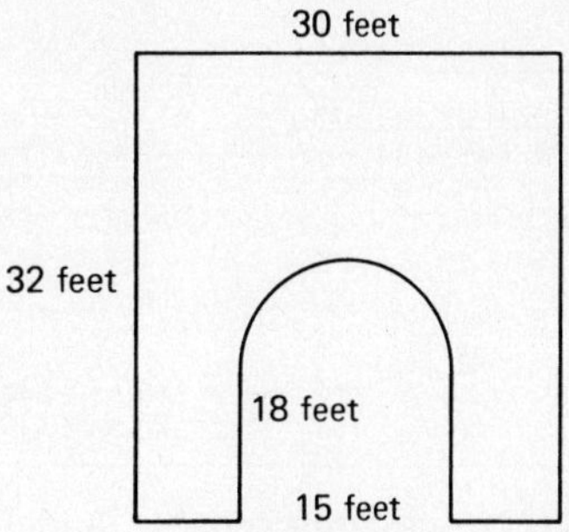

Figure E

S-12-15. What is the area of the garden?

S-12-16. Determine the inside perimeter of the sidewalk around the fountain illustrated by Figure F.

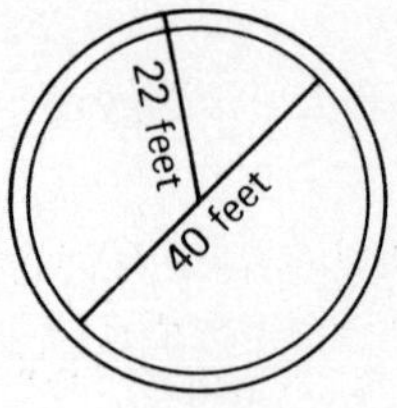

Figure F

S-12-17. What is the outside perimeter of the sidewalk?

S-12-18. What is the area covered by the sidewalk?

S-12-19. What is the perimeter of an ice-skating rink represented by Figure G?

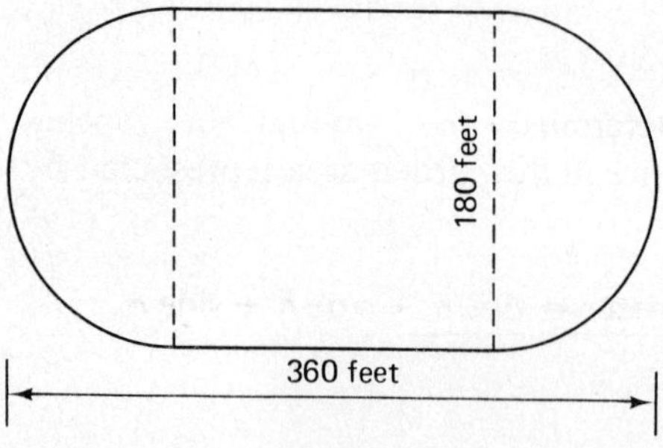

Figure G

S-12-20. What is the area of the ice-skating rink?

INTRODUCTION

The use of perimeters, circumferences, and areas of geometric figures is a daily experience for most people. However, geometric principles are seldom considered when a piece of bread is spread with peanut butter, a pie crust is rolled out, a wall is painted, or a tire size is selected. Yet in each case geometric problems are faced and solved either directly or indirectly.

The use of formulas to determine perimeters, circumferences, and areas can be most useful in saving both time and money at home or at work. Knowing the formulas for determining the areas of geometric figures is of great value when carpeting a room or buying a farm. Formulas for figuring perimeter are helpful when fencing land or framing a picture. Formulas for determining perimeters, circumferences, and areas can be of great service to craftsmen and tradesmen.

DEFINITIONS

Perimeter is the distance around a geometric plane. It is stated in linear measure such as inches, feet, miles, centimeters, and kilometers. The perimeter of a circle is called the *circumference* of the circle. *Area* is the amount of surface of a geometric plane. Generally the area of an object is calculated in square inches, square feet, square yards, square centimeters, square meters, etc.

CONTENT

The procedures for calculating perimeter, or circumference, and area can but be illustrated through the use of examples. Study each of the examples, noting the formulas used and the procedures followed.

Example 12-1

PROBLEM: Find the perimeter of the given square.

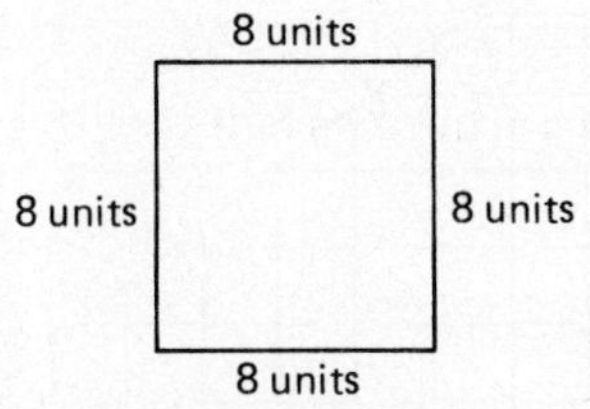

SOLUTION: The perimeter of the square can be determined by adding together the lengths of the four sides. Thus:

$$\text{Perimeter} = 8 \text{ units} + 8 \text{ units} + 8 \text{ units} + 8 \text{ units}$$
$$= 32 \text{ units}$$

The formula for determining the perimeter of a square is:

$$P = s + s + s + s$$

or

$$P = 4s$$

Example 12-2

PROBLEM: Find the perimeter of the given rectangle.

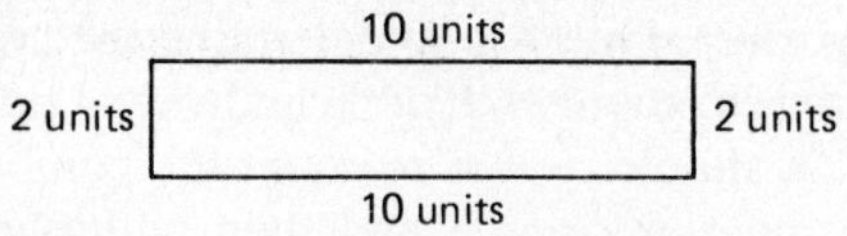

SOLUTION: The perimeter of the rectangle is the sum of the lengths of the four sides. Thus:

$$\text{Perimeter} = 10 \text{ units} + 2 \text{ units} + 10 \text{ units} + 2 \text{ units}$$
$$= 24 \text{ units}$$

The formula for determining the perimeter of a rectangle is:

$$\text{Perimeter} = \text{length} + \text{width} + \text{length} + \text{width}$$
$$P = l + w + l + w$$
$$= 2l + 2w$$
$$= 2(l + w)$$

Example 12-3

PROBLEM: Find the perimeter of the given triangle.

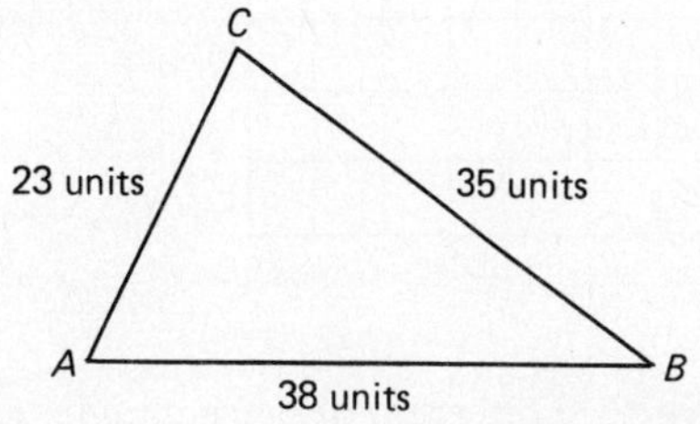

SOLUTION: The perimeter of the triangle is the sum of the lengths of the three sides. Thus:

$$\text{Perimeter} = 23 \text{ units} + 35 \text{ units} + 38 \text{ units}$$
$$= 96 \text{ units}$$

The formula for determining the perimeter of a triangle is:

$$\text{Perimeter} = \text{side } a + \text{side } b + \text{side } c$$
$$P = a + b + c$$

Example 12-4

PROBLEM: Find the area of the given square.

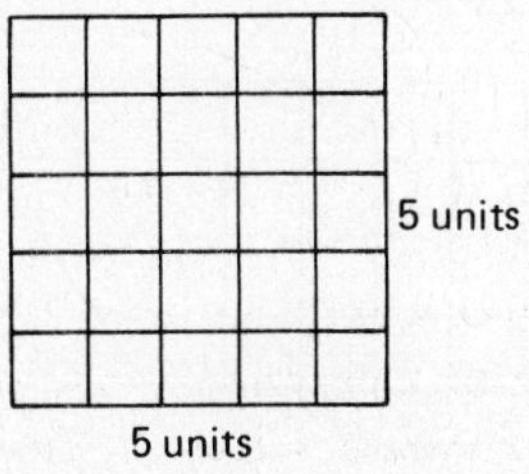

SOLUTION: One way to determine the area of the square is to count each of the individual smaller squares. A simpler way is to count the number of units in one row or side and then multiply this number by the number of rows. Thus:

$$\text{Area} = 5 \text{ units} \times 5 \text{ units}$$
$$= 25 \text{ square units}$$

The formula for determining the area of a square is:

$$\text{Area} = \text{side} \times \text{side}$$
$$A = s^2$$

It is important that area always be determined in like units. For example, units should be in square inches, square yards, square meters, etc.

Example 12-5

PROBLEM: Find the area of given rectangle.

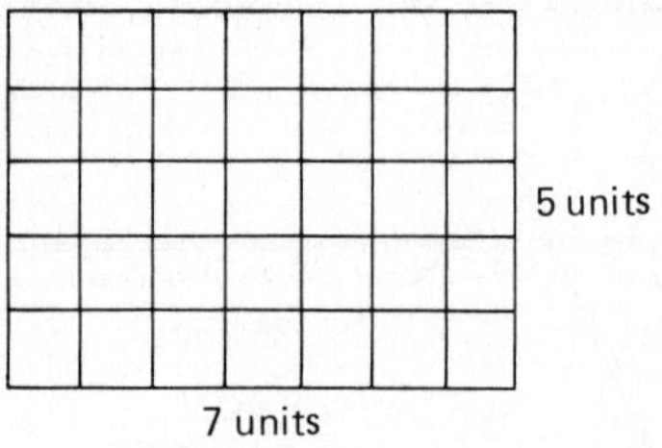

SOLUTION: The area of a rectangle can be determined by multiplying the number of units in one row (7) by the number of rows (5). Thus:

$$\text{Area} = 7 \text{ units} \times 5 \text{ units}$$
$$= 35 \text{ square units}$$

The formula for determining the area of a rectangle is:

$$\text{Area} = \text{length} \times \text{width}$$
$$A = lw$$

Example 12-6

PROBLEM: Find the area of the given parallelogram.

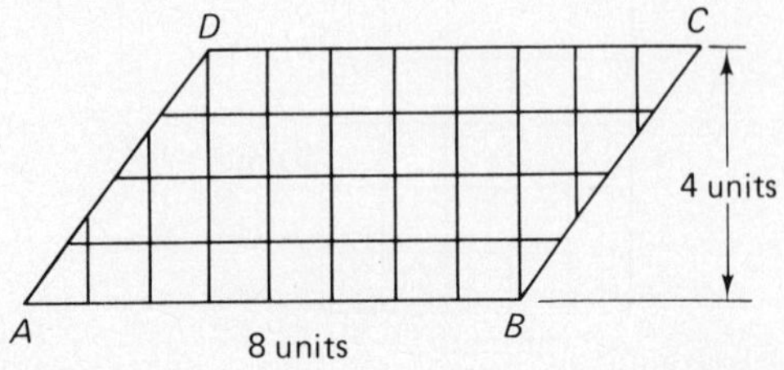

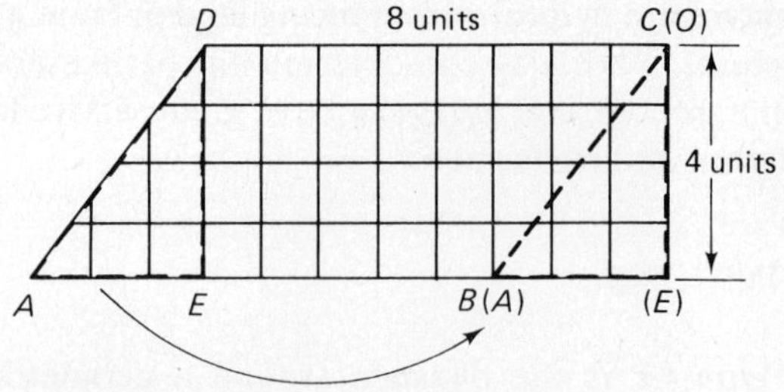

SOLUTION: One way to determine the area of the parallelogram is to move triangle *ADE* to the right of side *BC* to form rectangle *DCEE*, and then multiply length *DC* times width *DE*. A simpler way is to multiply the length *DC* times the height *DE*. Thus

$$\text{Area} = 8 \text{ units} \times 4 \text{ units}$$
$$= 32 \text{ square units}$$

The formula for determining the area of a parallelogram is:

$$\text{Area} = \text{length} \times \text{height (altitude)}$$
$$A = lh$$

Example 12-7

PROBLEM: Find the area of the given triangle.

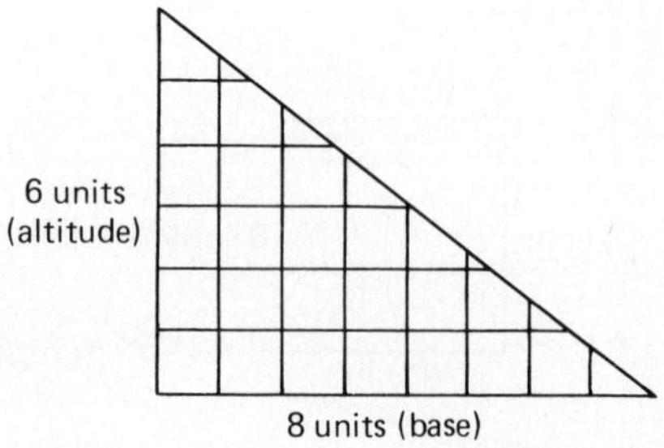

SOLUTION: One way to determine the area of the triangle is to form a rectangle as illustrated below.

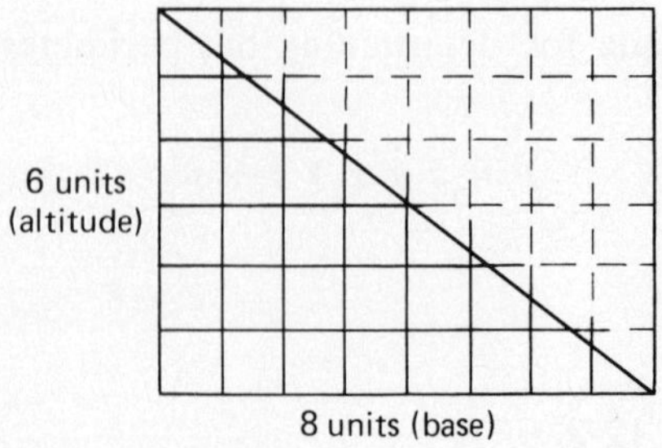

As shown in the sketch, the triangle is half of the area of the rectangle. Thus:

$$\text{Area} = \frac{8 \text{ units (base)} \times 6 \text{ units (altitude)}}{2}$$

$$= 24 \text{ square units}$$

The formula for determining the area of any triangle is:

$$\text{Area} = \frac{\text{base} \times \text{altitude}}{2}$$

$$A = \frac{ba}{2}$$

EXERCISE 12-1

Determine the perimeters, areas, and other items as called for in the following problems.

12-1. Find the perimeter of square A.

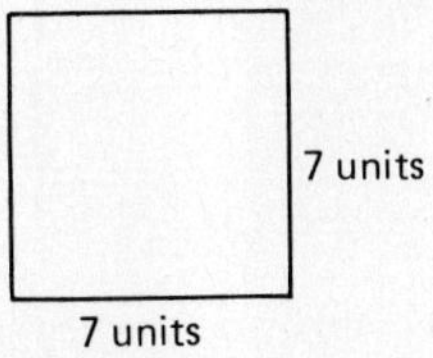

Square A

12-2. Find the area of square A.

12-3. Find the area of square B.

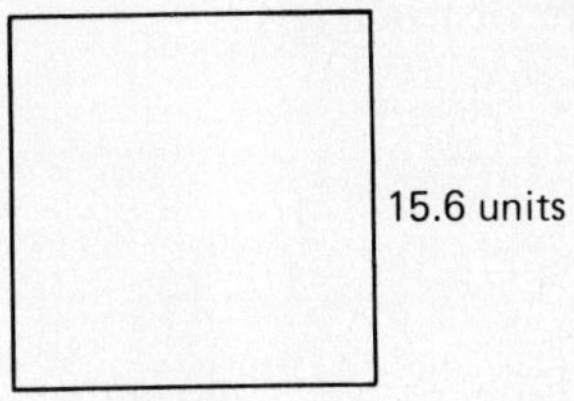

Square B

12-4. Find the perimeter of square B.

12-5. Find the area of rectangle C.

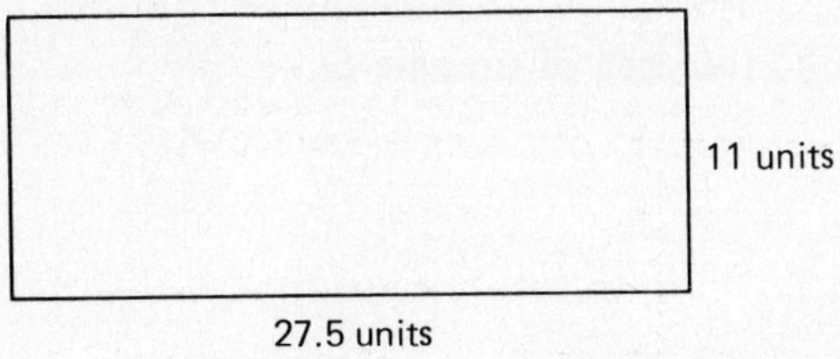

Rectangle C

12-6. Find the perimeter of rectangle C.

12-7. Find the perimeter of rectangle D.

12-8. Find the area of rectangle D.

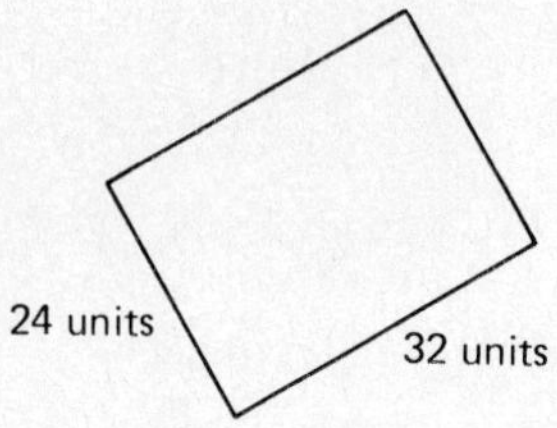

Rectangle D

12-9. Find the area of right triangle E.

12-10. Find the perimeter of right triangle E.

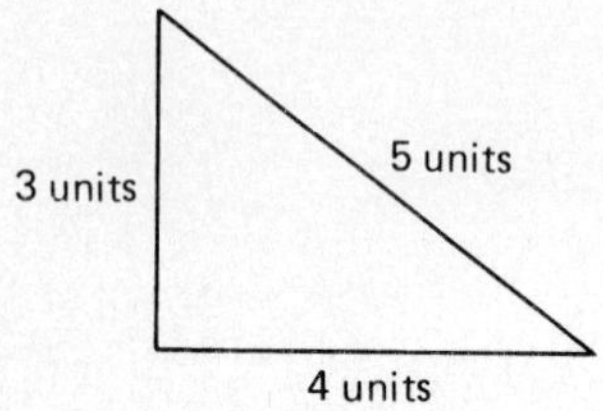

Triangle E

12-11. Find the area of triangle F.

12-12. Find the perimeter of triangle F.

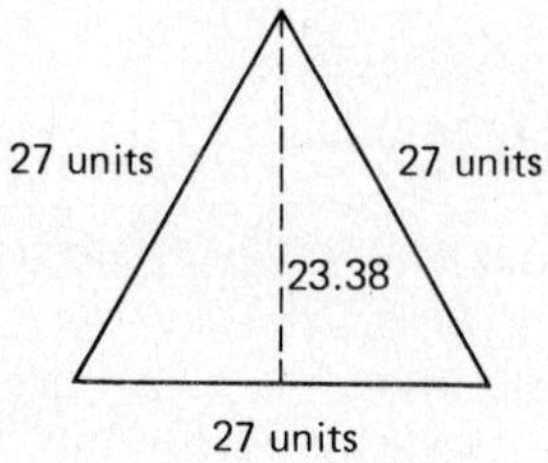

Triangle F

12-13. Find the perimeter of triangle G.

12-14. Find the area of triangle G.

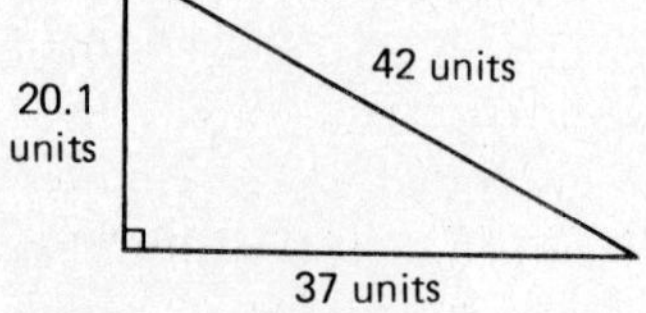

Triangle G

12-15. What is the altitude of triangle H?

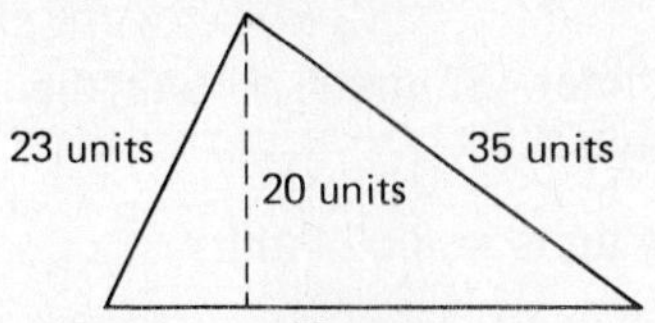

Triangle H

12-16. What is the perimeter of triangle H?

12-17. What is the area of triangle H?

12-18. What is the altitude of triangle I?

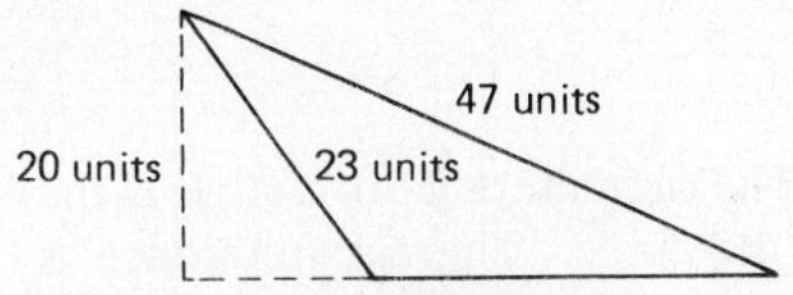

Triangle I

12-19. What is the perimeter of triangle I?

12-20. What is the area of triangle I?

DEFINITIONS

A *circle* is a set of points which are the same distance from a fixed point. The fixed point is called the *center* of the circle. The perimeter or distance around a circle is called the *circumference* of the circle. The following illustration shows a circle, its center, and its circumference.

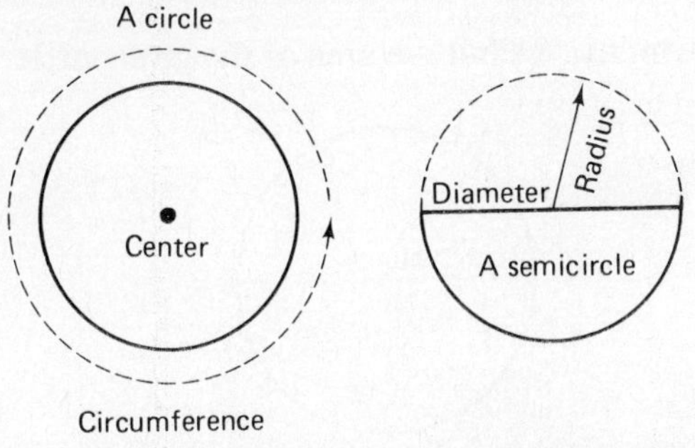

The *diameter* of a circle is a line drawn from one side of a circle, through the center, to the opposite side of the circle. The diameter divides the circle into two *semicircles.* Any line drawn from the center of a circle to the edge of the circle is called a *radius.* The radius is half of the diameter of a circle. Observe the illustration of a diameter, radius, and semicircle.

CONTENT

The *circumference* of a circle can be determined by multiplying the diameter of the circle times 3.14. The special number, 3.14, is called *pi* (pronounced "pie") and its symbol is the Greek letter π. The approximate numerical value for π, 3.14, is the same for all circles. This number is the ratio of the circumference of a circle to its diameter. A more exact value for π is 3.141592653. For calculation purposes, $\pi = 3.14$ or 3.142 or 3.1416 is used, depending on the accuracy desired. π was first determined by the ancient Egyptians as they developed a formula for finding the circumference of a circle.

The *area* of a circle can be determined by multiplying the radius of the circle *times* the radius of the circle *times* π.

Study the following examples for determining the circumference and area of a circle or semicircle.

Example 12-8

PROBLEM: Find the circumference of the given circle.

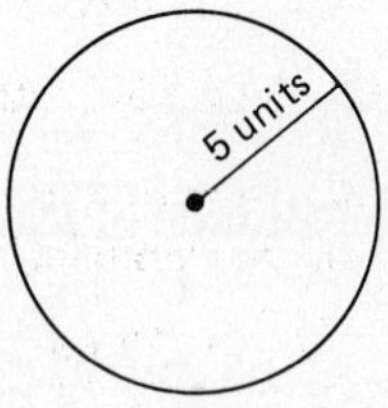

SOLUTION: The circumference of a circle is the distance around the circle. The circumference is found by multiplying the diameter (2 times the radius) by π, Thus:

$$\text{Circumference} = 2(5 \text{ units}) \times 3.14$$
$$= 31.4 \text{ units}$$

The formula for determining the circumference of a circle is:

$$\text{Circumference} = 2 \text{ times the radius times } \pi$$
$$C = 2r\pi$$

or

$$\text{Circumference} = \text{diameter times } \pi$$
$$C = d\pi$$

Example 12-9

PROBLEM: Find the circumference of the given semicircle.

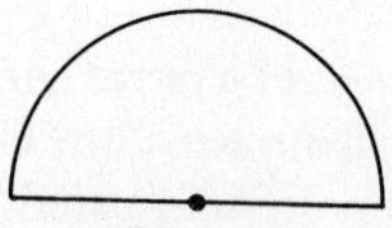

SOLUTION: The circumference of a semicircle is half the circumference of the whole circle plus the length of the diameter.

First, determine the circumference of the whole circle.

$$3.14 \times 32 \text{ units} = 100.48 \text{ units}$$

Second, divide the circumference (100.48 units) by 2.

$$\frac{100.48}{2} = 50.24 \text{ units}$$

Third, add the diameter (32 units) to half the circumference (50.24 units).

$$32 \text{ units} + 50.24 \text{ units} = 82.24 \text{ units}$$

The formula for determining the circumference of a semicircle is:

$$\text{Circumference of a semicircle} = \frac{\pi \text{ times diameter}}{2} + \text{diameter}$$

$$C = \frac{\pi d}{2} + d$$

Example 12-10

PROBLEM: Find the area of the given circle.

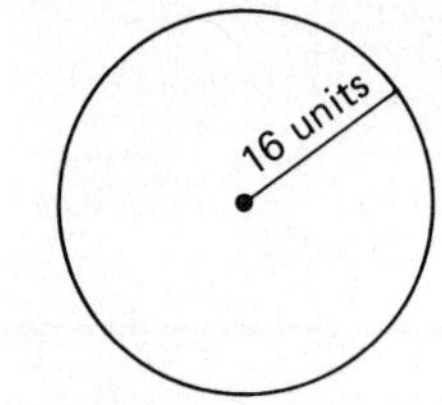

SOLUTION: The area of a circle is determined by multiplying π *times* the radius *times* the radius. Thus:

$$\text{Area} = 3.14 \times 16 \text{ units} \times 16 \text{ units}$$
$$= 803.84 \text{ square units}$$

The formula for determining the area of a circle using the radius is:

$$\text{Area} = \pi \text{ times radius times radius}$$
$$A = \pi r^2$$

Example 12-11

PROBLEM: Find the area of the given circle.

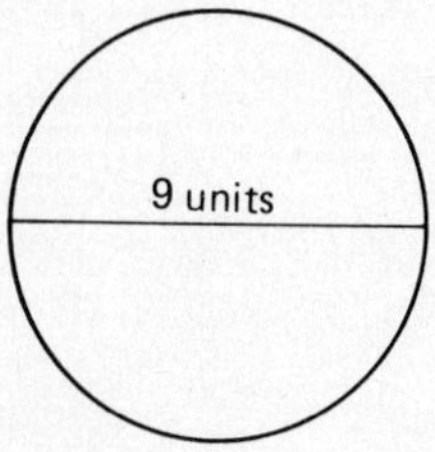

SOLUTION: The area of a circle, given the diameter, is determined by dividing the diameter by

2 to obtain the radius, and then multiplying π *times* the radius *times* the radius.

First, determine the radius.

$$9 \text{ units} \div 2 = 4.5 \text{ units}$$

Second, use the formula for determining area using the radius.

$$\text{Area} = 3.14 \times 4.5 \text{ units} \times 4.5 \text{ units}$$
$$= 63.585 \text{ square units}$$

The formula for determining the area of a circle using the diameter is:

$$\text{Area} = \pi \text{ times} \left(\frac{\text{diameter}}{2}\right)^2$$
$$A = \pi\left(\frac{d}{2}\right)^2$$
$$= \frac{\pi d^2}{4}$$

Example 12-12

PROBLEM: Find the area of the given semicircle.

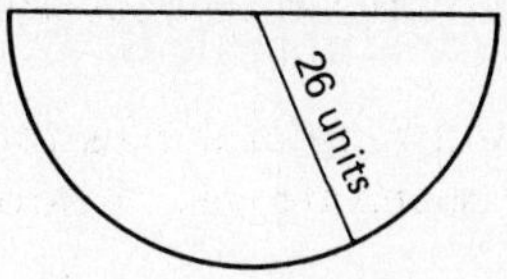

SOLUTION: The area of a semicircle is half the area of the whole circle.

First, determine the area of the whole circle.

$$\text{Area} = 3.14 \times 26 \text{ units} \times 26 \text{ units}$$
$$= 2122.64 \text{ square units}$$

Second, divide the total area of the circle by 2 to determine the area of the semicircle.

$$\frac{1}{2} \text{ area of a circle} = \frac{2122.64}{2} \text{ square units}$$
$$= 1061.32 \text{ square units}$$

The formula for determining the area of a semicircle is:

$$\text{Area of a semicircle} = \frac{\pi r^2}{2}$$

EXERCISE 12-2

Determine the area or circumference of the circles or semicircles in the following problems.

12-21. Find the diameter of circle A.

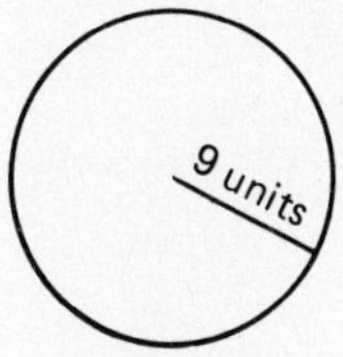

Circle A

12-22. Find the circumference of circle A.

12-23. Find the area of circle A.

12-24. Find the radius of circle B.

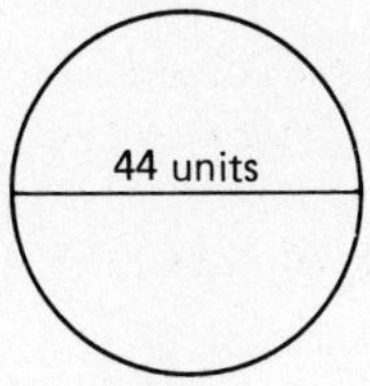

Circle B

12-25. Find the area of circle B.

12-26. Find the circumference of circle B.

12-27. Find the radius of circle C.

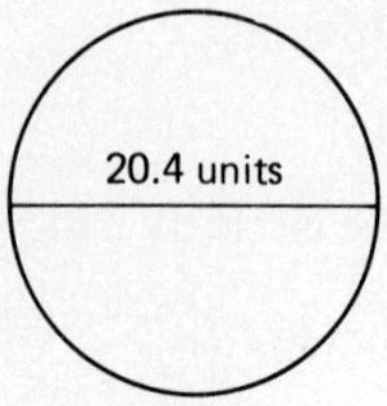

Circle C

12-28. Find the circumference of circle C.

12-29. Find the area of circle C.

12-30. Find the diameter of circle D.

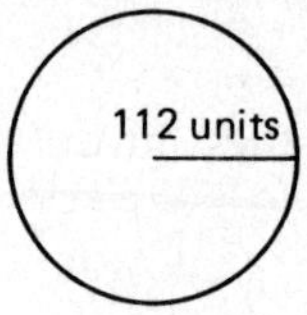

Circle D

12-31. Find the area of circle D.

12-32. Find the circumference of circle D.

12-33. Determine the radius of semicircle E.

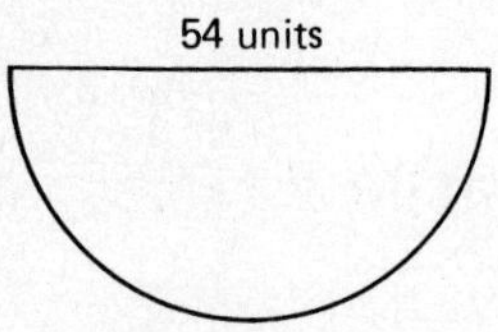

Semicircle E

12-34. Find the circumference of semicircle E.

12-35. Find the area of semicircle E.

12-36. Determine the diameter of semicircle F.

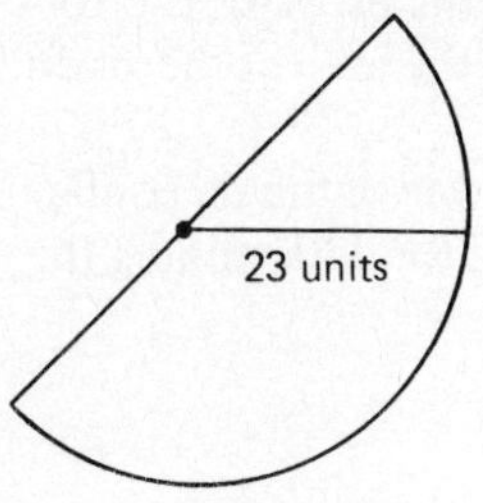

Semicircle F

12-37. Find the area of semicircle F.

12-38. Find the circumference of semicircle F.

CONTENT

The basic formulas for determining the perimeter and area of a simple geometric figure can also be used in more complex geometric figures. This is accomplished by dividing or changing the more complex figures into simple figures. The following examples will illustrate how some of the changes are made and the formulas that are involved.

Example 12-13

PROBLEM: Find the area of the given figure.

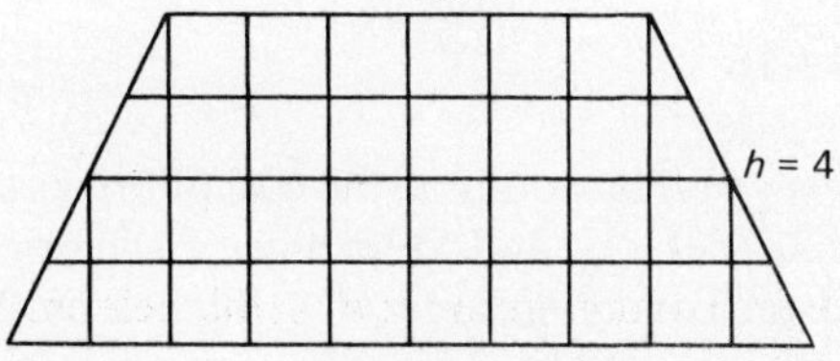

SOLUTION: The given figure is called a trapezoid. The upper base, b, and the lower base, B, are parallel to each other and the figure is closed. One way to determine the area of the trapezoid is to cut off the small triangles from each end of the trapezoid and reattach them to the upper corners to form a rectangle. See the illustration below.

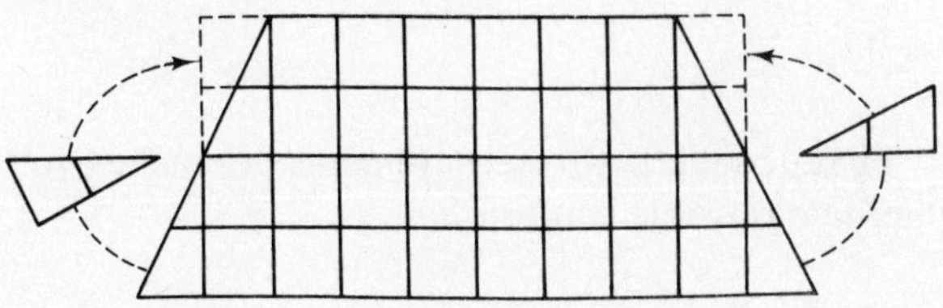

Note that the length of the newly formed rectangle is 8 units, or the average of the lengths of the upper base (6) and the lower base (10).

The height of the rectangle is 4 units. Thus:

$$\text{Area} = \frac{10 \text{ units} + 6 \text{ units}}{2} \times 4 \text{ units}$$

$$= \frac{16}{2} \text{ units} \times 4 \text{ units}$$

$$= 8 \text{ units} \times 4 \text{ units}$$

$$= 32 \text{ square units}$$

The formula for determining the area of a trapezoid is:

$$\text{Area} = \frac{\text{upper base} + \text{lower base}}{2} \times \text{height}$$

$$A = \left(\frac{B + b}{2}\right) h$$

Example 12-14

PROBLEM: Find the area of the given figure.

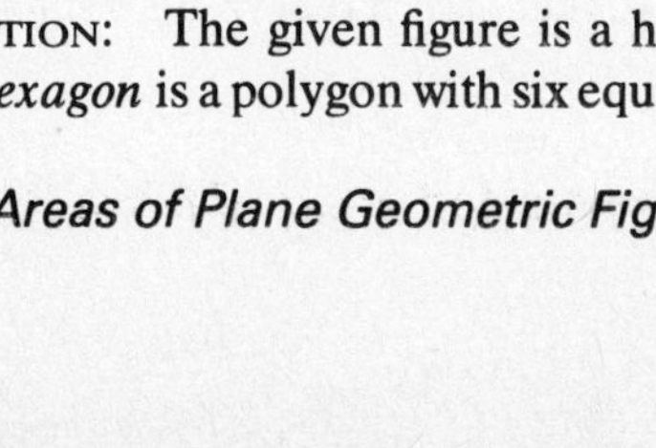

SOLUTION: The given figure is a hexagon. A *regular hexagon* is a polygon with six equal sides and angles, as shown in the sketch. Note that it can be divided into six equilateral triangles. Use the following procedure to determine the area of the given hexagon.

First, determine the area of one of the six equilateral triangles. Use the formula for determining the area of a triangle $\left(A = \frac{bh}{2}\right)$.

$$A = \frac{8 \text{ units} \times 6.9 \text{ units}}{2}$$

$$= \frac{55.2 \text{ square units}}{2} = 27.6 \text{ square units}$$

Second, multiply the area of one triangle (27.6 square units) by the number of triangles (6) to determine the area of the hexagon.

$$A = 27.6 \text{ square units} \times 6 = 165.6 \text{ square units}$$

The formula for determining the area of a hexagon is:

$$\text{Area} = 6\left(\frac{\text{base} \times \text{height}}{2}\right)$$

$$A = 6\left(\frac{bh}{2}\right)$$

Example 12-15

PROBLEM: Find the cost of carpeting for the given floor plan. The cost of carpeting is $12.00 per yard.

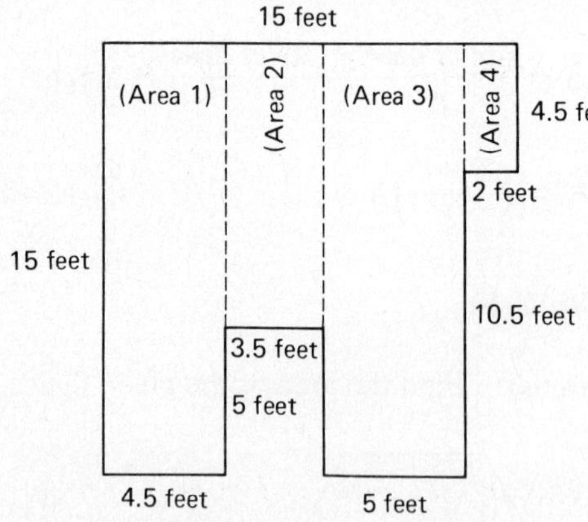

SOLUTION: First, divide the floor plan into rectangles as indicated by the broken lines.

Second, find the area of each of the rectangles.

Area 1 = 15 feet × 4.5 feet = 67.5 square feet

Area 2 = 10 feet × 3.5 feet = 35.0 square feet

Area 3 = 15 feet × 5 feet = 75.0 square feet

Area 4 = 4.5 feet × 2 feet = 9.0 square feet

Third, total the square footage of the four areas.

Area 1 =	67.5 square feet
Area 2 =	35.0 square feet
Area 3 =	75.0 square feet
+ Area 4 =	9.0 square feet
Total area =	186.5 square feet

Fourth, since 9 square feet = 1 square yard, divide 186.5 square feet by 9 to determine the number of square yards to be covered.

```
     20.7
 9 )186.50
    18
    06 5
     6 3
       2
```

The amount of carpet needed is 20.7 square yards.

Fifth, round off 20.7 square yards to the next full yard. Thus, 20.7 square yards becomes 21 square yards needed.

Sixth, determine the cost of the carpeting by multiplying the number of square yards (21) by the cost per square yard ($12).

```
     21    square yards
 × $12    cost per yard
     42
    21
$252.00   for 21 square yards
```

Example 12-16

PROBLEM: What would be the cost of bricks to build a wall 38 feet long and 4 feet high. Each brick with its mortar covers an area of $8\frac{1}{2}$ inches by $2\frac{3}{4}$ inches and its cost is $.28.

SOLUTION: Draw a sketch of the wall and the brick to gain a better understanding of what you are to do.

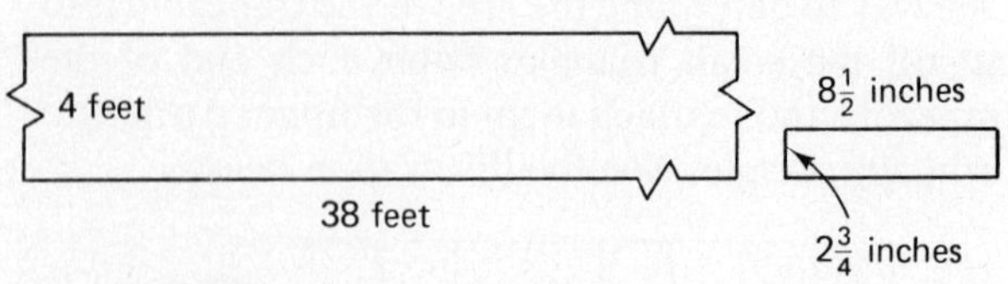

First, convert all measurements to inches to eliminate possible confusion.

38	feet	4	feet
× 12	inches per foot	× 12	inches per foot
76		48	inches
38			
456	inches		

Second, determine the area of the wall using the formula, $A = lw$.

456	inches (length)
× 48	inches (width)
3,648	
18,24	
21,888	square inches

Third, convert the mixed fractions to decimal fractions to simplify the process.

$8\frac{1}{2}$ inches = 8.5 inches

$2\frac{3}{4}$ inches = 2.75 inches

Fourth, determine the area of one brick using the formula, $A = lw$.

8.5	inches (length)
2.75	inches (width)
425	
595	
170	
23.375	square inches

Fifth, determine the number of bricks needed by dividing the area of the wall (21,888 square inches) by the area of one brick (23.375 square inches).

936.38 bricks

23.375)21,888.000.00

Sixth, change the decimal fraction to the next complete whole number. Thus, 936.36 bricks = 937 bricks.

Seventh, determine the total cost of the bricks by multiplying the total number of bricks (937) by the cost per brick ($.28)

937	bricks
× $.28	cost per brick
7496	
1874	
$262.36	total cost of bricks for the wall

EXERCISE 12-3

Complete the following problems which deal with perimeters, circumferences, and areas in order to further develop your skills in working with these kinds of problems.

12-39. What kind of a geometric figure is Figure A?

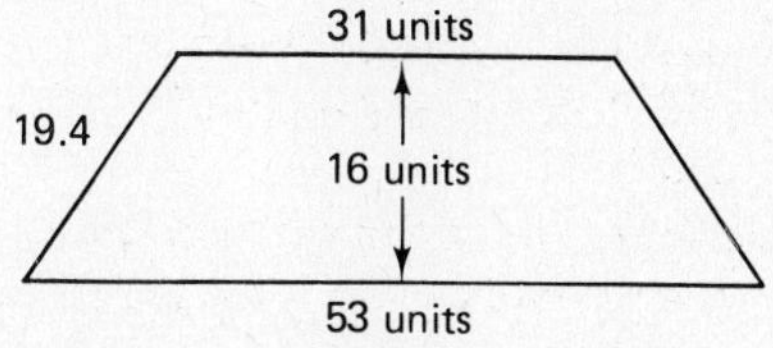

Figure A

12-40. Find the perimeter of Figure A.

12-41. Find the area of Figure A.

12-42. Find the perimeter of Figure B.

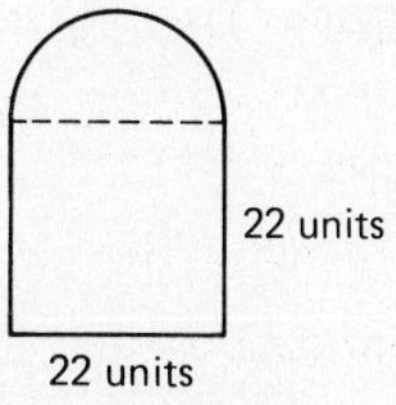

Figure B

12-43. Find the area of Figure B.

12-44. What kind of geometric figure is Figure C?

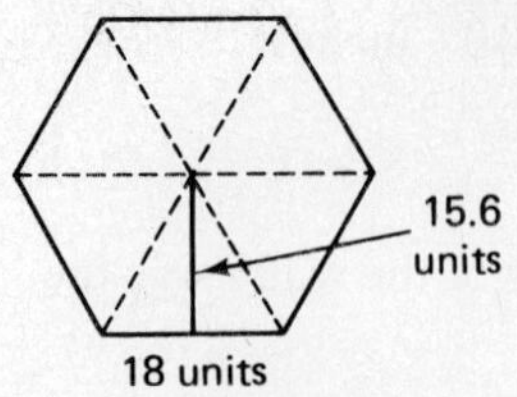

Figure C

12-45. Find the perimeter of one of the triangles of Figure C.

12-46. Find the perimeter of Figure C.

12-47. Find the area of one of the triangles of Figure C.

12-48. Find the area of Figure C.

12-49. Find the area of Figure D.

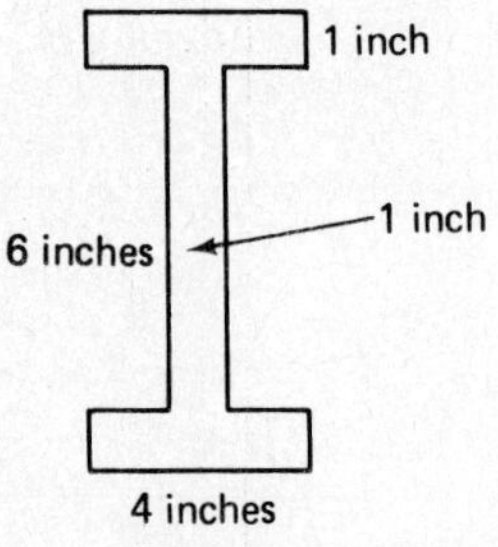

Figure D

12-50. Find the perimeter of Figure D.

12-51. Find the perimeter of the triangle in Figure E.

12-52. Find the perimeter of Figure E.

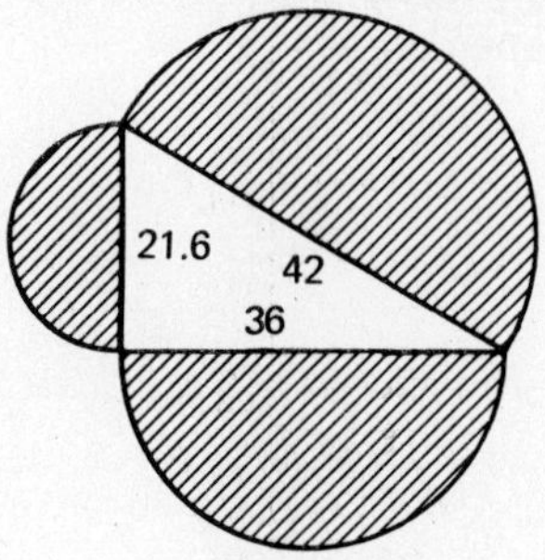

Figure E

12-53. Find the area of the triangle in Figure E.

12-54. Find the area of Figure E.

12-55. Find the area of the shaded portion of Figure E.

12-56. Find the perimeter of Figure F.

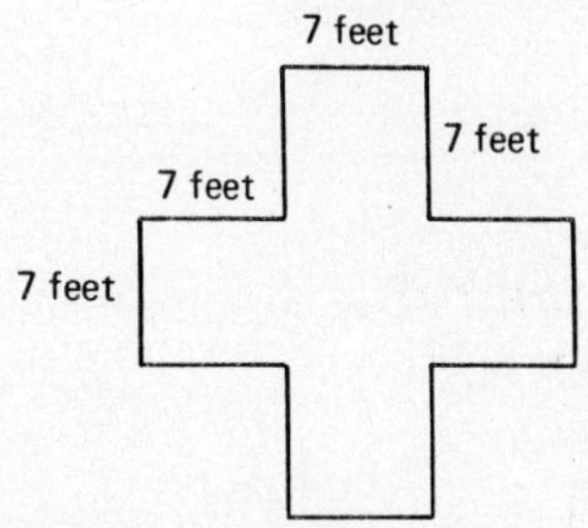

Figure F

12-57. Find the area of Figure F.

12-58. What would be the cost of painting Figure F if the cost of paint and labor is $.08 per square foot?

12-59. Find the area of Figure G.

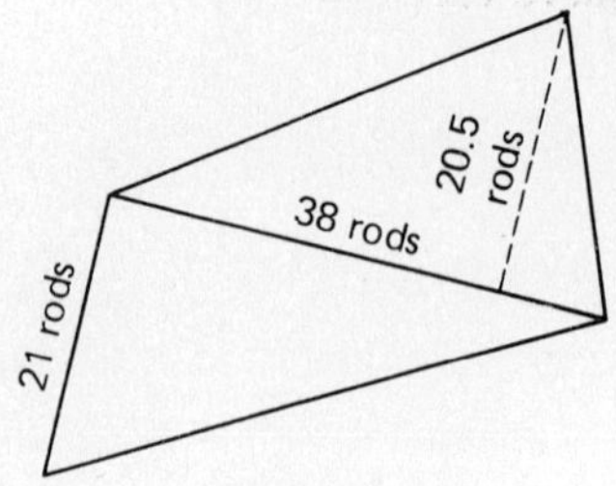

Figure G

12-60. If Figure G represents a small wheat field, how many bushels would be harvested per acre if the total number of bushels for the entire area was 182? (160 square rods equals one acre.)

12-61. What is the cost of sodding the playing area of the football field represented by Figure H at the cost of $1.25 per square yard?

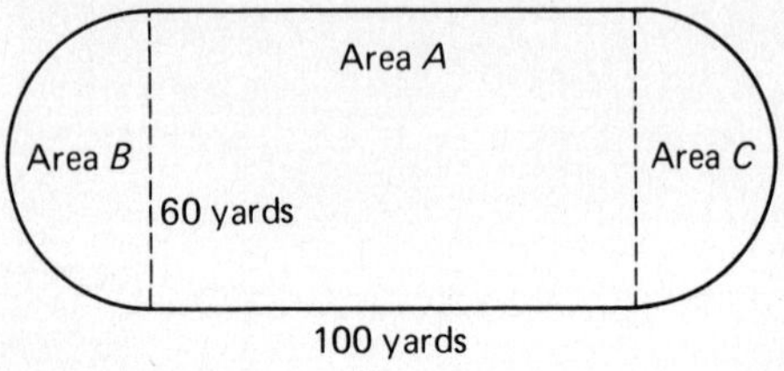

Figure H

12-62. What is the cost of sodding the bowl ends of the field (Figure H, areas *B* and *C*) at $.95 per square yard?

12-63. How many feet of fencing would be needed to fence in the entire area of Figure H.

12-64. How many feet of baseboard molding would be needed for the floor plan represented by Figure I?

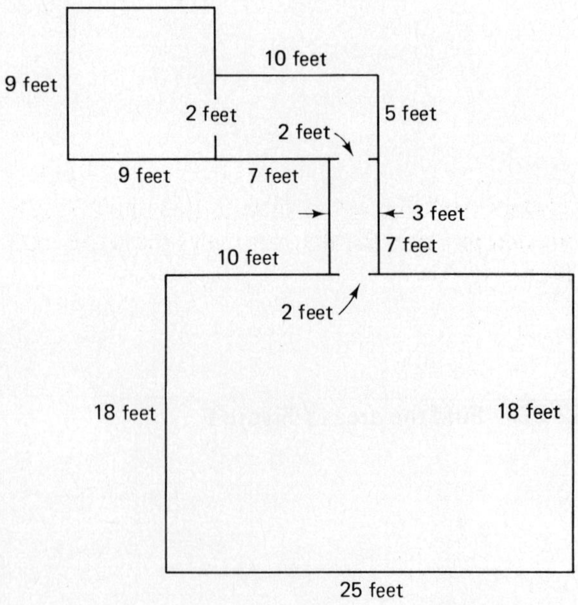

Figure I

12-65. What is the total area represented by Figure I?

12-66. What is the cost of carpeting the rooms represented by Figure I at $18.00 per square yard? (1 square yard = 9 square feet.)

12-67. What is the wall area of a room represented by Figure J?

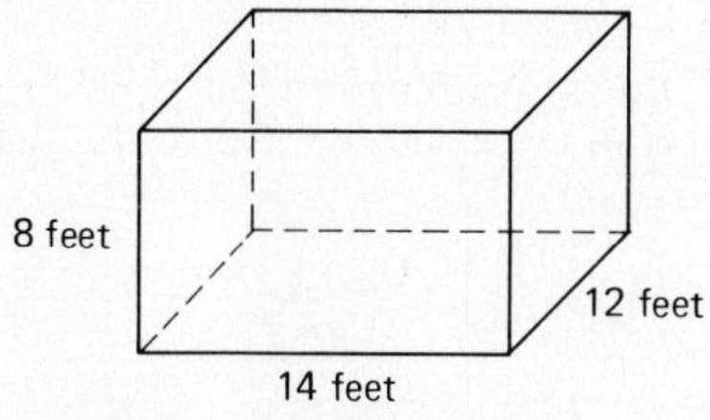

Figure J

12-68. What is the area of the ceiling of the room?

12-69. How many gallons of paint are needed to paint the room (walls and ceiling) twice if 550 square feet can be painted with 1 gallon of paint?

12-70. What would be the cost of painting the room (Figure J) if the paint costs $11.95 per gallon and $3.25 per quart?

12-71. How many 2-inch by 2-inch tiles would be needed to cover the floor of a bathroom that is $7\frac{1}{2}$ feet by $8\frac{1}{2}$ feet?

12-72. A landscape technician wishes to sod an area illustrated by Figure K. How much sod will be needed?

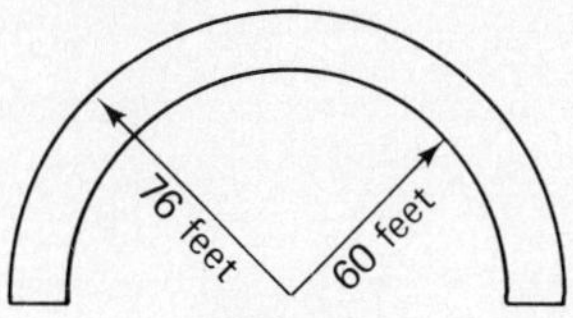

Figure K

12-73. A carpenter has accepted the job of paneling a wall 12 feet by 72 feet. How many panels will he need if each panel is 4 feet by 8 feet?

12-74. A nurseryman determines that he needs 36 square feet of ground space to grow a spruce tree. How many spruce trees could he grow on a piece of land 200 yards by 300 yards?

12-75. A swimming pool company wishes to advertise the swimming area of one of its large pools. What is the swimming area of a round pool which has a 35-foot diameter?

THINK TIME

Formulas for determining the perimeter, circumference, or area of a plane geometric figure are often difficult to remember. In order to refresh your memory from time to time, you may want to construct a chart for the purpose of recall. Complete the chart given below. When you finish, you can cover portions of the chart with a blank piece of paper to check your recall.

Sketch	*Name of Figure*	*Formula for Perimeter or Circumference*	*Formula for Area*
1. 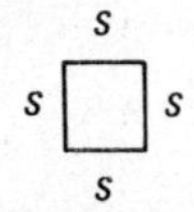			
2.	Rectangle		
3.		$P = a + b + c$	
4. 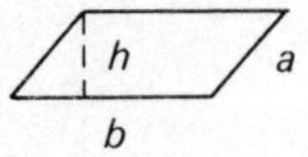			
5.			$A = \pi r^2$
6.		$C = d + \frac{1}{2}\pi d$ or $C = 2r + \pi r$	

Sketch	Name of Figure	Formula for Perimeter or Circumference	Formula for Area
7. 			
8.			$A = 6\left(\frac{bh}{2}\right)$

PROCEDURES TO REMEMBER

1. When determining the area of a plane geometric figure, use only like units.
2. Convert mixed fractions to decimal fractions, where possible, when determining the area of a plane geometric figure.
3. Divide or change complex plane geometric figures into simple figures, where possible, and apply basic formulas for determining perimeter, circumference, or area.

CHAPTER SUMMARY

1. The following definitions are essential to the measurement of plane geometric figures.
   (a) *Perimeter* is the distance around a plane geometric figure and is stated in linear measure such as inches, miles, centimeters, and meters.
   (b) *Circumference* is the perimeter of a circle.
   (c) *Area* is the amount of surface of a plane geometric figure and is measured in square inches, square yards, square meters, etc.
   (d) A *circle* is a set of points which are the same distance from a fixed point called the *center*.
   (e) The *radius* of a circle is a line drawn from the center of the circle to its edge.
   (f) The *diameter* of a circle is a line drawn from one side of a circle, through the center, to the opposite side.
   (g) A *semicircle* is half of a circle; it is created by the division of a circle into two equal parts by a diameter.
   (h) *Pi* is a special number, approximately equal to 3.14, which is the ratio of the circumference of a circle to its diameter. Pi is symbolized by the Greek letter, π. For calculation purposes, $\pi = 3.14$ or 3.142 or 3.1416 is used, depending on the accuracy desired.
2. The following formulas are essential to the measurement of plane geometric figures.

Figure	Perimeter (Circumference)	Area
Square	$P = 4s$	$A = s^2$
Rectangle	$P = 2l + 2w$ $P = 2(l + w)$	$A = lw$
Triangle	$P = a + b + c$	$A = \frac{1}{2}bh$
Parallelogram	$P = 2a + 2b$ $P = 2(a + b)$	$A = bh$
Circle	$C = d\pi$;	$A = \pi r^2$; $A = \frac{\pi d^2}{4}$
Semicircle	$C = d + \frac{1}{2}\pi d$; $C = 2r + \pi r$	$A = \frac{\pi r^2}{2}$
Trapezoid	$P = 2a + B + b$	$A = \left(\frac{B + b}{2}\right)h$
Hexagon	$P = 6s$	$A = 6\left(\frac{bh}{2}\right)$

CHAPTER TEST

T-12-1. What kind of a geometric figure is Figure A?

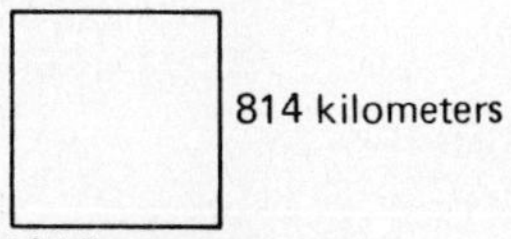

Figure A

T-12-2. Determine the area of Figure A.

T-12-3. Find the perimeter of Figure A.

T-12-4. What kind of a geometric figure is Figure B?

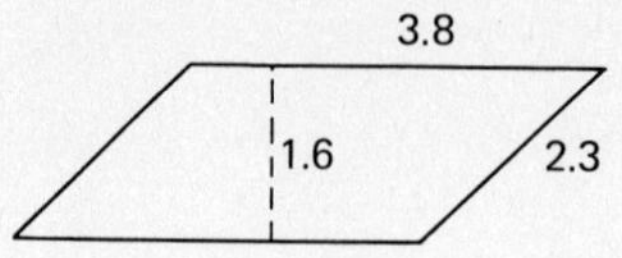

Figure B

T-12-5. What is the altitude of Figure B?

T-12-6. What is the perimeter of Figure B?

T-12-7. What is the area of Figure B?

T-12-8. What is the altitude of Figure C?

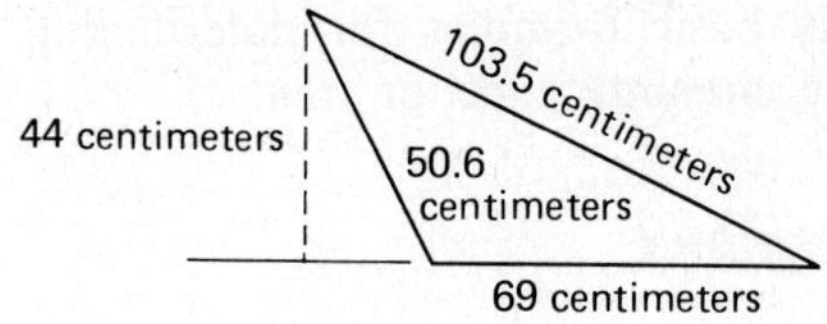

Figure C

T-12-9. What is the perimeter of Figure C?

T-12-10. What is the area of Figure C?

T-12-11. What is the radius of Figure D?

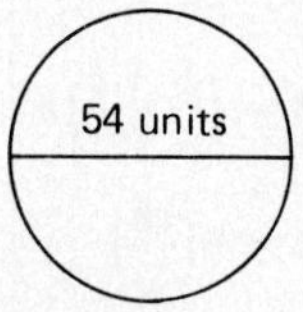

Figure D

T-12-12. What is the area of Figure D?

T-12-13. What is the circumference of Figure D?

T-12-14. What kind of a geometric figure is Figure E?

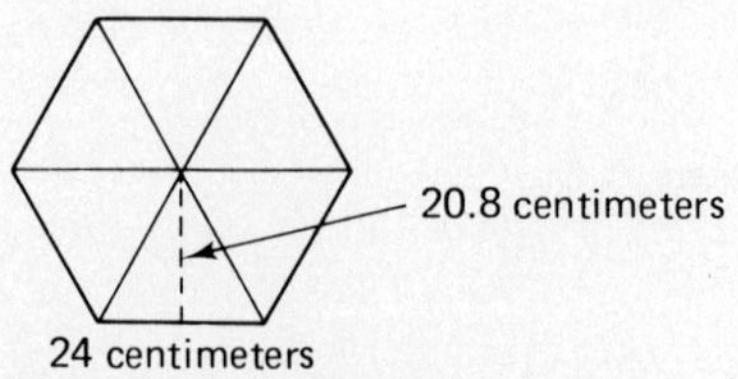

Figure E

T-12-15. What is the perimeter of Figure E?

T-12-16. Determine the area of Figure E.

T-12-17. What is the cost of carpeting the area around the swimming pool illustrated by Figure F? The cost of carpeting is \$8.95 per square yard. (9 square feet = 1 square yard.)

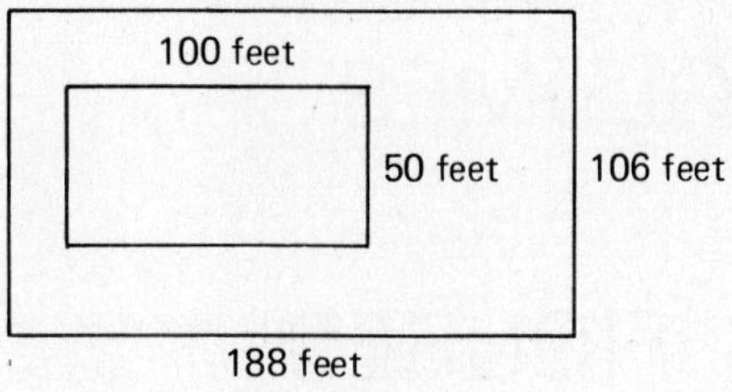

Figure F

T-12-18. Two heating ducts are joined together as illustrated by Figure G. What will the diameter of the single opening (c) have to be if the area of the opening is to be equal to the area of openings $a + b$?

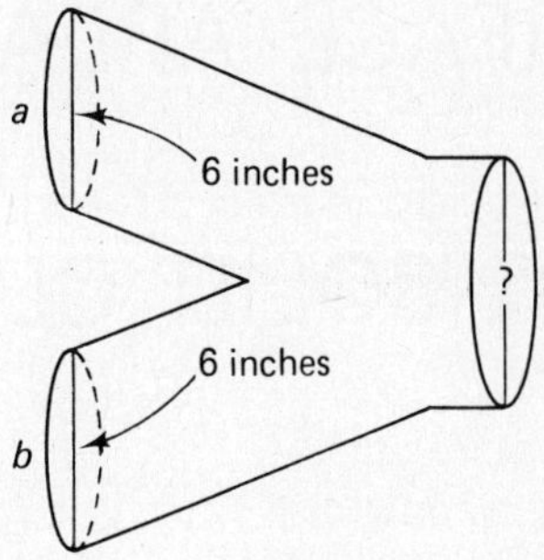

Figure G

T-12-19. How many sections of acoustical ceiling tile will be needed to cover the ceiling illustrated by Figure H? The tile sections are 2 feet by 4 feet.

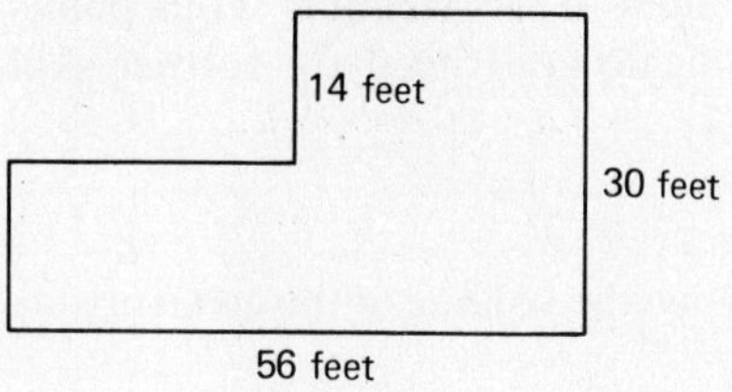

Figure H

T-12-20. How many gallons of paint are needed to paint the walls and ceiling of the storage room illustrated by Figure I? Each gallon of paint covers 520 square feet.

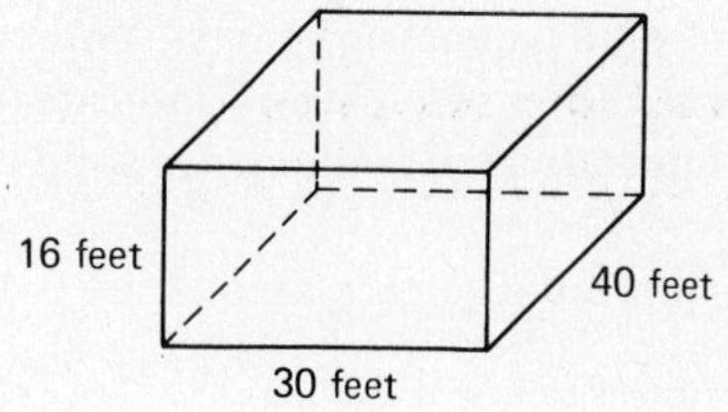

Figure I

CHAPTER

13

SURFACE AREAS AND VOLUMES OF BASIC SOLID GEOMETRIC FIGURES

OBJECTIVES

1. To develop the skills necessary to determine the lateral and total surface areas of prisms, pyramids, cylinders, cones, and spheres.
2. To develop the skills necessary to determine the volumes of prisms, pyramids, cylinders, cones, and spheres.
3. To solve practical problems of industry and daily living which require the skills of determining lateral and total surface areas and volumes of solid geometric figures.

SELF-TEST

The test which follows will evaluate your ability to determine the lateral surface areas, total surface areas, and volumes of solid geometric figures. Your successful completion of this self-test will indicate your ability to determine areas and volumes of solid figures and your readiness to develop your abilities beyond what is taught in this chapter. Your points of failure will indicate your need for further skill development.

S-13-1. Determine the total surface area of a 15-inch cube.

S-13-2. Determine the volume of the given prism.

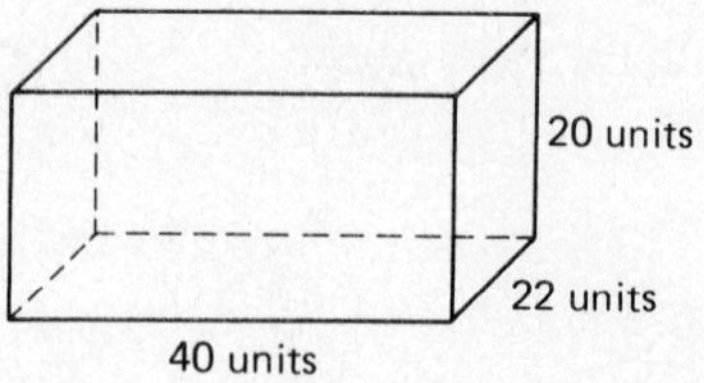

S-13-3. Determine the lateral surface area of a rectangular solid which is 3 centimeters wide × 4 centimeters long × 30 centimeters high.

S-13-4. Determine the lateral surface area of the given cylinder.

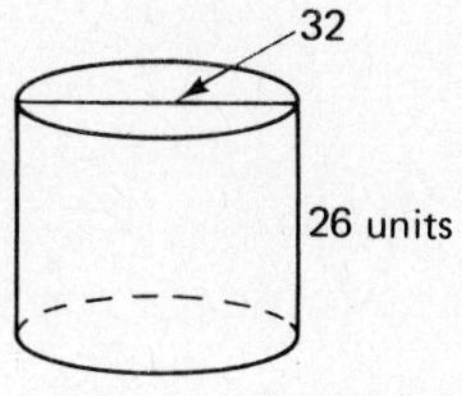

S-13-5. Determine the volume of the given cone.

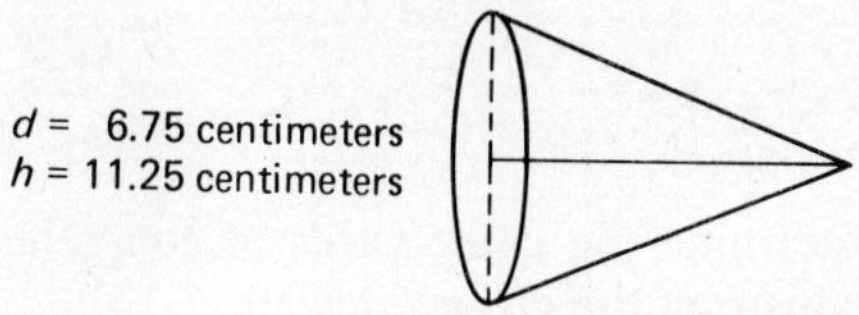

S-13-6. Determine the volume of the given cylinder.

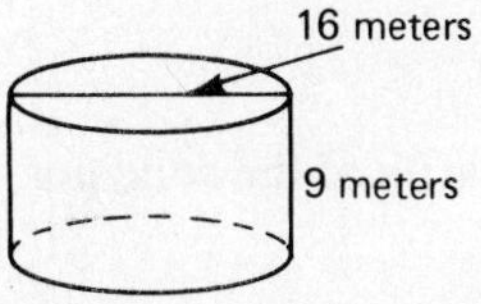

S-13-7. Determine the lateral surface area of the given cylinders.

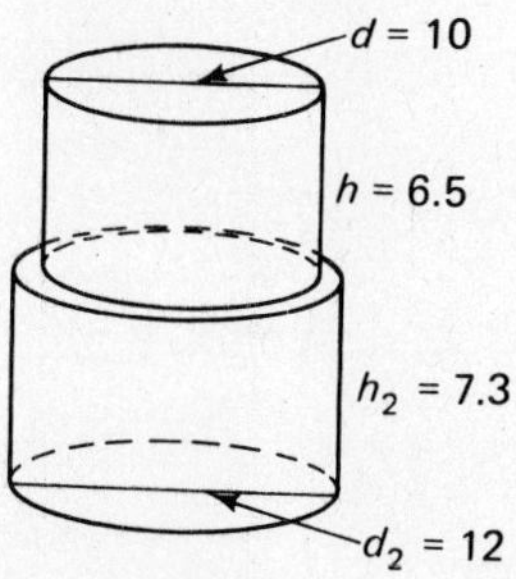

S-13-8. Determine the total surface area of the given cylinder.

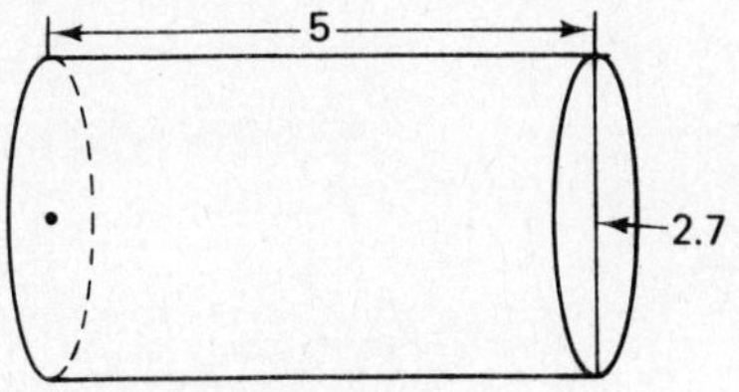

S-13-9. Determine the volume of a sphere which has a 24-inch diameter.

S-13-10. Determine the volume of the given pyramid.

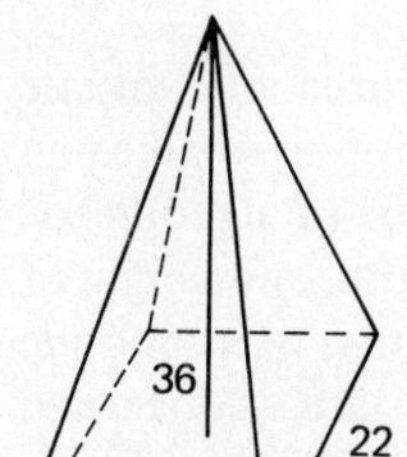

S-13-11. How many gallons will the given tank hold? (1 cubic foot = 7.5 gallons.)

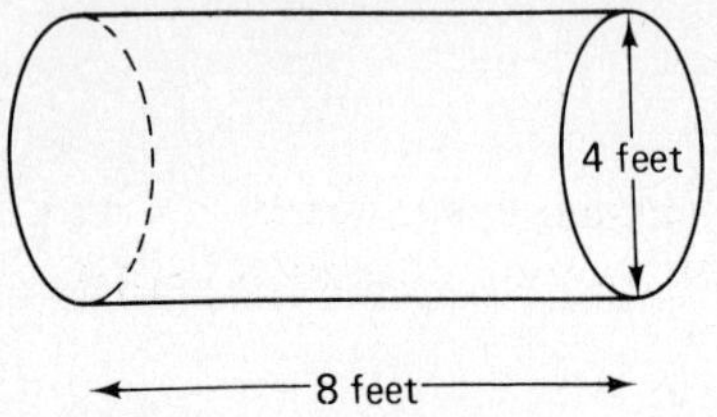

S-13-12. How many square feet of sheet metal will be necessary to construct the duct illustrated? (Add 15% for seams.)

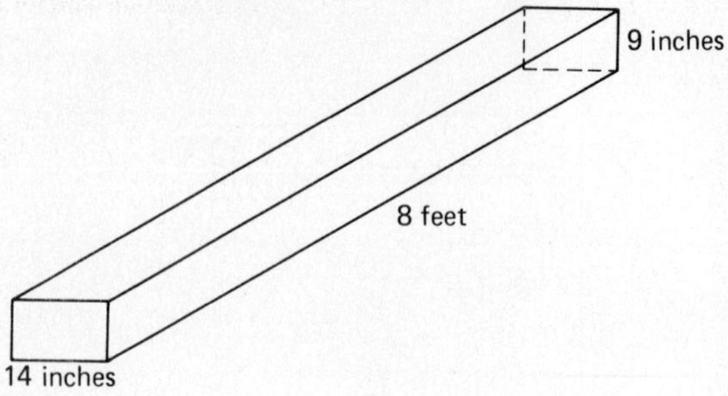

S-13-13. Determine the capacity of the cone portion of the given funnel.

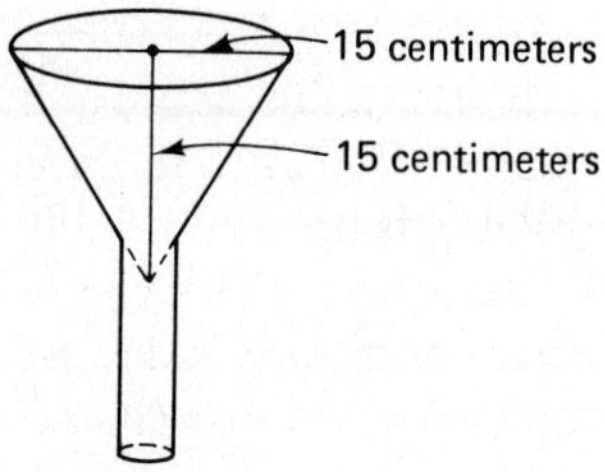

S-13-14. Determine the cubic yards of concrete necessary to construct the given sidewalk.

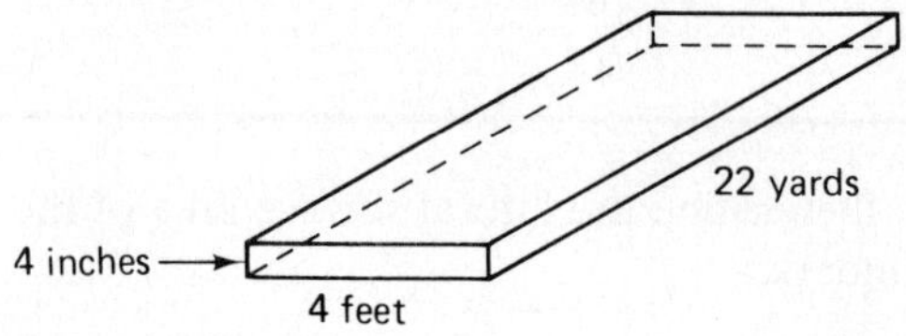

S-13-15. Determine the cost of painting a 50-foot-diameter spherical water tower at $.75 per square yard.

INTRODUCTION

Every solid object has surface area and volume whether it is as large as the earth or as small as an atom. The surface area and volume of a geometric solid are determined by length, width, and height or a combination of these dimensions. Automobile manufacturers and consumers are concerned about the cubic inches or volume of compression of an engine. A thrifty shopper carefully notes the volume of a container as well as its cost. An air-conditioning salesman measures the volume of an area to be air-conditioned. Most occupations require working knowledge of the skills necessary to determine the surface area and volume of a geometric solid.

DEFINITIONS

The formulas for determining the surface area and volume of a geometric solid are closely related to formulas used to determine the perimeter and area of geometric plane figures. The ability to work with geometric plane figure formulas will be of distinct value in determining the area and volume of solid geometric figures.

A *prism* is a solid geometric figure with parallel edges and uniform cross sections. The figures below are prisms.

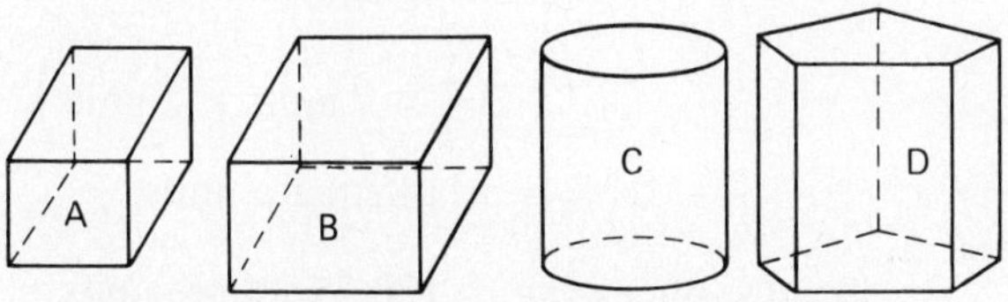

A *face* is any plane surface of a solid figure. A *lateral face* of a prism is a side of a geometric figure. The *bases* of a prism are the top face and bottom face, generally called the top and bottom of the solid. The *lateral edge* of a prism is the place where two sides of a solid come together. Observe Figure E below.

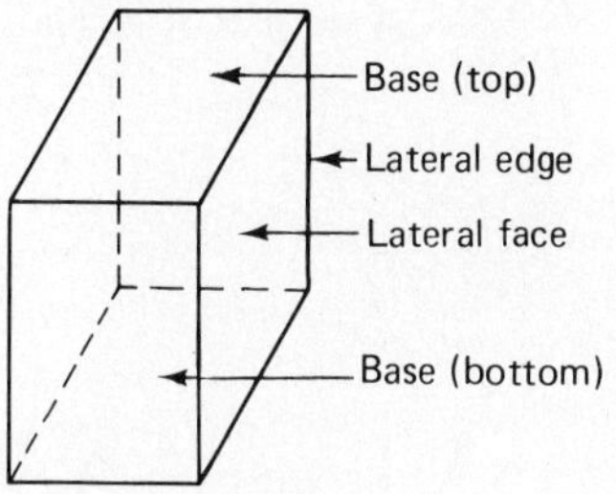

Figure E

All prisms are classified by the shape of their base. A *right prism* has lateral faces which are perpendicular to its base. A prism whose sides are not perpendicular to its base is called an *oblique prism.*

Figure A above is a *right square prism* because it has a square for a base. Figures B and E are *right rectangular prisms* because they have rectangular bases. Figure D is a *right pentagonal prism* because of its pentagon-shaped base. Figure C represents a special type of prism called a *cylinder.*

The *lateral surface area* of a solid geometric figure is the area or surface of the sides of the figure, excluding the area of the top and bottom. The shaded portion of Figure F indicates the lateral surface.

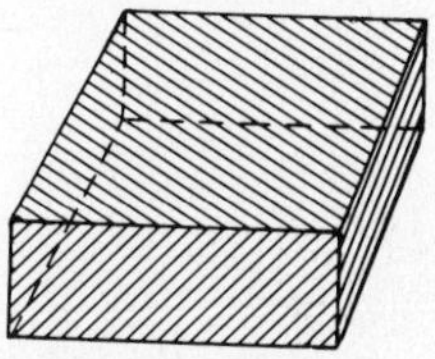

Figure F

The *total surface area* of a geometric solid includes the area of the top and bottom of the figure as well as the area of the sides.

CONTENT

To determine the lateral surface of a prism, multiply the perimeter of the prism by its height.

Example 13-1

PROBLEM: Determine the lateral surface (area) of Figure G.

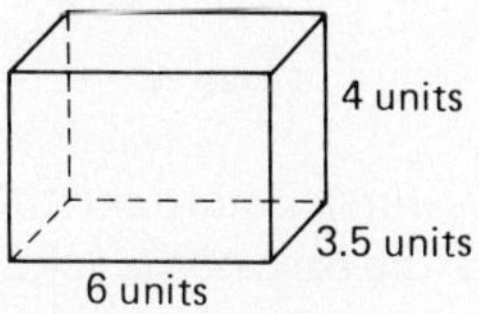

Figure G

SOLUTION: First, determine the nature of Figure G. Because the base of the solid figure is a rectangle and its faces are perpendicular to its base, Figure G is a right rectangular prism.

Second, determine the lateral area. This can be done by finding the area of each of the four sides and then adding these areas together. Note the following illustration.

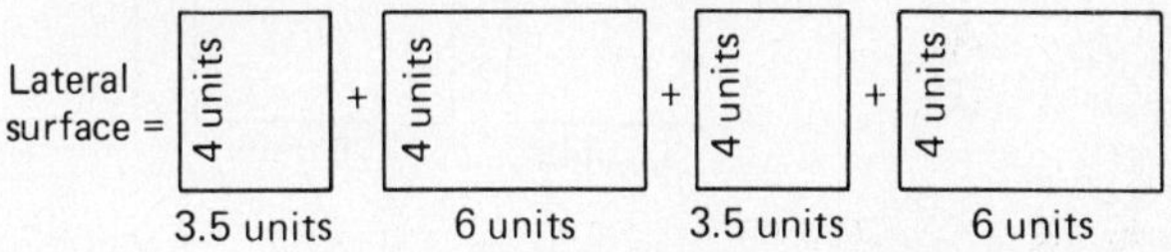

Lateral surface = { 4 units × 3.5 units = 14 square units
4 units × 6 units = 24 square units
4 units × 3.5 units = 14 square units
4 units × 6 units = 24 square units }

Lateral surface = 76 square units

A simpler method for determining lateral surface is to use the formula:

Lateral surface (area) = perimeter $(2l + 2w) \times$ height (h)

$$A = p \times h$$

Example 13-2

PROBLEM: Find the total surface area of Figure H.

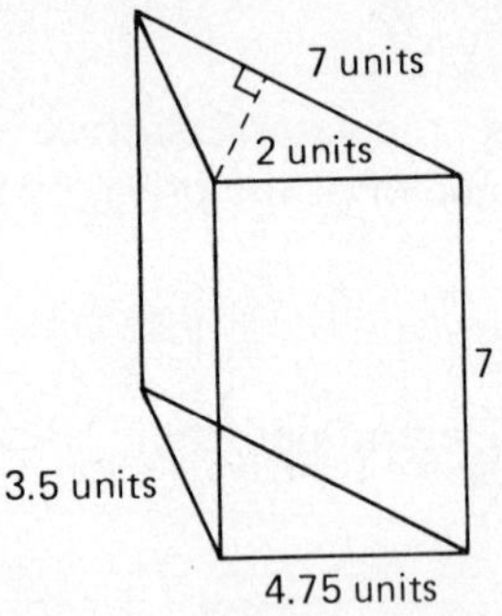

Figure H

SOLUTION: First, determine the nature of Figure H. Because the base of the solid figure is a right triangle and its faces are perpendicular to its base, the figure is a right triangular prism.

Second, determine the surface area. The total surface area may be determined by finding the area of each of the sides, the top, and the bottom and then adding these areas together. Note the illustration that follows.

$$\text{Total surface area} = \begin{cases} 7 \text{ units} \times 3.5 \text{ units} \\ \quad = 24.5 \text{ square units} \\ 7 \text{ units} \times 4.75 \text{ units} \\ \quad = 33.25 \text{ square units} \\ 7 \text{ units} \times 7 \text{ units} \\ \quad = 49 \text{ square units} \\ 2\left(\frac{1}{2} \times 7 \text{ units} \times 2 \text{ units}\right) \\ \quad = 14 \text{ square units} \end{cases}$$

$$\text{Total surface area} = 120.75 \text{ square units}$$

The formula for determining the total surface area of a right triangular prism is:

$$\text{Total surface area} = \text{perimeter } (2l + 2w) \times \text{height} + \frac{2(\text{base} \times \text{height})}{2}$$

$$A = p \times h + bh$$

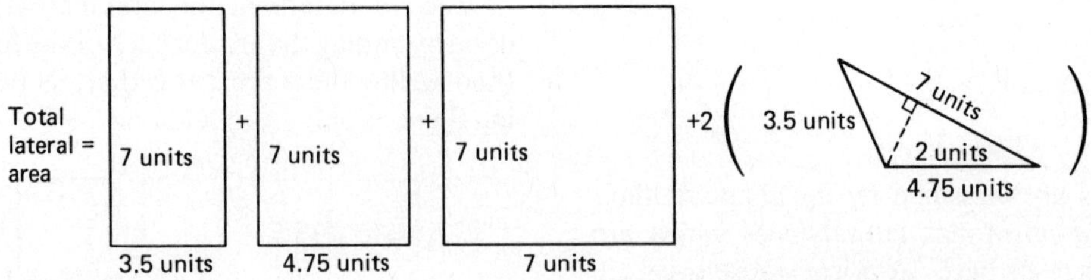

EXERCISE 13-1

Determine the lateral and total surface areas as called for in the following problems:

13-1. What type of prism is Figure I?

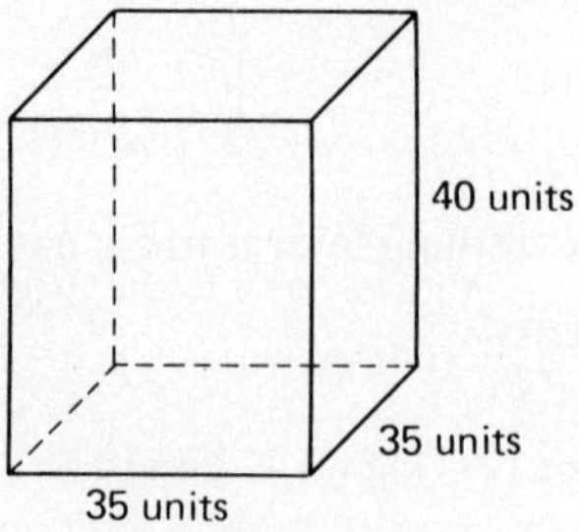

Figure I

13-2. What is the lateral area of Figure I?

13-3. What is the total surface area of Figure I?

13-4. What type of prism is Figure J?

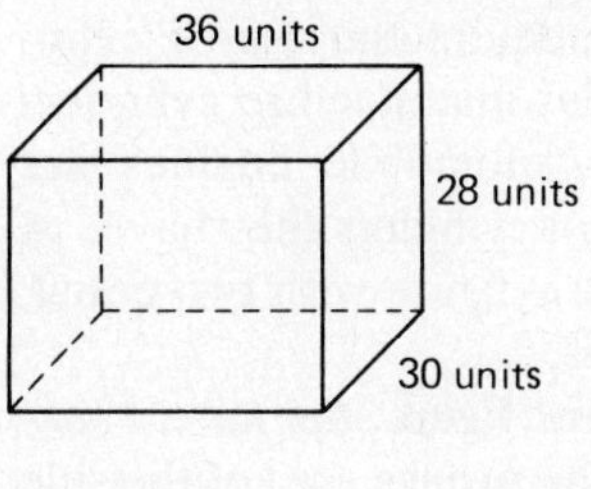

Figure J

13-5. What is the lateral area of Figure J?

13-6. What is the total surface area of Figure J?

13-7. What type of prism is Figure K?

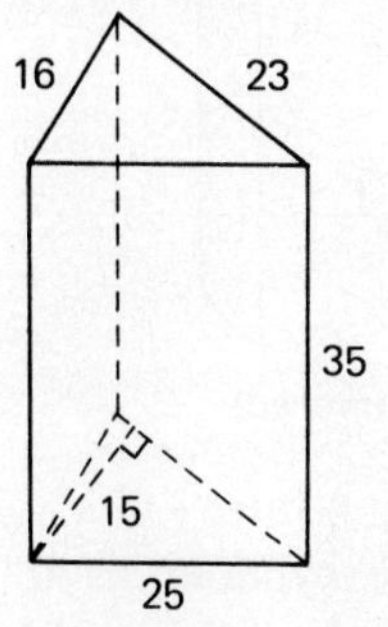

Figure K

13-8. Determine the lateral area of Figure K.

13-9. What is the total surface area of Figure K?

13-10. What type of prism is Figure L?

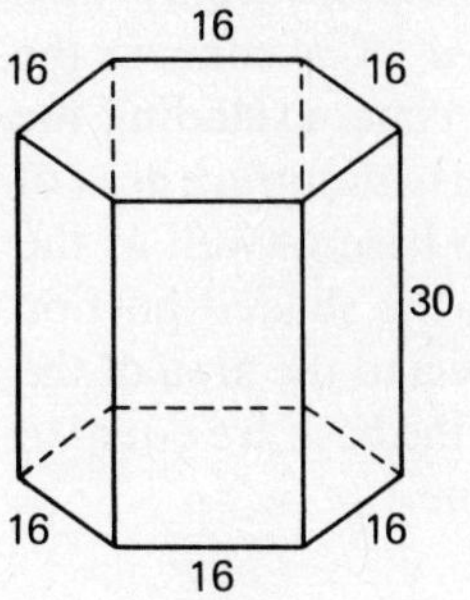

Figure L

13-11. What is the lateral area of Figure L?

13-12. What is the total surface area of Figure L?

DEFINITIONS

Numerous industries require the use of cylinders. Packaging companies manufacture cylinders for use as containers. Automobile engines are described by the number of cylinders they have. A *cylinder* is a solid geometric figure with two equal, parallel, and circular bases.

Like any solid geometric figure, the *lateral surface area* of a cylinder is the surface area of the side of the cylinder, excluding the area of the top and bottom. The shaded portion of Figure M indicates the lateral surface area. The *total surface area* of a cylinder includes the area of the top and the bottom of the figure as well as the area of the side. Thus, in Figure M, the area of the shaded portion plus the area of the top and bottom are equal to the total lateral surface of the cylinder.

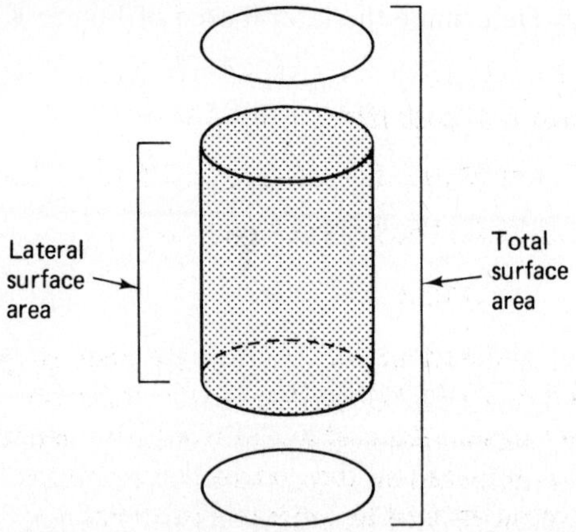

Figure M

A *cone* is a solid geometric figure which has one circular base and whose side tapers evenly to a point. The *lateral surface area* of a cone is the surface area of the side of the cone, excluding the bottom or base of the cone. The *total surface area* of a cone includes the area of the base as well as the area of the side. In Figure N, the shaded portion indicates that the lateral surface and the area of the shaded portion plus the area of the base are equal to the total surface area of the cone.

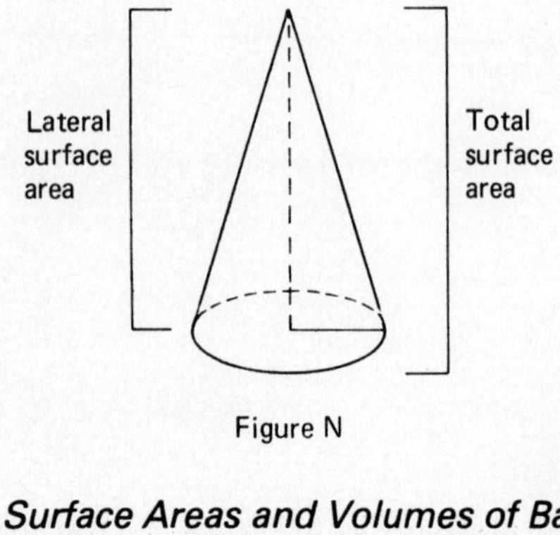

Figure N

CONTENT

The *lateral surface area* of a cylinder is determined by multiplying the diameter of the base (top or bottom) times π times the height. The *total surface area* of a cylinder is determined by multiplying the diameter of the base times π times the height plus $\frac{1}{2}$ the radius squared times π times 2.

Example 13-3

PROBLEM: Determine the total surface area of the cylinder represented by Figure O which has a diameter of 12 meters and a height of 20 meters.

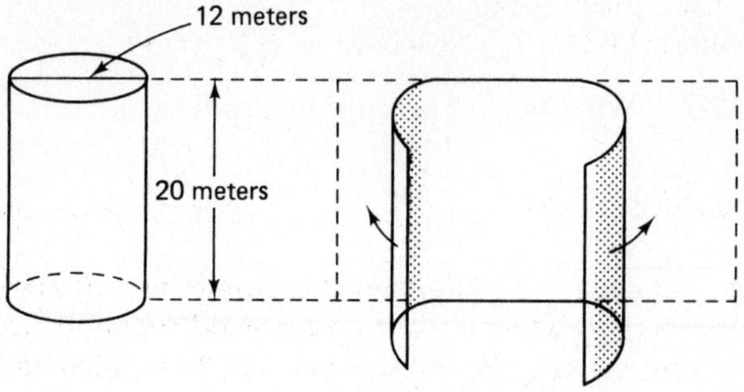

Figure O Figure O (unrolled)

SOLUTION: First, observe the cylinder as it could be "unrolled" to form a rectangle. If this could be done for each cylinder, the lateral surface area of a cylinder could be determined by multiplying the length times the width of the newly formed rectangle.

Second, determine the lateral surface area by multiplying the diameter (12 meters) times π (3.142) times the height (20 meters).

$$\text{Lateral surface area} = 12 \text{ meters} \times 3.142 \times 20 \text{ meters}$$
$$= 754.08 \text{ square meters}$$

Third, determine the area of the bases (top and bottom). Since the bases are two equal circles, the formula for determining the area of a circle can be used. Thus, we multiply $\frac{1}{2}$ the diameter (12 meters) (or the radius) times $\frac{1}{2}$ the diameter (12 meters) (or the radius) times π (3.142) times the number of bases (2).

$$\text{Area of bases} = \frac{1}{2}(12 \text{ m}) \times \frac{1}{2}(12 \text{ m}) \times 3.142 \times 2$$
$$= 226.224 \text{ m}^2$$

Fourth, determine the total surface area of the cylinder. To do this, add the lateral surface area and the surface area of the bases.

Total surface area = lateral surface area + area of bases

$$= 754.08 \text{ m}^2 + 226.224 \text{ m}^2$$

$$= 980.304 \text{ m}^2$$

Note the two formulas which were used to solve this problem:

Lateral surface area of a cylinder $= d\pi h$ or $2r\pi h$

Total surface area of a cylinder $= d\pi h + 2\left[\left(\frac{1}{2}d\right)^2 \pi\right]$

$$= 2r\pi h + 2\pi r^2$$

Example 13-4

PROBLEM: Determine the square feet of plastic material needed to make a wind-tunnel cone, represented by Figure P, which is 10 feet high and has a radius of 2.5 feet.

SOLUTION: First, observe the drawing of the cone, Figure P. Note that the height, h, is 10 feet, the radius of the base, r, is 2.5 feet, and the length of the slant height, s, is unknown.

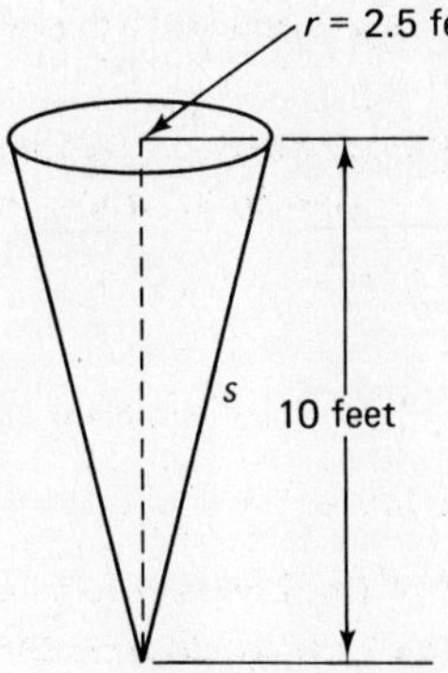

Figure P

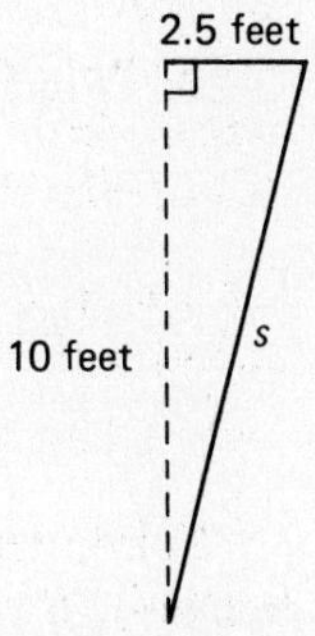

Figure P (abbreviated)

Second, determine the length of the slant height, s. Note in Figure P (abbreviated) that a right triangle has been formed from the cone. Thus, the formula for determining the hypotenuse of a right triangle can be used to determine the length of slant height s.

$$\text{Slant height} = \sqrt{(\text{radius})^2 + (\text{height})^2}$$

$$= \sqrt{(2.5)^2 + (10)^2}$$

$$= \sqrt{6.25 + 100}$$

$$= \sqrt{106.25}$$

$$= 10.3 \text{ feet (to the nearest tenth)}$$

Third, determine the lateral surface area of the cone.

Lateral surface area $= \text{radius} \times \text{slant height} \times \pi$

$$= 2.5 \text{ feet} \times 10.3 \text{ feet} \times 3.142$$

$$= 80.967 \text{ square feet}$$

Fourth, determine the total surface area of the cone.

Total surface area = lateral surface area + πr^2

$$= 80.967 \text{ ft}^2 + 3.142 \times 2.5 \text{ ft}^2 \times 2.5 \text{ ft}^2$$

$$= 80.967 \text{ ft}^2 + 19.6375 \text{ ft}^2$$

$$= 100.60 \text{ ft}^2$$

Thus, a minimum of 101 square feet of plastic material would be needed to fabricate the wind-tunnel cone. More material would be needed for seams and waste in the construction process. Note the formulas used to solve this problem:

Lateral surface area of a cone $= r\pi \times \text{slant height}$

Total surface area of a cone $= \pi \times \text{slant height} + \pi r^2$

EXERCISE 13-2

Determine the lateral and total surface areas as called for in the following problems. Use 3.142 as the value for π.

13-13. What type of solid figure is Figure Q?

13-14. Find the lateral surface area of Figure Q.

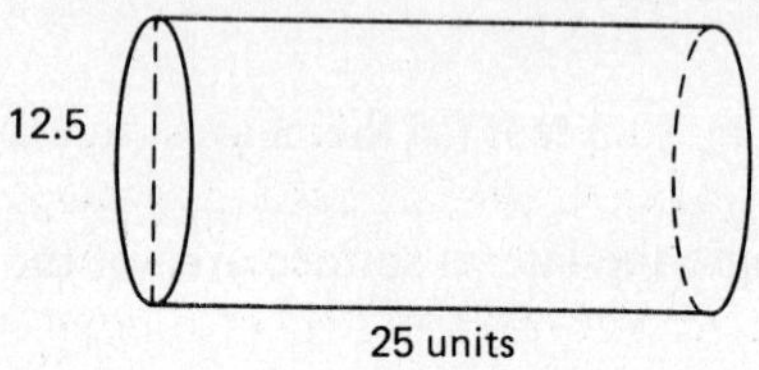

Figure Q

13-15. Find the total surface area of Figure Q.

13-16. What type of solid figure is Figure R?

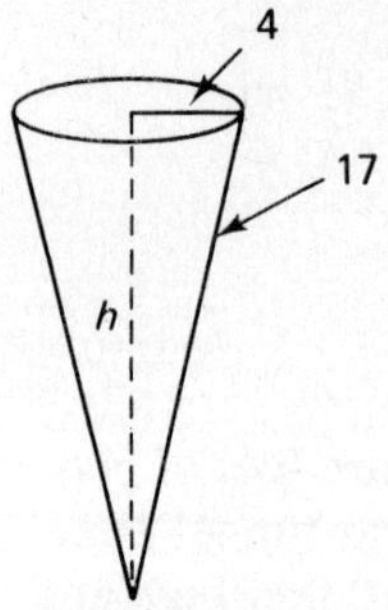

Figure R

13-17. Find the diameter of Figure R.

13-18. Find the height of Figure R.

13-19. Find the lateral surface area of Figure R.

13-20. Find the total surface area of Figure R.

13-21. What is the radius of Figure S?

13-22. What is the slant height of Figure S?

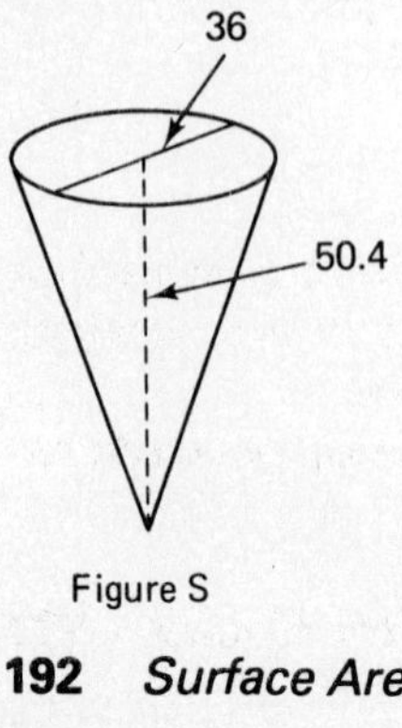

Figure S

13-23. What is the lateral surface area of Figure S?

13-24. What is the total surface area of Figure S?

DEFINITIONS

Another type of solid figure, similar in shape to the cone, is called a *pyramid.* The names of pyramids are derived from the shape of the base of the pyramid. A *right pyramid* is a pyramid whose base is a regular polygon and whose sides or faces form equal isosceles triangles. Study the illustrations of three different types of right pyramids. Note the components and nomenclature of each.

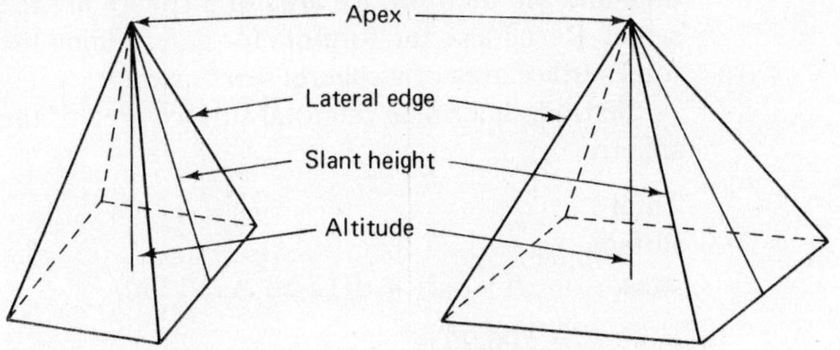

Figure T – Square pyramid

Figure U – Rectangular pyramid

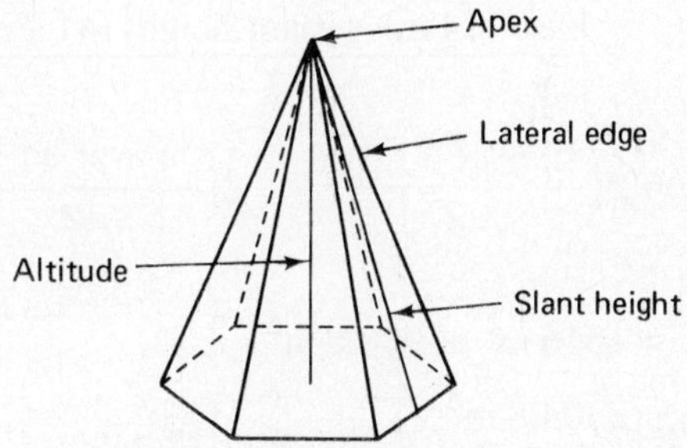

Figure V – Hexagonal pyramid

A *sphere* is a solid figure in which all points on the surface of the solid are the same distance from the center point. A unique property of the sphere is that it has a greater volume in proportion to its surface area than any other solid geometric figure. A *hemisphere* is a half sphere. Study the illustrations of the sphere and hemisphere. Note the components and nomenclature of each.

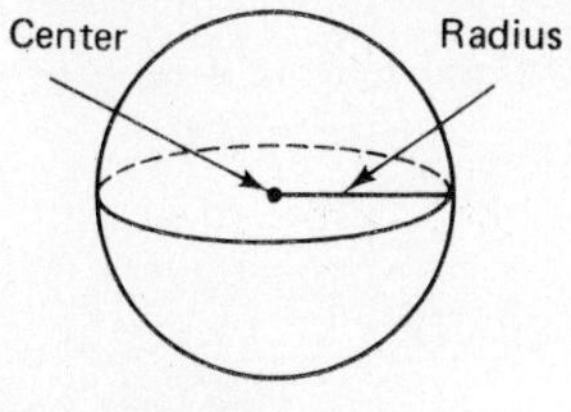

Figure W – Sphere

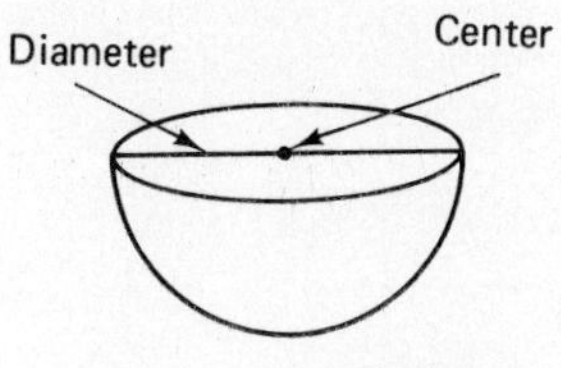

Figure X – Hemisphere

CONTENT

The *lateral surface area* of a pyramid is determined by adding together the areas of each of the face triangles. The *total surface area* of a pyramid is determined by adding together the lateral surface area and the base polygon surface area. The *total surface area* of a sphere is determined by formula $4\pi r^2$ or πd^2. The *lateral surface area* of a hemisphere is determined by the formula $2\pi r^2$ or $\frac{\pi d^2}{2}$. The *total surface area* of a hemisphere is determined by the formula $3\pi r^2$ or $\frac{1.5\pi d^2}{2}$.

Example 13-5

PROBLEM: Find the lateral and total surface areas of the pyramid illustrated by Figure Y.

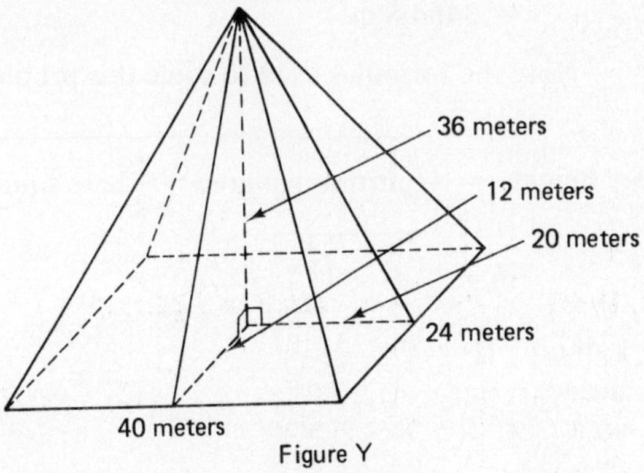

Figure Y

SOLUTION: First, observe that the figure is a rectangular pyramid because its base is a rectangle and all faces are isosceles triangles.

Second, determine the slant height of two different faces.

Slant height

$= \sqrt{(12)^2 + (36)^2}$

$= \sqrt{144 + 1296}$

$= \sqrt{1440}$

$= 37.9$ meters

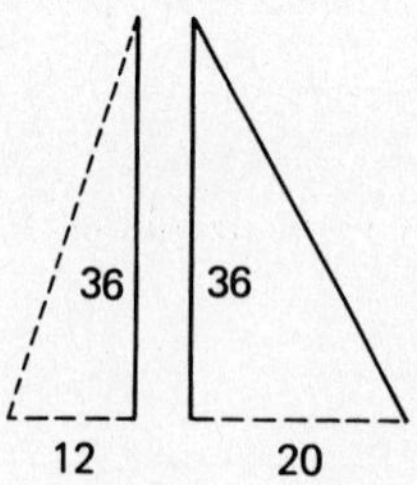

Slant height

$= \sqrt{(36)^2 + (20)^2}$

$= \sqrt{1296 + 400}$

$= \sqrt{1696}$

$= 41.2$ meters

Third, determine the lateral surface area of the pyramid. To do this, use the formula for determining the surface area of a triangle. Thus the lateral surface area of a pyramid can be found by adding together the surface areas of the four sides or triangles.

Lateral surface area

$$= \frac{\text{base } (40 \text{ m}) \times \text{slant height } (37.9 \text{ m})}{2}$$

$$+ \frac{\text{base } (24 \text{ m}) \times \text{slant height } (41.2 \text{ m})}{2}$$

$$= \overset{1}{\cancel{2}}\left(\frac{40 \text{ m} \times 37.9 \text{ m}}{\cancel{2}}\right) + \overset{1}{\cancel{2}}\left(\frac{24 \text{ m} \times 41.2 \text{ m}}{\underset{1}{\cancel{2}}}\right)$$

$$= 1516 \text{ m}^2 + 988.8 \text{ m}^2$$

$$= 2504.8 \text{ m}^2$$

Fourth, determine the total surface area of the pyramid.

Total surface area $=$ lateral surface area $+$ base surface area

$$= 2504.8 \text{ m}^2 + 960 \text{ m}^2$$

$$= 3464.8 \text{ m}^2$$

Note the formulas used to solve this problem:

$$\text{Slant height} = \sqrt{\text{altitude squared} + \frac{1}{2}\text{base squared}}$$

$$sh = \sqrt{a^2 + \left(\frac{b}{2}\right)^2}$$

$$\text{Lateral surface area} = 2\left(\frac{1}{2}\text{ base} \times \text{slant side } 1\right)$$

$$+ 2\left(\frac{1}{2}\text{ base} \times \text{slant side } 2\right)$$

$$= 2\left(\frac{b}{2}sh_1\right) + 2\left(\frac{b}{2}sh_2\right)$$

Total surface area $=$ lateral surface area $+$ base surface area

$$TSA = lsa + bsa$$

Example 13-6

PROBLEM: Find the total surface area of the sphere illustrated by Figure Z.

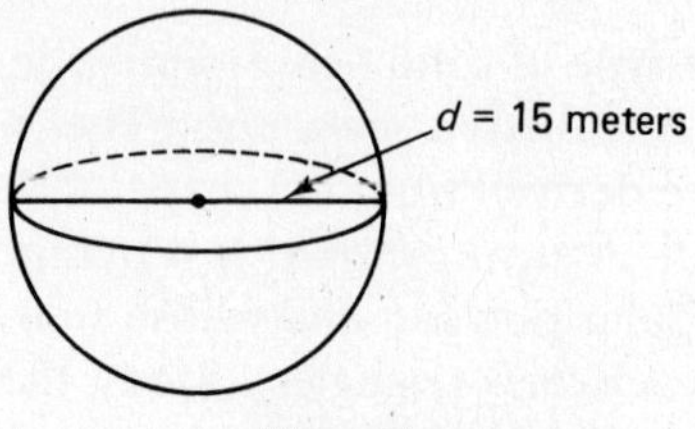

Figure Z

SOLUTION: First, note that the lateral surface area and the total surface area of a sphere are the same. Recall also the formula for determining the total surface area of a sphere: $4\pi r^2$ or πd^2.

Second, determine the total surface area of the sphere.

Total surface area $= \pi(3.142) \times d(15 \text{ m}) \times d(15 \text{ m})$

$$= 706.95 \text{ m}^2$$

Note that the lateral surface area of a hemisphere is determined by the formula $\frac{\pi d^2}{2}$, or $2\pi r^2$. Thus the lateral surface area of this hemisphere is 353.48 square meters. The total surface area of a hemisphere is determined by the formula $3\pi r^2$, or $\frac{3}{4}\pi d^2$. Thus the total surface area for this hemisphere is 530.22 square meters.

EXERCISE 13-3

Determine the lateral and total surface areas as called for in each of the following problems:

13-25. What type of a solid figure is Figure AA?

13-26. Find the slant height of Figure AA.

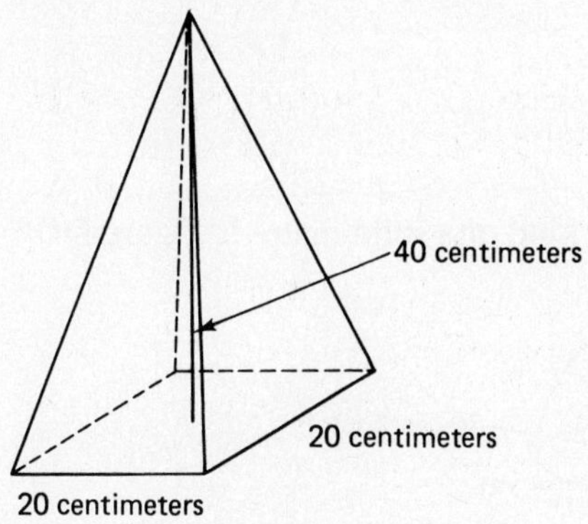

Figure AA

13-27. Find the lateral surface area of Figure AA.

13-28. Find the total surface area of Figure AA.

13-29. What is the total surface area of Figure BB?

13-30. What is the lateral surface area of a hemisphere of Figure BB?

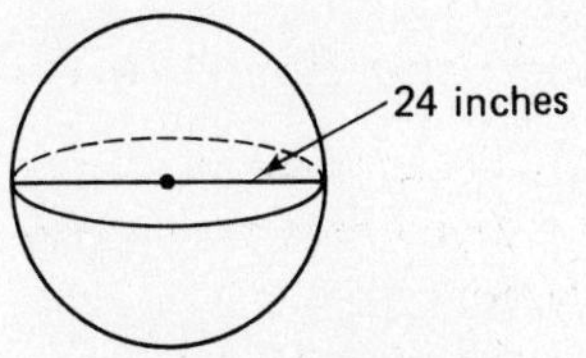

Figure BB

13-31. What kind of a solid figure is Figure CC?

13-32. What is the slant height on the 45-unit side of Figure CC?

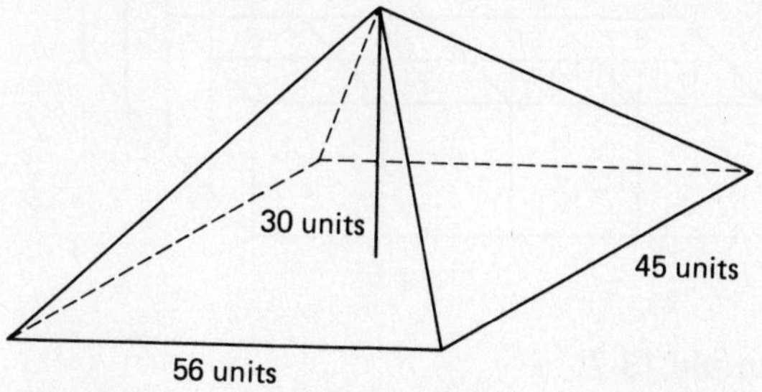

Figure CC

13-33. What is the slant height on the 56-unit side of Figure CC?

13-34. What is the lateral surface area of Figure CC?

13-35. What is the total surface area of Figure CC?

13-36. What kind of a solid figure is Figure DD?

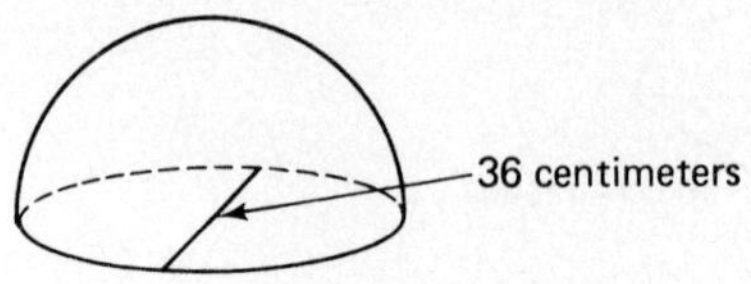

Figure DD

13-37. What is the radius of Figure DD?

13-38. What is the altitude of Figure DD?

13-39. Determine the lateral surface area of Figure DD.

13-40. Determine the total surface area of Figure DD.

DEFINITIONS

Volume is the amount of cubic space or content which a solid figure may contain. The volume of a prism or cylinder may be determined by multiplying the area of the base of the figure times the altitude or height.

For a better understanding of volume, study Figure EE below. The cubes in the first layer are numbered from 1 to 16. Remember that a cube is a solid figure in which all faces are squares and thus all edges are equal in length. If each cube is 1 cubic inch, the volume of the first layer would be 16 cubic inches. The volume of all three layers combined would be 16 cubic inches in a layer times the number of layers (3), or a total of 48 cubic inches.

The volume of a pyramid or cone is determined by multiplying the area of the base times the altitude or height times one third.

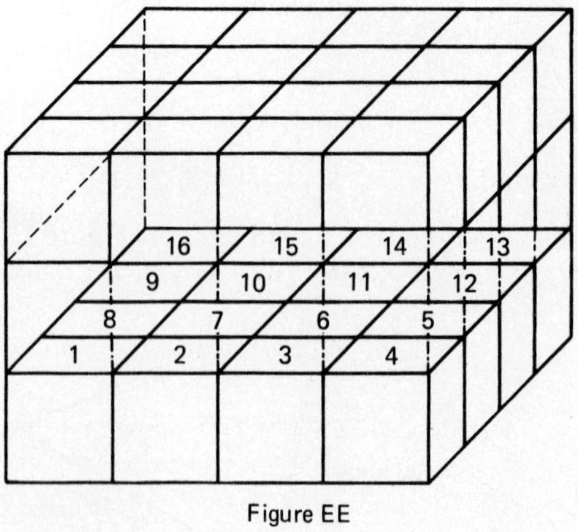

Figure EE

Example 13-7

PROBLEM: Find the volume of the solid illustrated by Figure FF.

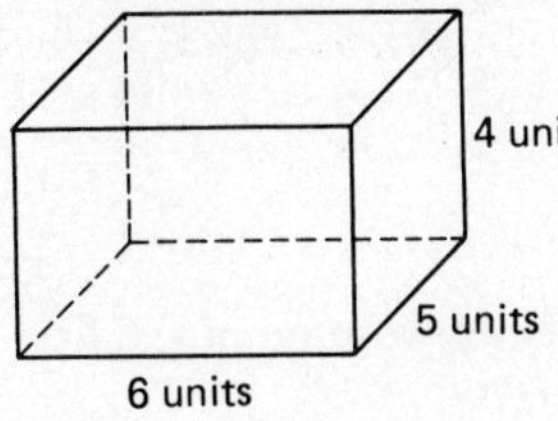

Figure FF

SOLUTION: Use the formula for determining the volume of a rectangular solid.

Volume = length times width times height.

Thus, in this problem:

$$\text{Volume} = 6 \text{ units} \times 5 \text{ units} \times 4 \text{ units}$$

$$= 120 \text{ cubic units}$$

Example 13-8

PROBLEM: Find the volume of the cylinder illustrated by Figure GG.

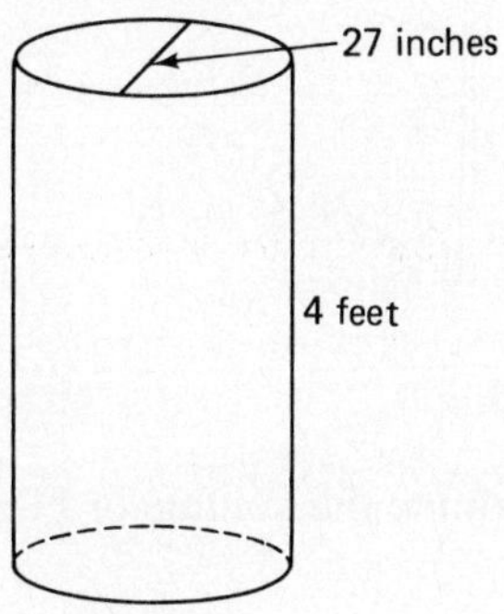

Figure GG

SOLUTION: First, convert mixed units (inches and feet) to like units (inches).

$$12 \text{ inches} = 1 \text{ foot}$$

$$12 \text{ inches} \times 4 = 48 \text{ inches}$$

Second, use the formula for determining the volume of a right circular cylinder ($V = \pi r^2 h$).

Volume = area of base times height or altitude

$$= \pi \times \frac{1}{2}d \times \frac{1}{2}d \times \text{altitude or height}$$

$$= 3.142 \times 13.5 \text{ in.} \times 13.5 \text{ in.} \times 48 \text{ in.}$$

$$= 27{,}468.22 \text{ in.}^3$$

Example 13-9

PROBLEM: Find the volume of the cone illustrated by Figure HH.

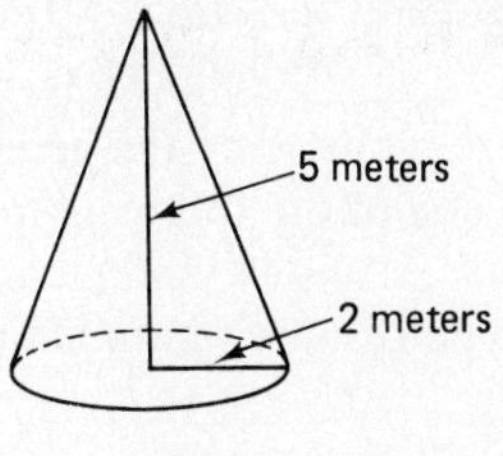

Figure HH

SOLUTION: Use the formula for determining the volume of a cone $\left(V = \frac{\pi r^2 h}{3}\right)$.

Volume = area of base × altitude × one third

$$= \pi \times r \times r \times \text{altitude} \times \frac{1}{3}$$

$$= 3.142 \times 2 \times 2 \times 5 \times \frac{1}{3}$$

$$= \frac{3.142 \times 20 \text{ cubic meters}}{3}$$

$$= 20.94 \text{ cubic meters}$$

Note that the formula for the volume of a pyramid is $V = \frac{1}{3} \times$ area of the base × the height.

Example 13-10

PROBLEM: Determine the volume of an 8-inch sphere.

SOLUTION: Use the formula for determining the volume of a sphere.

Volume = 4 × π × the radius cubed divided by 3

$$= \frac{4 \times 3.142 \times 4 \times 4 \times 4}{3}$$

$$= 268.12 \text{ cubic inches}$$

The formula for determining the volume of a hemisphere is:

$$V = \frac{2\pi r^3}{3}$$

EXERCISE 13-4

Determine the volume of the geometric figures as indicated in each of the following problems:

13-41. What type of solid figure is Figure II?

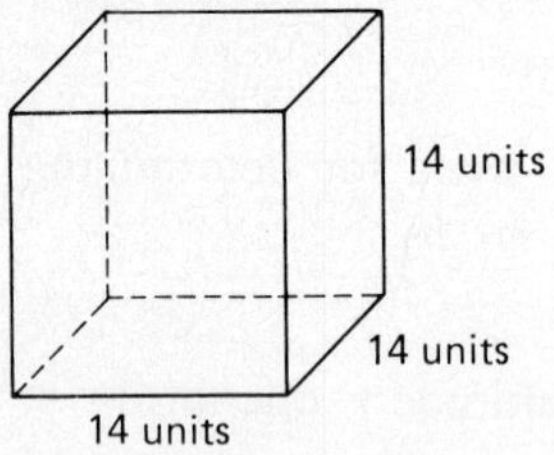

Figure II

13-42. What is the volume of the first layer of the solid in Figure II?

13-43. What is the volume of the solid in Figure II?

13-44. What kind of a solid is Figure JJ?

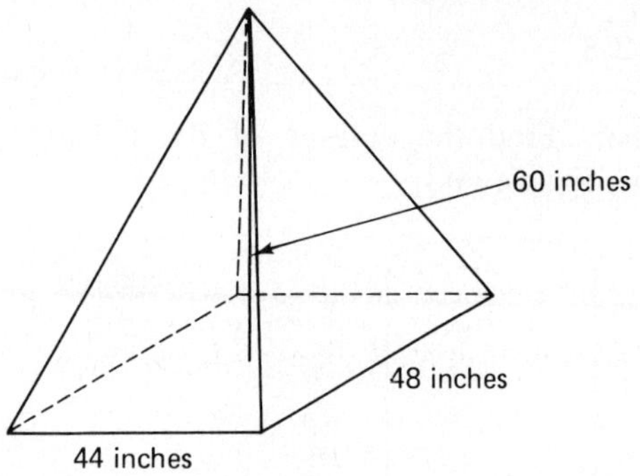

Figure JJ

13-45. Determine the area of the base of Figure JJ.

13-46. Determine the volume of Figure JJ.

13-47. What kind of a solid is Figure KK?

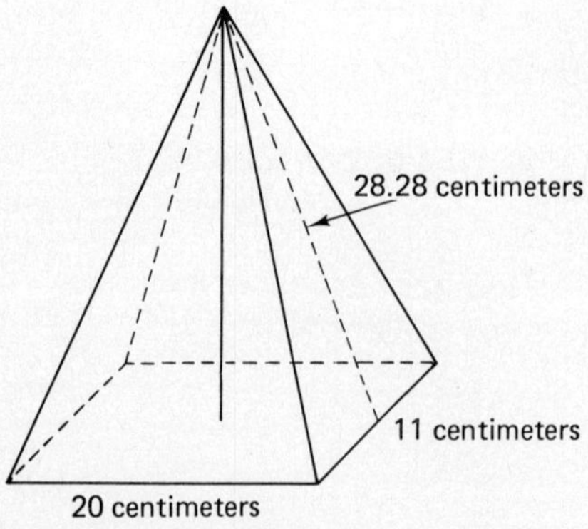

Figure KK

13-48. Determine the altitude of Figure KK.

13-49. Determine the area of the base of Figure KK.

13-50. Determine the volume of Figure KK.

13-51. What kind of a solid is Figure LL?

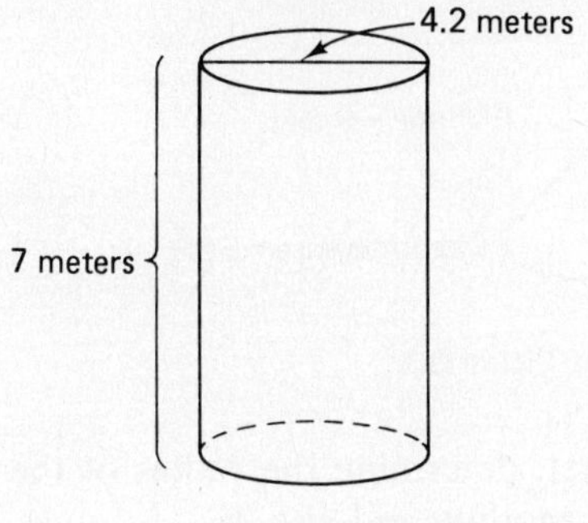

Figure LL

13-52. Find the area of the base of Figure LL.

13-53. Find the volume of Figure LL.

13-54. What kind of a solid is Figure MM?

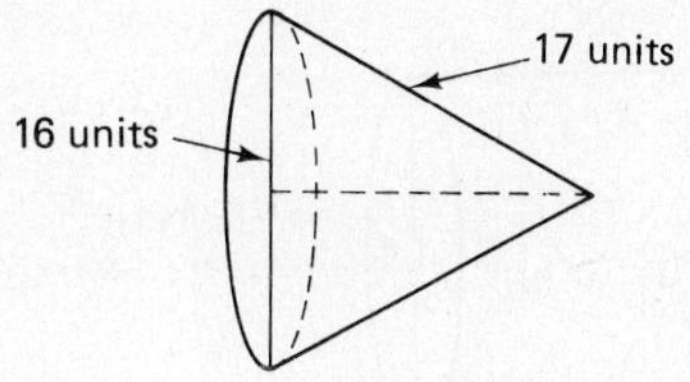

Figure MM

13-55. What is the area of the base of Figure MM?

13-56. Determine the volume of Figure MM.

13-57. What kind of a solid is Figure NN?

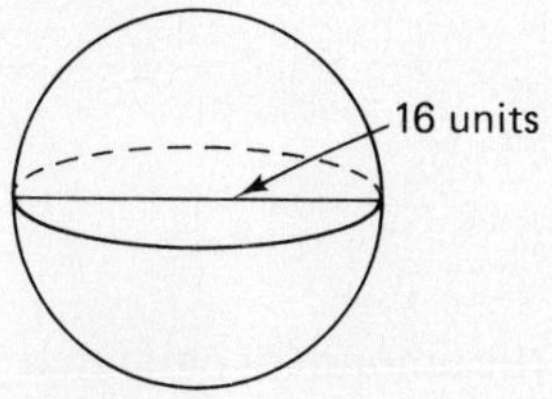

Figure NN

13-58. What is the volume of Figure NN?

13-59. What kind of a solid is Figure OO?

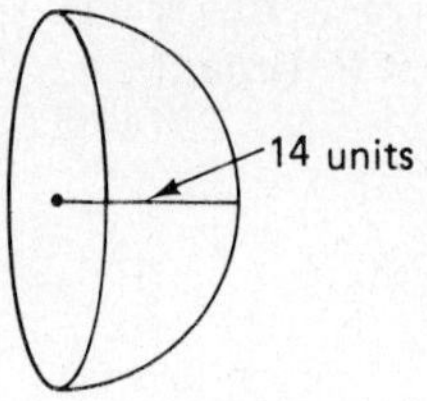

Figure OO

13-60. What is the volume of Figure OO?

CONTENT

The following examples deal with problems of industry. Note in each case that the formulas for determining the volumes of solids are used. Sometimes these formulas are used in combination.

Example 13-11

PROBLEM: Determine the number of gallons of gasoline that can be stored in the cylindrical tank illustrated by Figure PP.

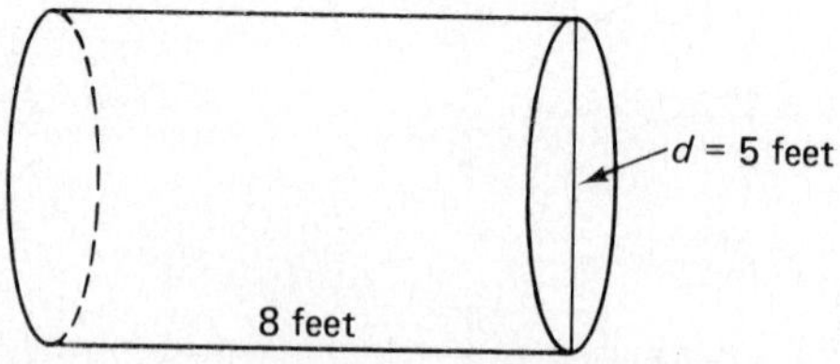

Figure PP

SOLUTION: First, determine the volume of the cylinder using the formula, $V = \pi r^2 h$.

$$V = 3.142 \times 2.5 \times 2.5 \times 8$$
$$V \doteq 157.1 \text{ cubic feet}$$

Second, determine the number of gallons using the formula, 1 cubic foot = 7.5 gallons.

$$\frac{157.1 \text{ ~~cubic feet~~}}{1} \times \frac{7.5 \text{ gallons}}{1 \text{ ~~cubic foot~~}} = 1178.2 \text{ gallons}$$

Example 13-12

PROBLEM: Determine the number of tons of coal in a pile whose circumference is 62.8 feet and height is 9 feet. Observe Figure QQ.

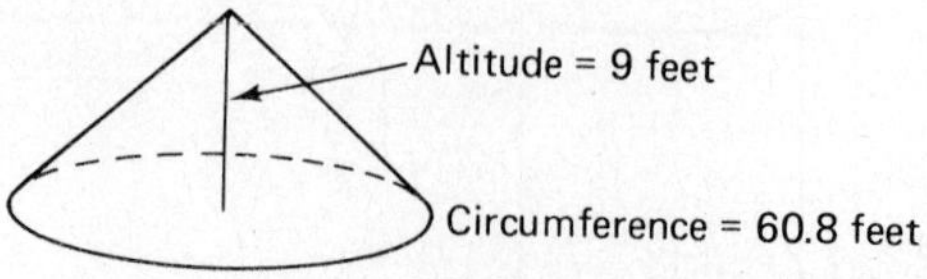

Figure QQ

SOLUTION: First, determine the radius of the base of the pile. Modify and use the formula, circumference = diameter times π.

$$\frac{\text{circumference}}{\pi} = \text{diameter}$$

$$\frac{60.8 \text{ feet}}{\pi} = \text{diameter}$$

$$20 \text{ feet} = \text{diameter}$$

$$10 \text{ feet} = \text{radius}$$

Second, determine the volume of the coal by using the formula for the volume of a cone,

$$V = \frac{\pi r^2 h}{3}.$$

$$= \frac{3.14 \times 10 \text{ feet} \times 10 \text{ feet} \times 9 \text{ feet}}{3}$$

$$= 942 \text{ cubic feet}$$

Third, determine the number of tons of coal by using the formula, 1 cubic foot = .0235 tons of coal.

$$\frac{942 \text{ ft}^3}{1} \times \frac{.0235 \text{ tons of coal}}{1 \text{ ft}^3} = 22.137 \text{ tons of coal}$$

Example 13-13

PROBLEM: Determine the amount of paint that would be needed to paint the water tower illustrated in Figure RR.

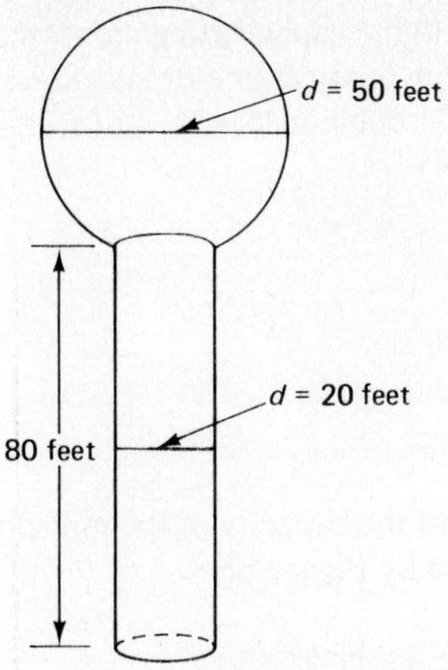

Figure RR

SOLUTION: First, determine the total surface area of the sphere using the formula πd^2.

$$\text{Total surface area} = 3.14 \times 50 \text{ feet} \times 50 \text{ feet}$$
$$= 7850 \text{ square feet}$$

Second, determine the lateral surface area of the cylinder using the formula $d\pi h$.

$$\text{Lateral surface area} = 20 \text{ feet} \times 3.14 \times 80 \text{ feet}$$
$$= 5024 \text{ square feet}$$

Third, determine the surface area which is inside the cylinder using the formula πr^2.

$$\text{Base area} = 3.14 \times 10 \text{ feet} \times 10 \text{ feet}$$
$$= 314 \text{ square feet}$$

Fourth, determine the total surface area to be painted.

$$\text{Total area of sphere} = 7850 \text{ square feet}$$
$$+ \text{ Total area of cylinder} = 5024 \text{ square feet}$$

$$\text{Total area} = 12{,}874 \text{ square feet}$$
$$- \text{ Area of base} = 314 \text{ square feet}$$

$$\text{Area to be painted} = 12{,}560 \text{ square feet}$$

Fifth, determine the amount of paint needed by using the formula, 1 gallon of paint covers 550 square feet.

$$\frac{12{,}560 \cancel{\text{ft}^2}}{1} \times \frac{1 \text{ gal}}{550 \cancel{\text{ft}^2}} = 22.84 \text{ gal}$$

EXERCISE 13-5

Determine the surface areas and volumes of the solid figures as indicated in the following problems:

13-61. Determine how much fabric would be needed to cover a cylindrical wastebasket which is 18 inches high and 8 inches in diameter.

13-62. Determine the total amount of steel needed to make a tank 6.75 meters long and 6 meters in diameter. (Allow 5% for seams.)

13-63. What is the volume of the tank in problem 13-62?

13-64. What is the weight of a brass solid which is 3 yards long, 1 inch thick, and $\frac{1}{2}$ foot wide? (Brass weighs 512 pounds per cubic foot.)

13-65. Find the capacity in gallons of a gas tank that is a rectangular solid 12 inches × 14.5 inches × 48 inches. (1 gallon = 231 cubic inches.)

13-66. Find the capacity in gallons of a cylindrical gas tank 12 inches in diameter and 48 inches long. (1 gallon = 231 cubic inches.)

13-67. How many gallons of crude oil will a mile section of the Alaskan pipeline contain if the inside diameter is 48 inches? (1 cubic foot = 7.5 gallons.)

13-68. Find the capacity of the mill-dust collector as illustrated by Figure SS.

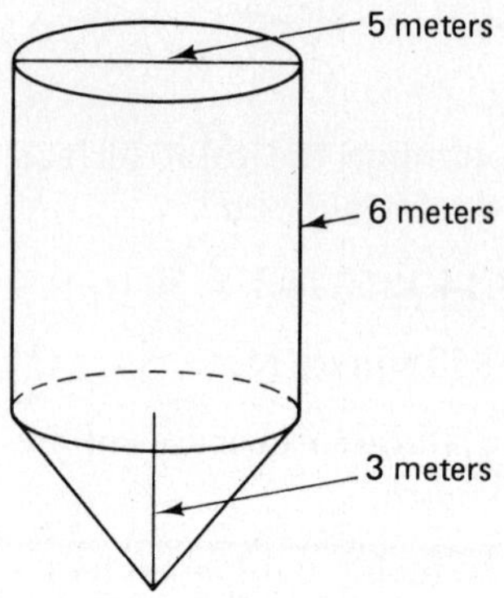

Figure SS

13-69. A stainless steel cylindrical milk tanker is 38 feet long and 5 feet in diameter. What is the amount of stainless steel needed to construct a tanker, allowing 12% for seams and waste?

13-70. Determine the capacity in gallons of the milk tanker in problem 13-69. (1 cubic foot = 7.5 gallons.)

13-71. A swimming pool is to be 50 meters long and 20 meters wide and will have an average depth of 3 meters. How many cubic meters must be excavated if an additional meter must be excavated for each side and the bottom?

13-72. How many square meters must be tiled if the inside of the pool (problem 13-71) is to be tiled?

13-73. A new house is built in the form of a hemisphere which has a radius of 7 meters. If the air in the house is to be circulated every 90 seconds, how many cubic meters of air must be moved per hour?

THINK TIME

To provide yourself with a simple review guide for geometric figures and the formulas for determining their volumes and surface areas, complete the following chart from the given sketches.

SOLID GEOMETRIC FORMULAS

Sketch	*Name of Figure*	*Formula for Lateral Surface Area*	*Formula for Total Surface Area*	*Formula for Volume*
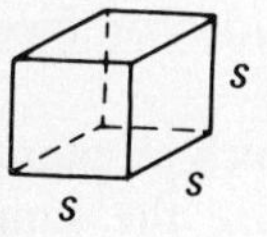				
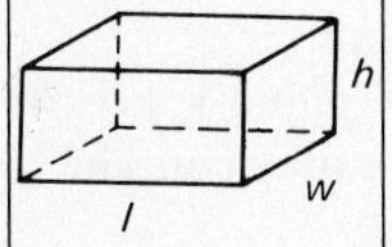				
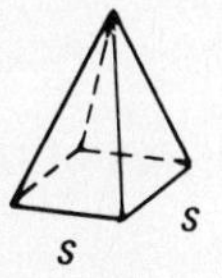				
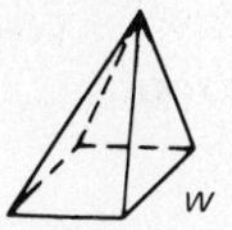				
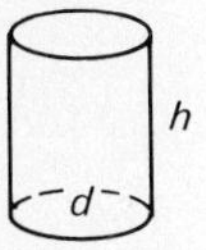				

PROCEDURES TO REMEMBER

1. When determining the area or volume of a solid geometric figure, use only like units.
2. Divide or change complex solid geometric figures into simple figures where possible, and apply basic formulas for determining lateral surface area; total surface area, or volume.

CHAPTER SUMMARY

The following definitions are essential to the measurement of solid geometric figures.

1. A *prism* is a solid geometric figure with parallel edges and uniform cross sections.
2. A *face* is any plane surface of a solid figure.
3. A *lateral face* of a prism is a side of a geometric figure.
4. The *bases* of a prism are the top face and the bottom face and are generally called the top and bottom of the solid.
5. The *lateral edge* of a prism is the place where two sides of a solid come together.
6. A *right prism* has lateral faces which are perpendicular to its base.
7. An *oblique prism* has lateral faces which are not perpendicular to its base.
8. A *right square prism* has a square for a base.
9. A *right rectangular prism* has a rectangle for a base.
10. A *right pentagonal prism* has a pentagon-shaped base.
11. The *lateral surface area* of a solid geometric figure is the surface area of the sides of the figure, excluding the area of the top and bottom.
12. The *total surface area* of a solid geometric figure is the area of the top and bottom of the figure as well as the area of the sides.
13. A *right pyramid* is a pyramid whose base is a regular polygon and whose sides or faces form equal isosceles triangles.
14. A *sphere* is a solid figure in which all points on the surface of the solid are the same distance from the center point.
15. A *hemisphere* is a half sphere.
16. *Volume* is the amount of cubic space or content which a solid figure may contain.

CHAPTER TEST

T-13-1. Determine the volume of a rectangular solid 2 centimeters × 5.5 centimeters × 8 centimeters.

T-13-2. Determine the lateral surface area of the given prism.

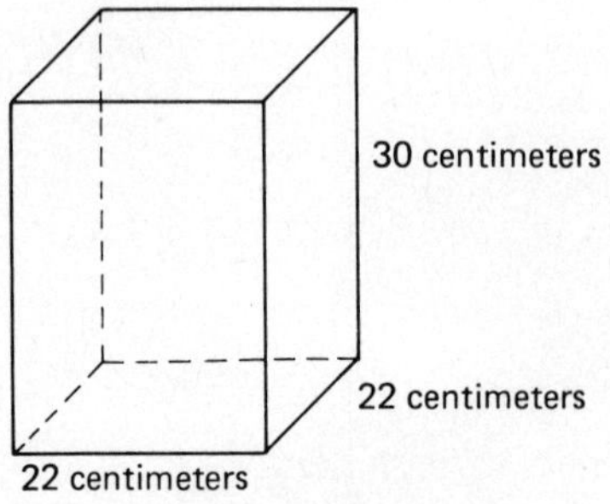

T-13-3. Determine the total surface area of a rectangular room 8 feet wide, 16 feet long, and 8 feet high.

T-13-4. Determine the lateral surface area of the given cylinder.

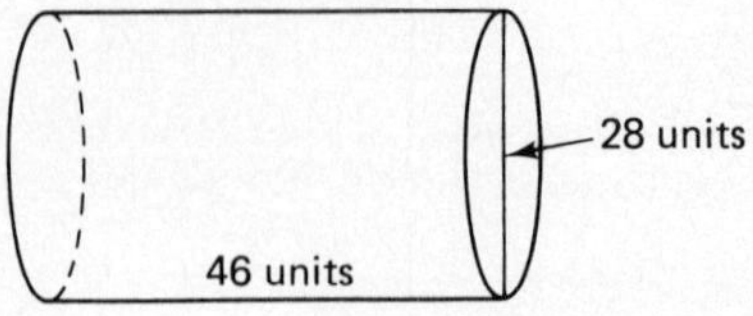

T-13-5. Determine the volume of the given cylinder.

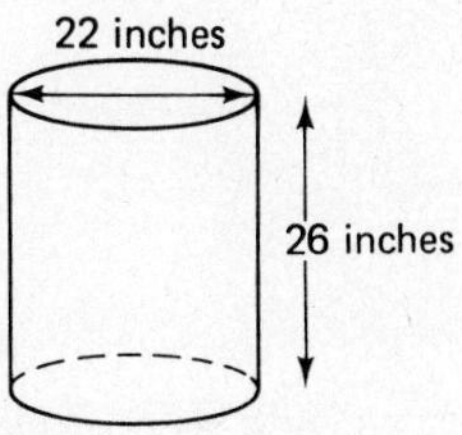

T-13-6. Determine the lateral surface area of the given cone.

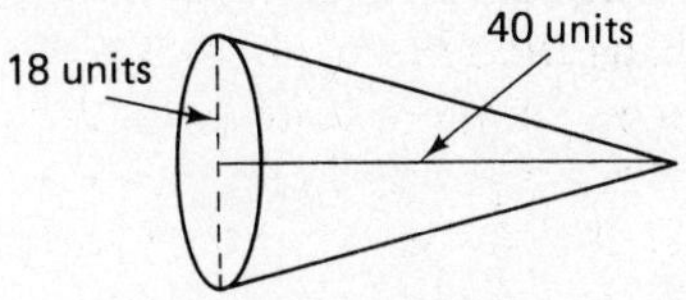

T-13-7. Determine the total surface area of the given cylinders.

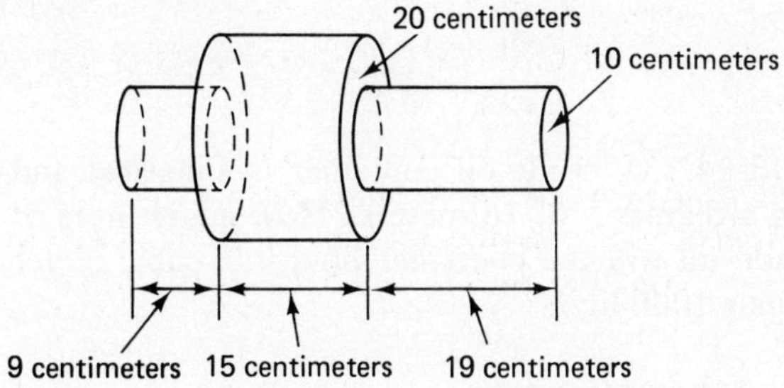

T-13-8. Determine the volume of the given cone.

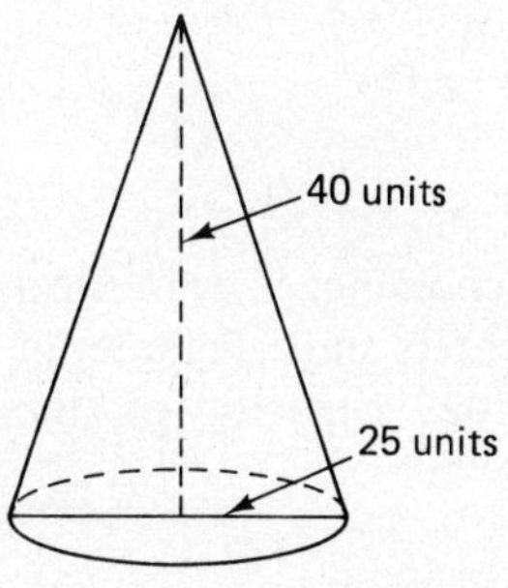

T-13-9. Determine the surface area of the given hemisphere.

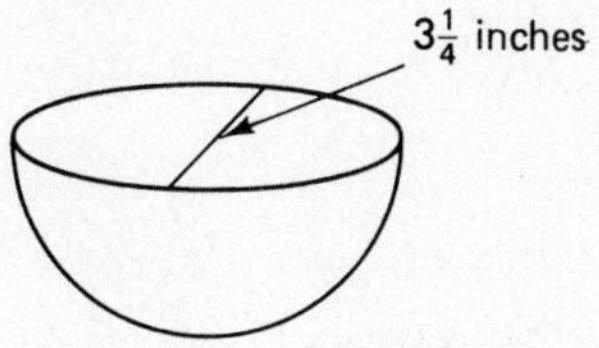

T-13-10. Determine the total surface area of the given pyramid.

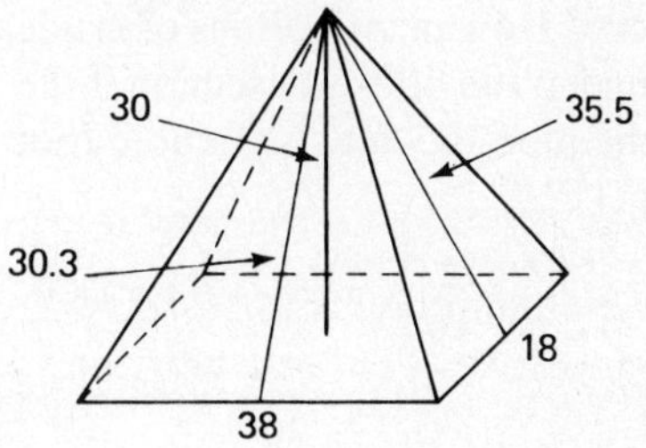

T-13-11. A stainless steel milk tanker is 36 feet long and 6 feet in diameter. How many gallons will it hold if 1 cubic foot is equal to 7.5 gallons?

T-13-12. Find the capacity of the given sawdust collector.

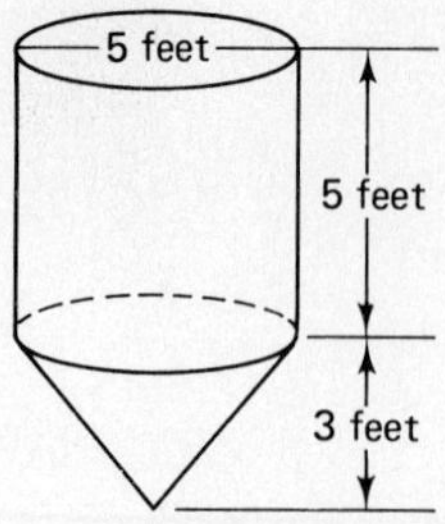

T-13-13. A tooth-powder container is cylindrical and has a length of 10 centimeters and a diameter of 4 centimeters. What is the capacity of the container?

T-13-14. A crude-oil container is spherical and has a diameter of 16 meters. How many liters of crude oil will the container hold if 1 cubic meter equals 1000 liters?

T-13-15. A pipeline is 400 feet long between shut-off valves in a refinery. How many gallons of crude oil would be contained in the 400-foot section if the inner diameter of the pipe is 3 feet? (1 cubic foot equals 7.5 gallons.)

CHAPTER
14

CONSTRUCTION OF SIMPLE GEOMETRIC FORMS

OBJECTIVES

To develop the skills necessary to construct the following with a compass and straight edge: circles, hexagons, squares, triangles, perpendiculars, parallel lines, angles (60°, 30°, 45°), tangents to circles, bisections (lines, angles), copies (lines, angles), and proportions.

SELF-TEST

This test will measure your ability to construct simple geometric forms. To complete this test you will need to use a compass and a straight edge. Satisfactory construction of the items called for in the problems will give evidence of your ability to construct simple geometric forms and your readiness to advance your skills in areas beyond the limits of this chapter. Problems you are not able to solve will indicate potential areas for needed instruction. When necessary make your constructions on a separate sheet of paper.

S-14-1. Bisect $\overline{XY}$.

S-14-2. Bisect $\angle ABC$.

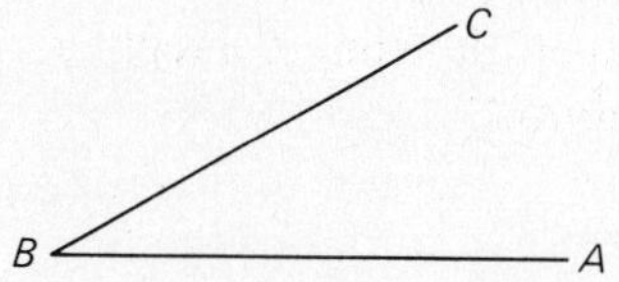

S-14-3. Copy $\angle FOB$.

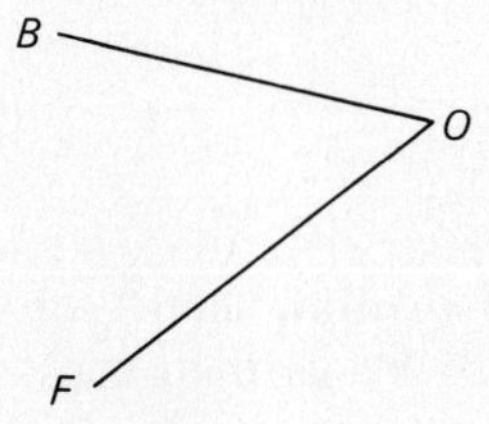

S-14-4. Divide $\overline{GJ}$ into three equal parts.

S-14-5. Construct a perpendicular on $\overline{JY}$ at point *L*.

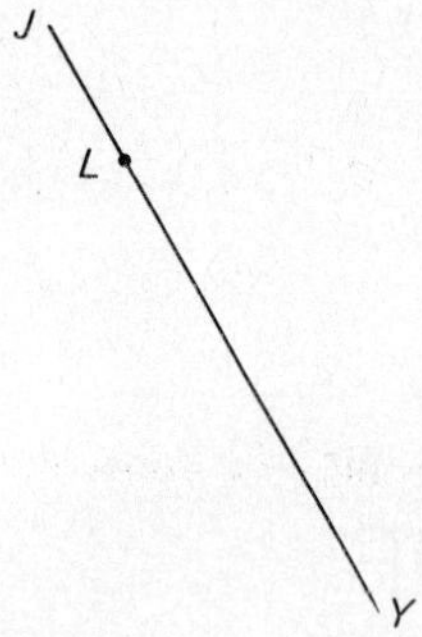

S-14-6. Construct a perpendicular to $\overline{KC}$ from point *M*.

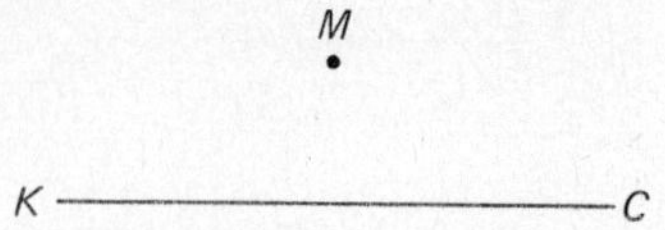

S-14-7. Construct a line parallel to $\overline{ND}$.

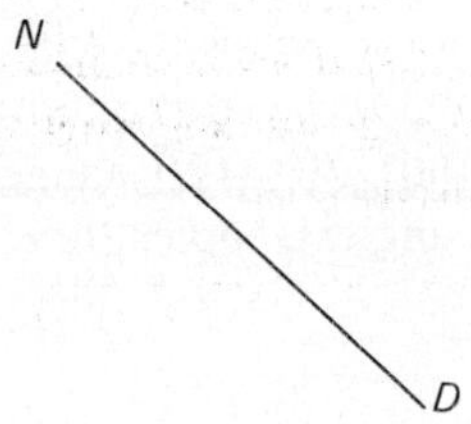

S-14-8. Construct tangents to the circle from point *P*.

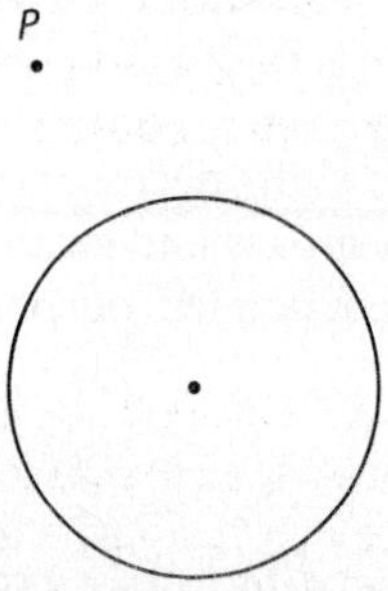

S-14-9. Construct a square with sides equal to $\overline{AB}$.

S-14-10. Construct a triangle from segments $\overline{AB}$, $\overline{CD}$, and $\overline{EF}$.

INTRODUCTION

It is important for tradesmen and technicians to be able to construct simple, yet accurate sketches of the work they are about to undertake. Often these sketches help to simplify the work and save valuable time. It is the purpose of this chapter to help you develop the skills necessary to construct these simple geometric forms.

DEFINITIONS

A *compass* is a tool used to draw circles or portions of circles. It may also be used to copy lengths of line segments. One type of compass is illustrated in Figure A.

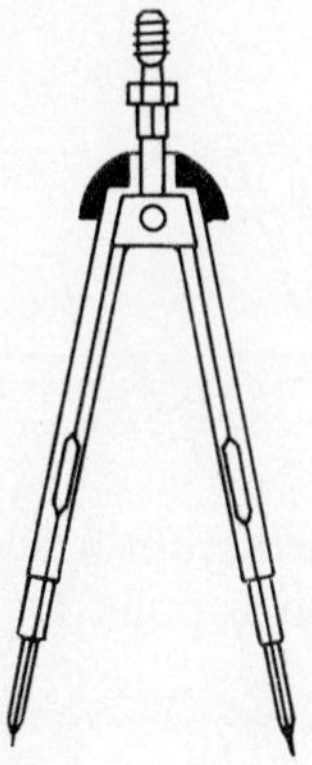

A *straight edge* is used to draw straight lines between two points. A ruler is generally used as a straight edge, but any material or instrument may be used if it has a straight edge. The opening between two points of the compass is called the *radius.* The letters O and P are frequently used as points of identification in geometric construction.

CONTENT

Constructions are learned best through practice. The first thing you will construct is a circle. Open the compass so that the metal tip and lead tip are approximately an inch or two apart. Place the metal tip on point P at the right. Draw or "swing" a circle with the compass. Point P is the center of the circle. Practice constructing circles similar to those given below.

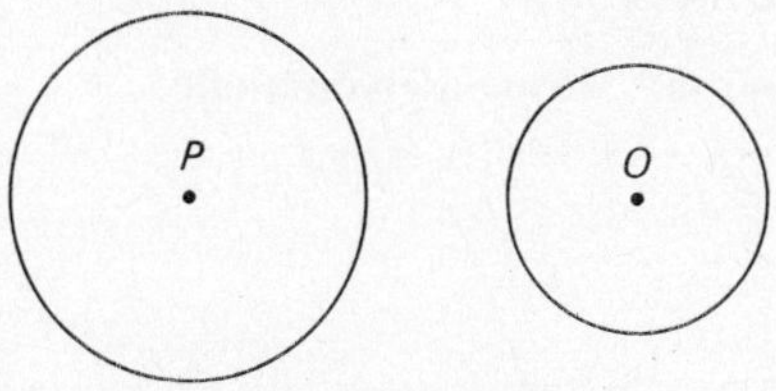

Refer to Figure B, and use your straight edge to draw a line from the center O to point A on the circle. This line, $\overline{OA}$, is the radius of the circle. Draw in radius $\overline{OB}$ and radius $\overline{OC}$. $\overline{OA}$, $\overline{OB}$, and $\overline{OC}$ are called *radii* of the circle. (Radii is the plural of radius.) Note that $\overline{OB}$ and $\overline{OC}$ form a straight line and also pass through the center point O. Thus $\overline{BC}$ is the *diameter* of the circle. The curved line AC in Figure B is called an *arc.* An *arc* is a portion of the circumference of a circle. The symbol for an arc is $\frown$, which is placed over appropriate letters: $\overset{\frown}{AC}$. A circle may have several arcs. $\overset{\frown}{AB}$ and $\overset{\frown}{BC}$ are also arcs of circle O.

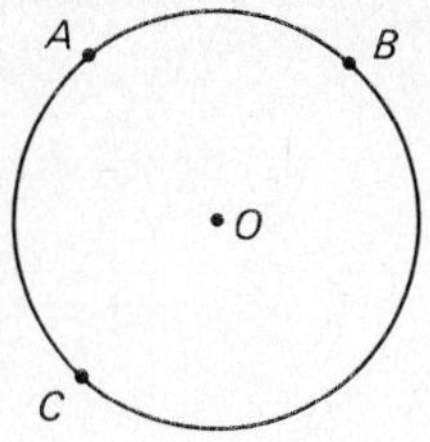

Figure B

Given the radius $\overline{LM}$, construct a circle using M as the center point.

Place two additional letters, J and S, on the circumference of the circle. Identify the three arcs of circle M. Identify the three radii of circle M.

To copy a line segment, both a compass and a straight edge are used. To copy line segment JG, illustrated below, first draw a "working line" which is longer than the given line segment. Use the straight edge to do this.

Next, open the compass to the length of line segment JG. Place the metal point on J and the lead point on G. Then place the metal point on J of the working line and "arc" the working line. At the point where the arc crosses the working line, identify point G.

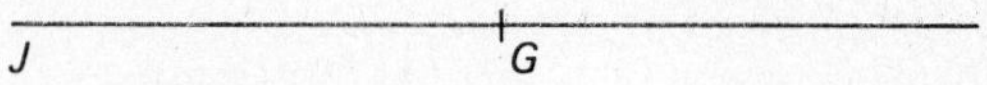

Copy the given lines below on the right side of the page.

Example 14-1

PROBLEM: Bisect a given line segment.

SOLUTION: To bisect a line means to divide it into two equal parts. First, open the compass to a position greater than half the length of the line segment. Place the metal tip at point K and swing an arc as shown.

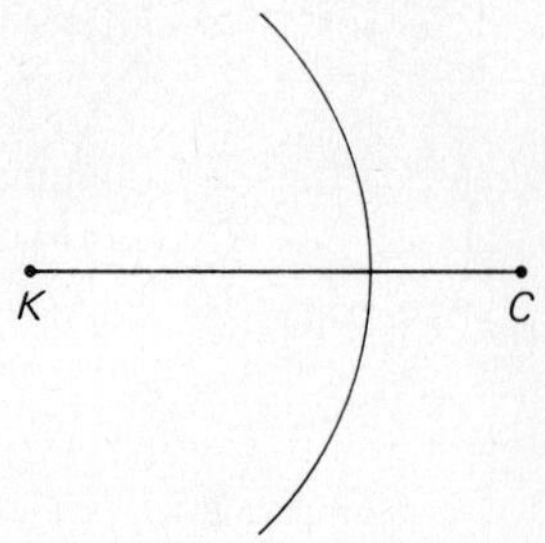

Second, keep the compass at the same opening and swing an arc from point *C*. Label the points where the arcs intersect as *X* and *Y*.

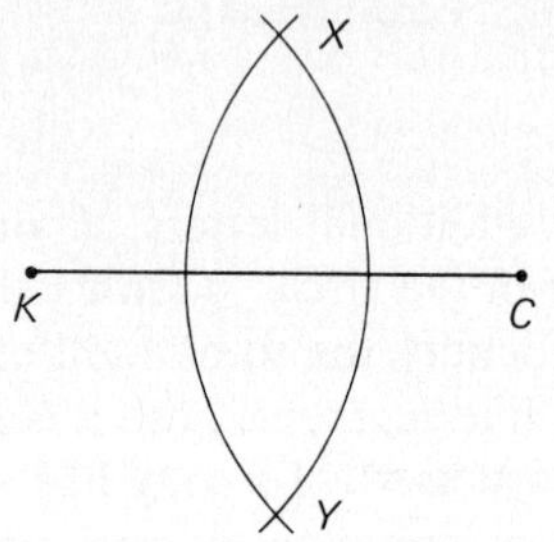

Third, sketch a line from the point of intersection *X* to the point of intersection *Y*. Where line *XY* intersects line *KC*, label the point *Z*. Point *Z* is the midpoint of line *KC*. Thus $\overline{KZ} = \overline{ZC}$.

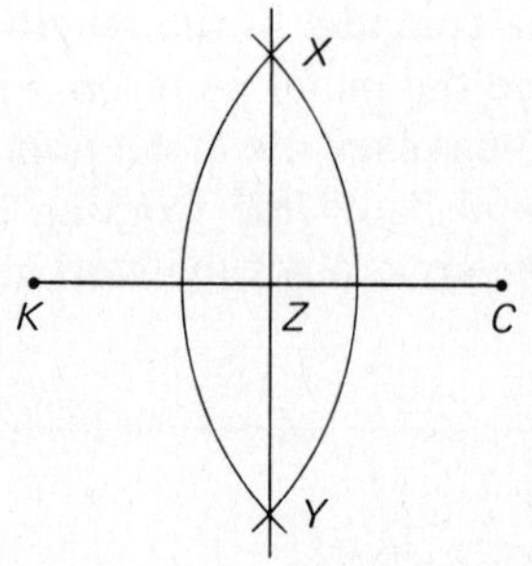

Example 14-2

PROBLEM: Bisect a given angle.

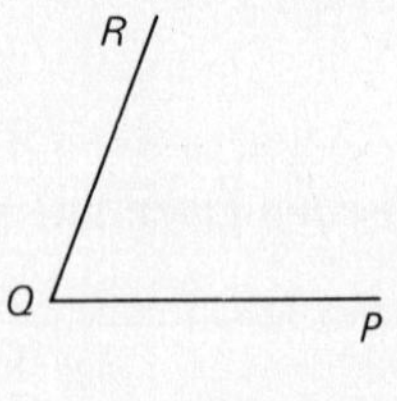

SOLUTION: To bisect an angle means divide it into two equal parts. First, place the metal tip of your compass at the vertex of the angle (point *Q*) and swing an arc through the rays of the angle cresting points *A* and *B*. The compass opening should be $\frac{1}{2}$ to $\frac{3}{4}$ the length of $\overrightarrow{RQ}$. Label points *A* and *B*.

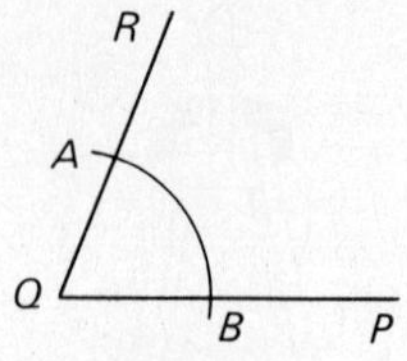

Second, swing arcs from *A* and *B* and label their point of intersection as point *D*.

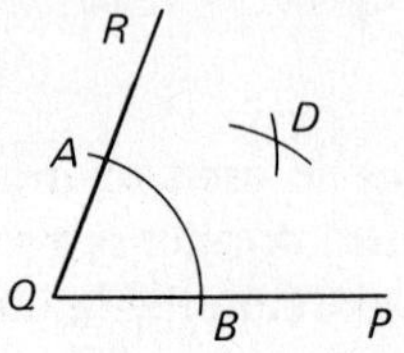

Third, draw line *QD*. Line *QD* bisects angle *PQR* to form two equal angles.

Angle *PQD* = angle *RQD* or $\angle PQD = \angle RQD$

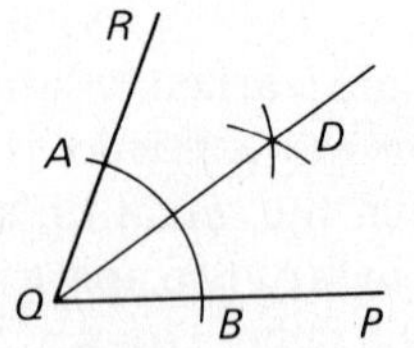

Example 14-3

PROBLEM: Copy a given angle.

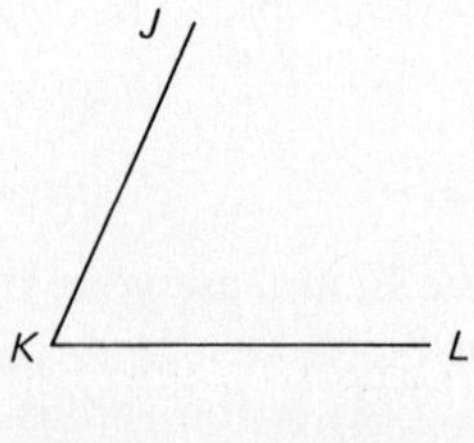

SOLUTION: To copy an angle means to duplicate it or produce another angle which is equal or congruent to the given angle. First, draw a working line and label it *QR*. This line should be somewhat longer than line *KL* of $\angle JKL$.

Second, swing an arc from point *K* of $\angle JKL$ to form points *N* and *M*.

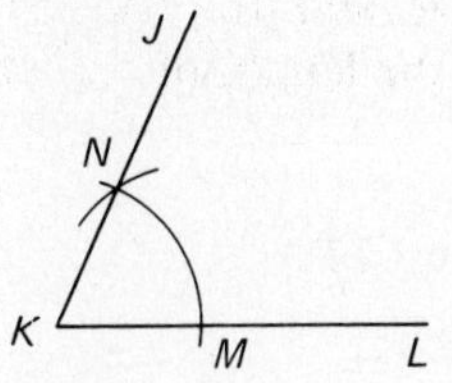

Third, swing a similar arc on the working line from point Q. Identify the point of intersection as point X.

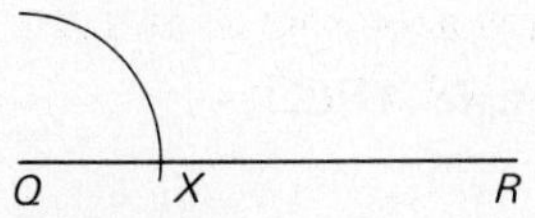

Fourth, set the compass so that the metal tip is on point M and the lead tip is on point N.

Fifth, place the metal tip on point X of $\overline{QR}$ and swing an arc so that it intersects the other arc. Label this point of intersection as point Y.

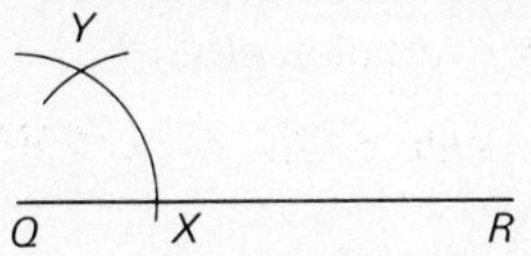

Sixth, draw a line from point Q through point Y to form line QP.

Angle JKL = angle PQR or $\angle JKL = \angle PQR$

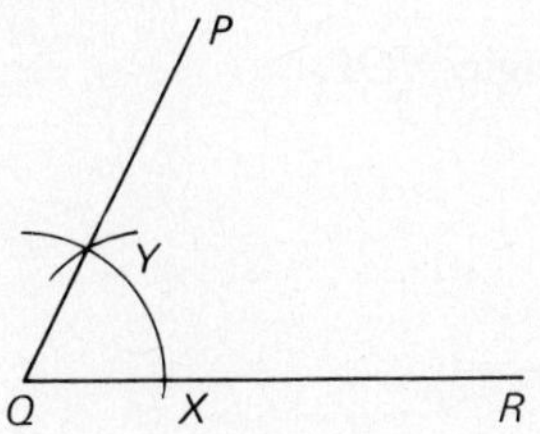

Example 14-4

PROBLEM: Divide a line into a given number of equal parts.

SOLUTION: To divide the given line SA into three equal parts, first draw a working line about 20° to 50° above line SA to form working angle MSA.

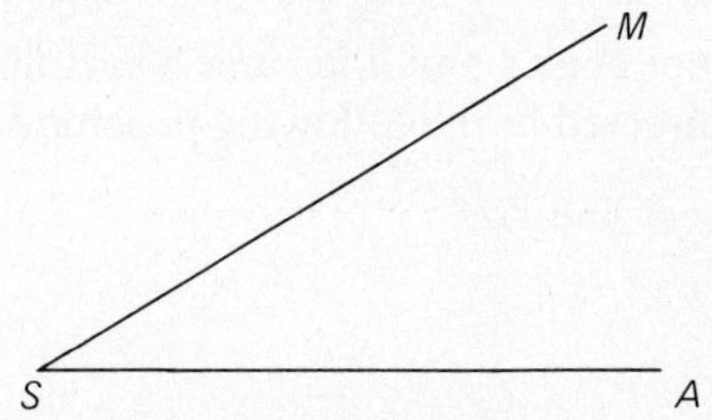

Second, divide the line SM into three equal parts by opening the compass and swinging it to form point E. From point E, swing the same arc to form point F. From point F, swing the same arc to form point G. Draw a line from point G to point A.

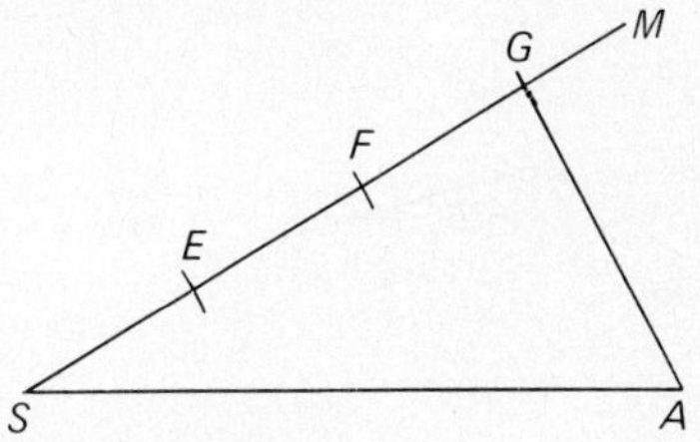

Third, using point F as a center, copy angle SGA and draw line FB.

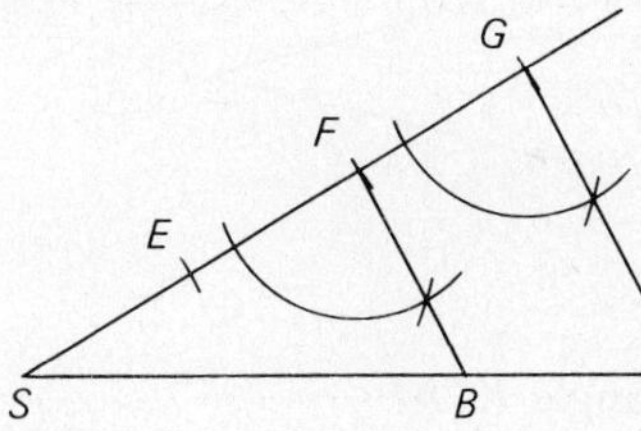

Fourth, using point E as a center, copy angle SGA and draw line EK.

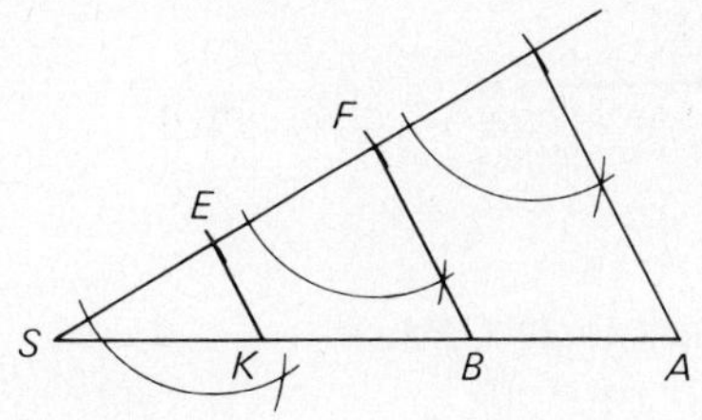

Thus, $\overline{SK} = \overline{KB} = \overline{BA}$.

The same process could be used to divide any straight line into any number of equal parts.

EXERCISE 14-1

Construct angles and lines and bisect lines and angles as directed in the following problems.

14-1. Bisect line *AB*.

14-2. Bisect line *CD*.

14-3. Bisect line *EF*.

14-4. Bisect angle *ABC*.

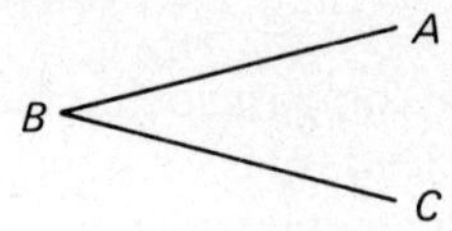

14-5. Bisect angle *EFG*.

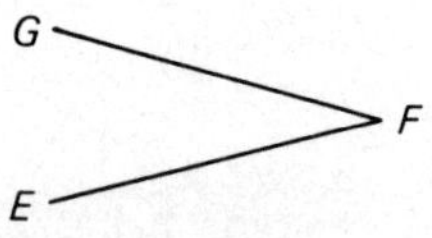

14-6. Bisect angle *HIJ*.

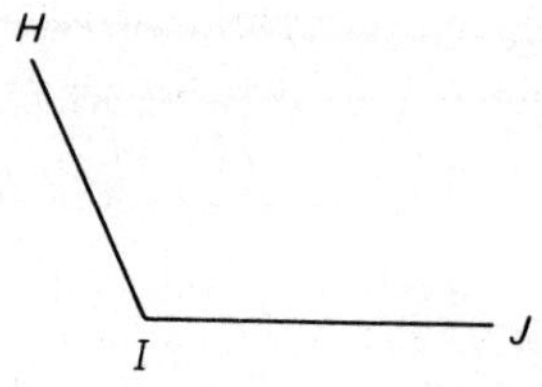

14-7. Copy angle *KLM*.

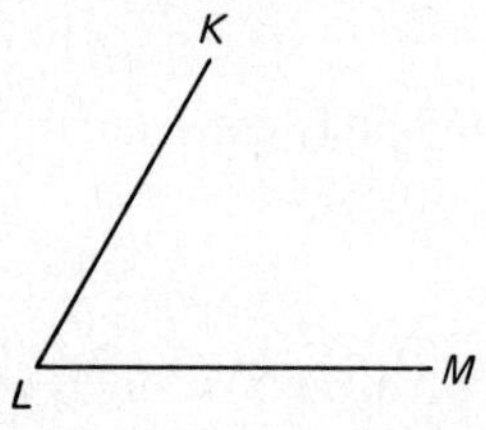

14-8. Copy angle *NOP*.

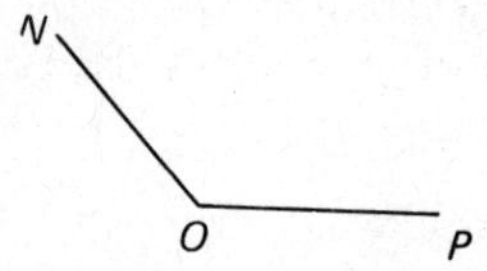

14-9. Copy angle *QRS*.

14-10. Divide line *AB* into three equal parts.

14-11. Divide line *CD* into five equal parts.

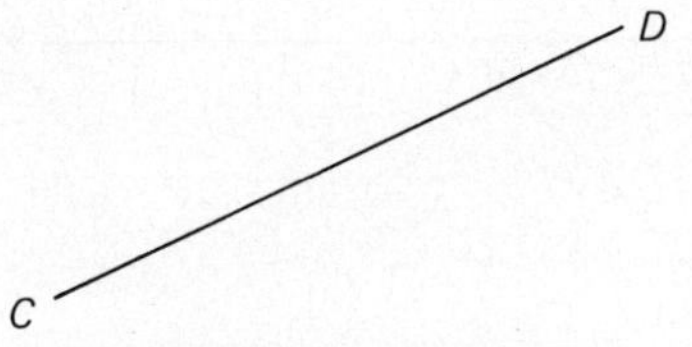

CONTENT

Generally walls are constructed straight up from the floor so that they form a 90° angle on each side of the wall. When this occurs, we say that the wall is perpendicular to the floor. Thus when one line is *perpendicular* to another, it forms two 90° angles. When two lines run side by side and never cross no matter how far they are extended, we say the lines are *parallel.* When a line touches a circle at only one point, we say that the line is *tangent* to the circle. In the next series of examples, you will see how to construct perpendicular, parallel, and tangent lines.

Example 14-5

PROBLEM: Construct a perpendicular line from a point on a line.

SOLUTION: First, place the compass point on point *P*, swing arcs on each side of point *P*, and label them as points *X* and *Y*.

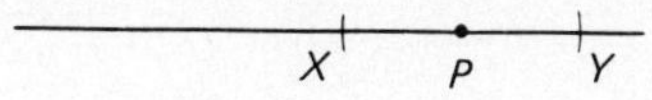

Second, open the compass and swing arcs above and below the line from points *X* and *Y* to form points *A* and *B*.

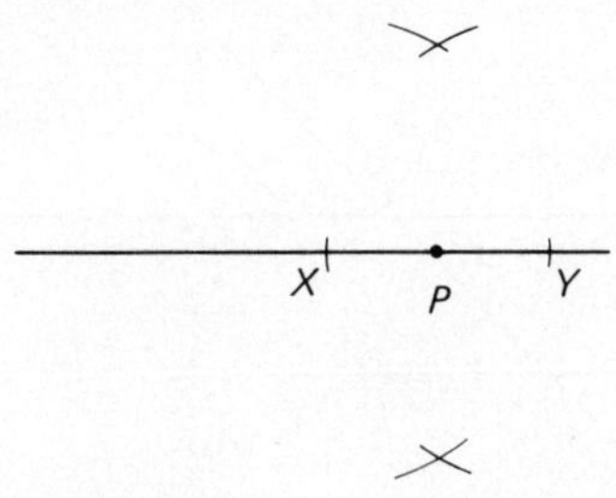

14-12. Divide line *EF* into five equal parts.

Third, draw $\overline{AP}$ using point *B* as a check point. $\overline{AP}$ will be perpendicular to the line at point *P*. Thus angles *XPA* and *APY* are 90°.

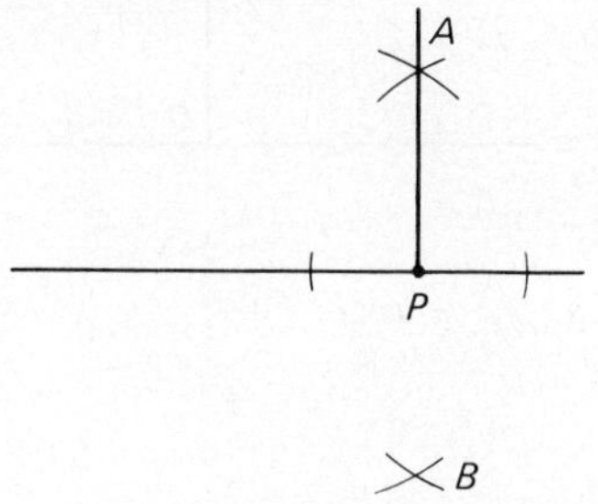

Example 14-6

PROBLEM: Construct a perpendicular to line *AB* from point *P*.

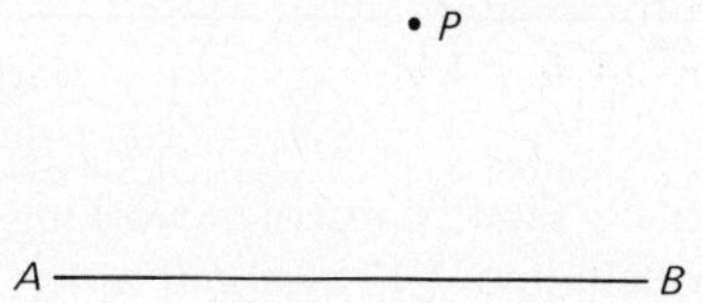

SOLUTION: First, open the compass slightly more than the distance from point *P* to $\overline{AB}$ and arc the line in two places. Label these points *J* and *K*.

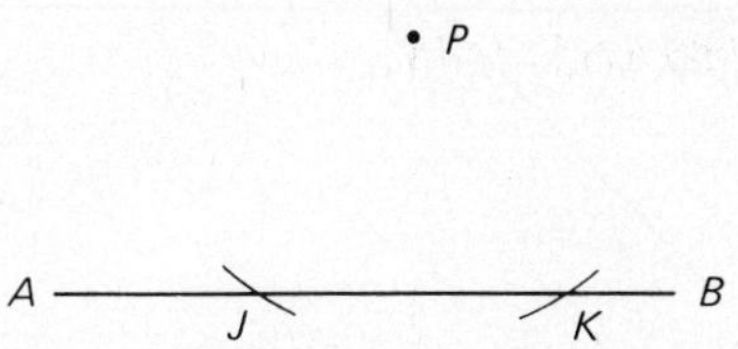

Second, using *J* and *K* as centers, swing arcs below $\overline{AB}$ and label the new point *L*.

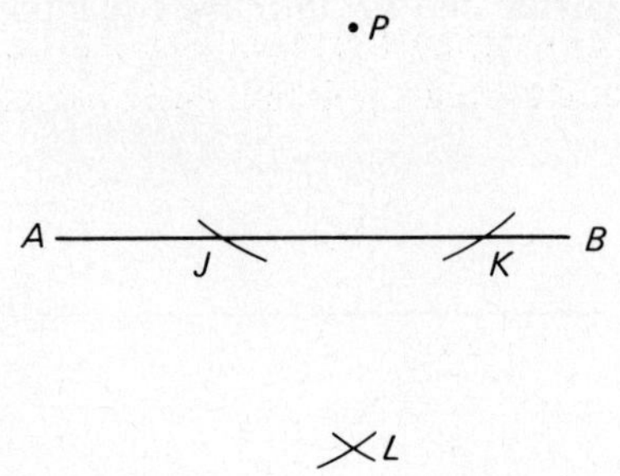

Third, draw the line segment from point P to $\overline{AB}$ using point L as the guide point to determine the line. Label the point of intersection as point Q. Thus $\overline{PQ}$ is perpendicular to $\overline{AB}$.

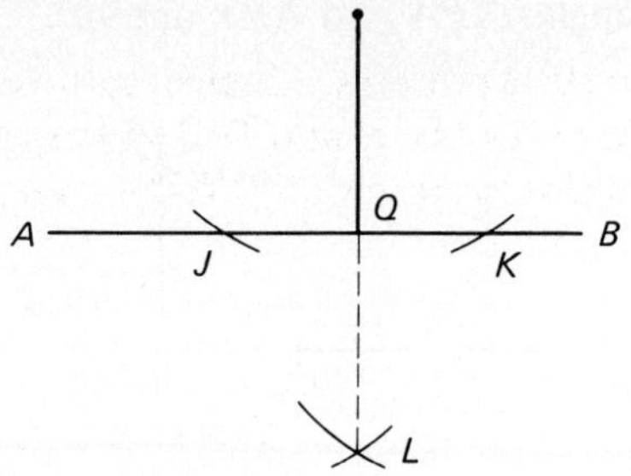

Example 14-7

PROBLEM: Construct a line parallel to a given line.

SOLUTION: First, construct working line JL which intersects line XY at about a 45° angle as shown.

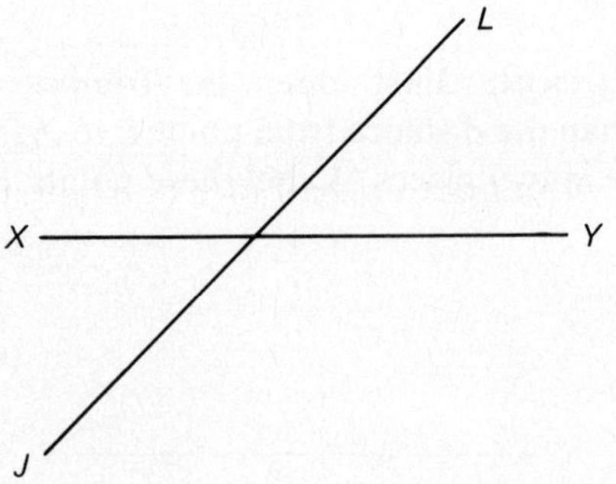

Second, swing an arc above $\overline{XY}$ at the intersection of $\overline{XY}$ and $\overline{JL}$.

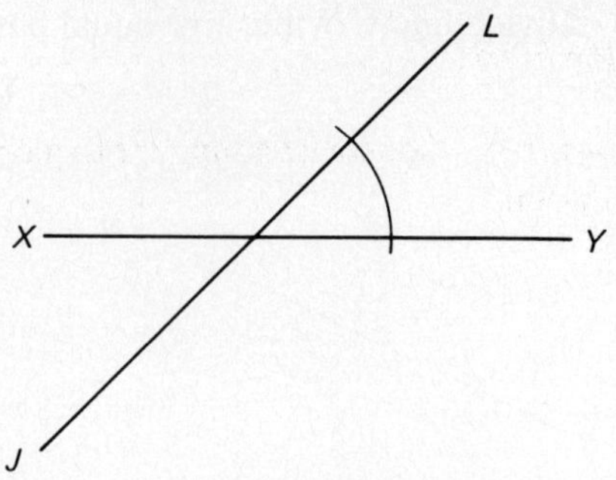

Third, swing an arc from a point on $\overline{JL}$ which might be identified as point K.

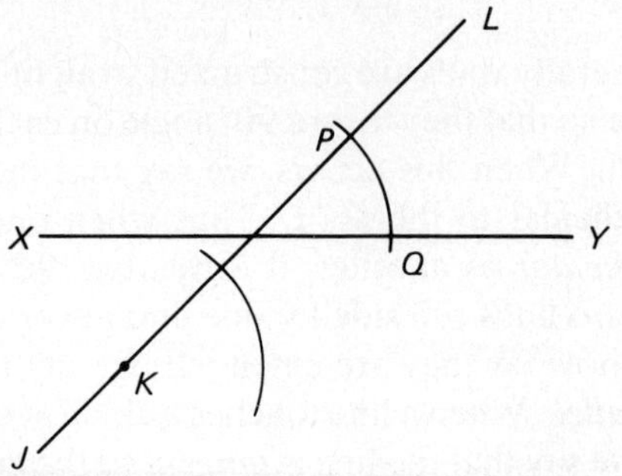

Fourth, open the compass a distance equal to PQ and swing an arc from the point where the previous arc intersected $\overline{JL}$ to intersect the previous arc at point W.

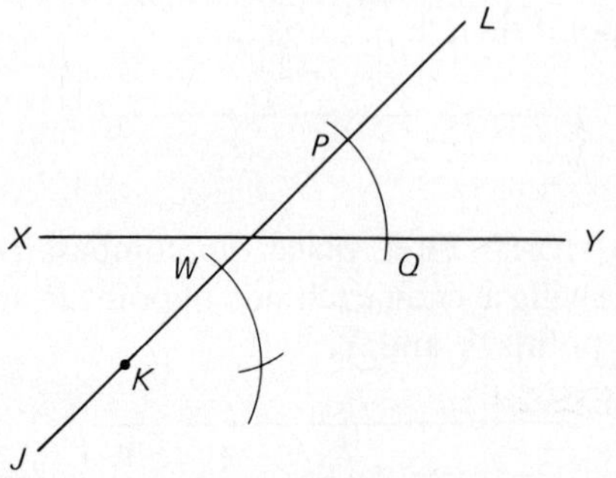

Fifth, draw line VT through points K and W. Thus $\overline{XY}$ is parallel to $\overline{VT}$, or $XY \parallel VT$.

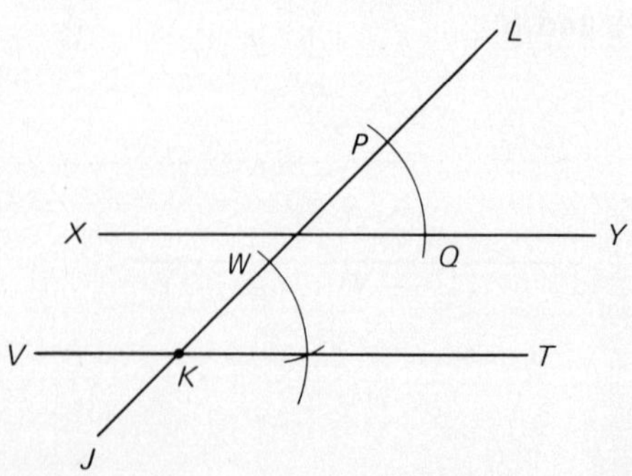

Example 14-8

PROBLEM: Construct a tangent to a circle from a given point.

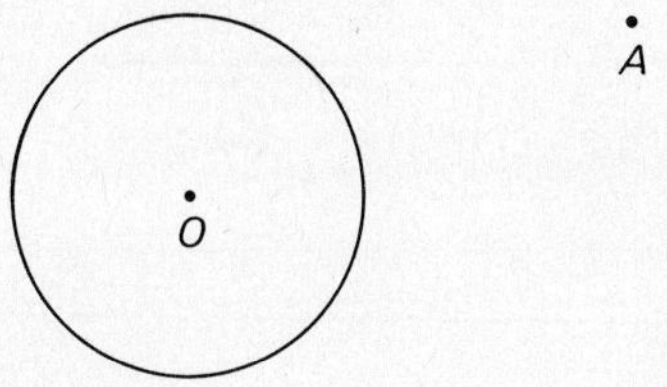

SOLUTION: First, draw a line from point A to the center of the circle, point O.

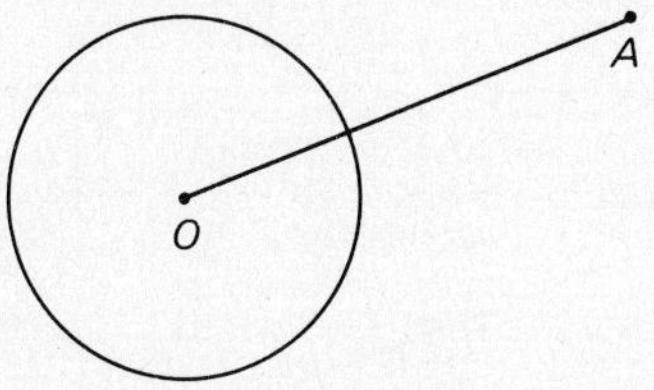

Second, construct a line that bisects AO as shown. Label the point of bisection as point K.

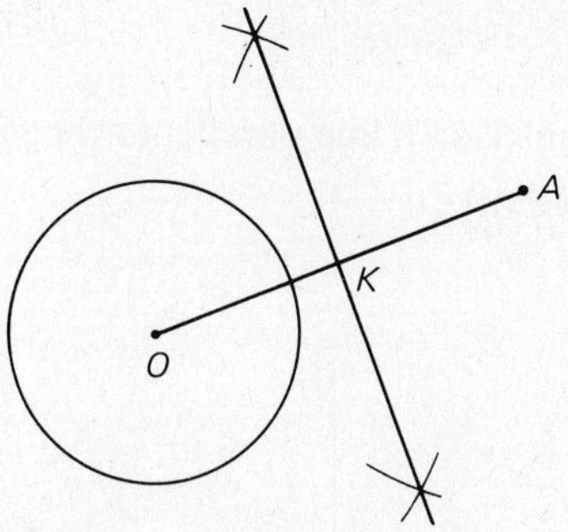

Third, swing an arc from point K to intersect point O and to create points E and D on the circle.

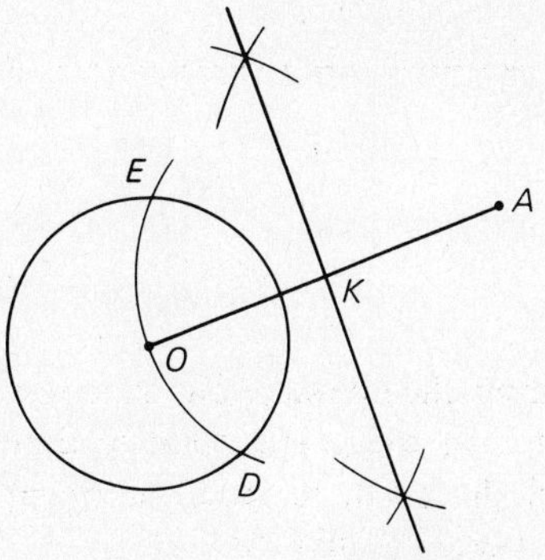

Fourth, draw rays AE and AD from point A to intersect at points E and D, respectively. Thus $\overrightarrow{AE}$ is a tangent to circle O and $\overrightarrow{AD}$ is a tangent to circle O.

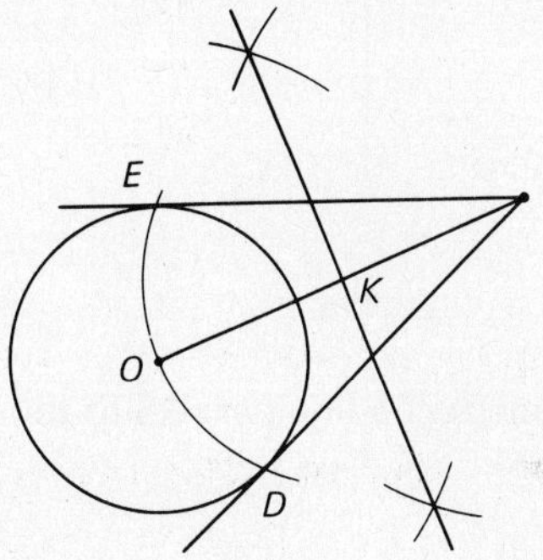

EXERCISE 14-2

Construct perpendiculars, parallels, and tangents as directed in the following problems.

14-13. Construct a perpendicular to point X.

14-14. Construct a perpendicular to point P.

14-15. Construct a perpendicular to point *K*.

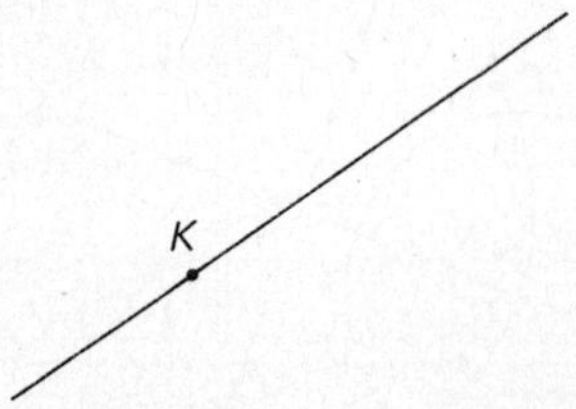

14-16. Construct a perpendicular to the line from point *L*.

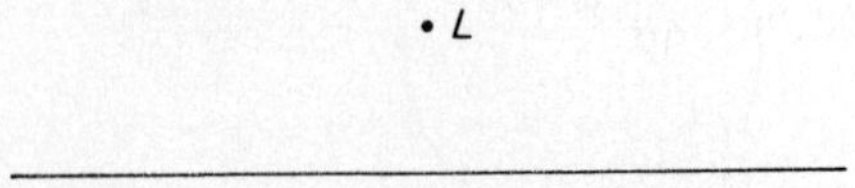

14-17. Construct a perpendicular to the line from point *J*.

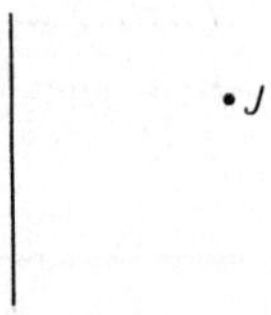

14-18. Construct a perpendicular to the line from point *K*.

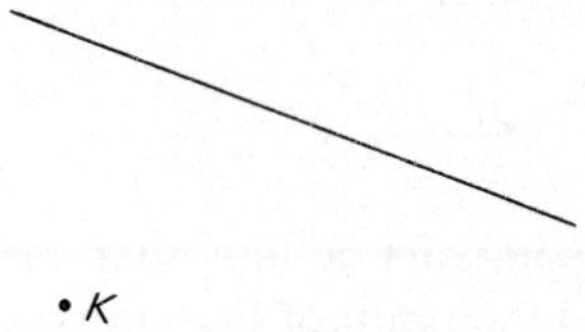

14-19. Construct a line parallel to the given line.

14-20. Construct a line parallel to the given line.

14-21. Construct a line parallel to the given line.

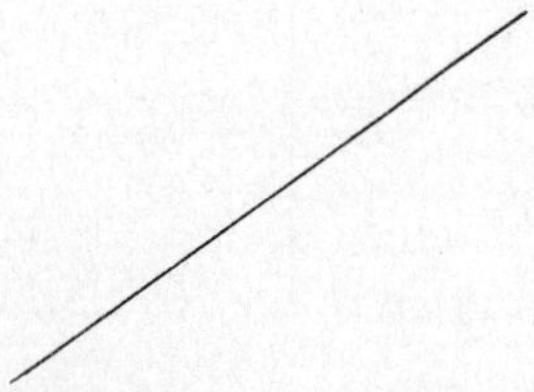

14-22. Construct tangents to circle *L* from point *M*.

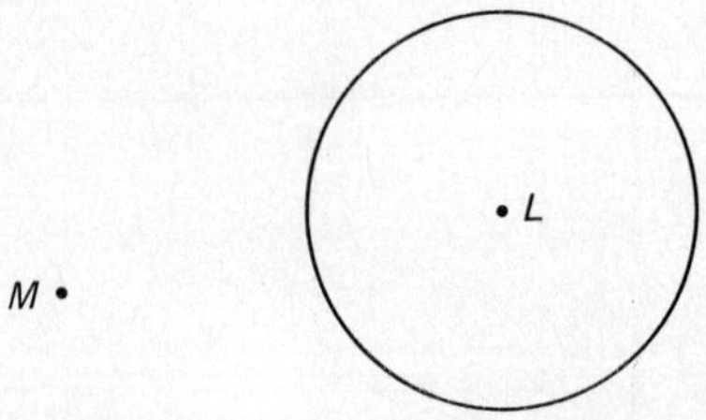

14-23. Construct tangents to circle X from point Y.

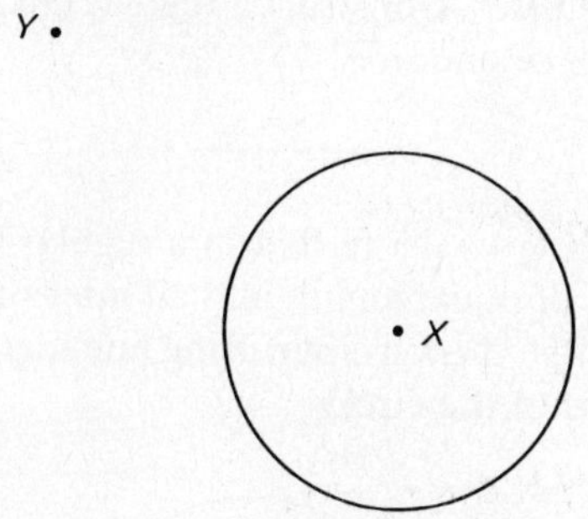

14-24. Construct tangents to circle S from point T.

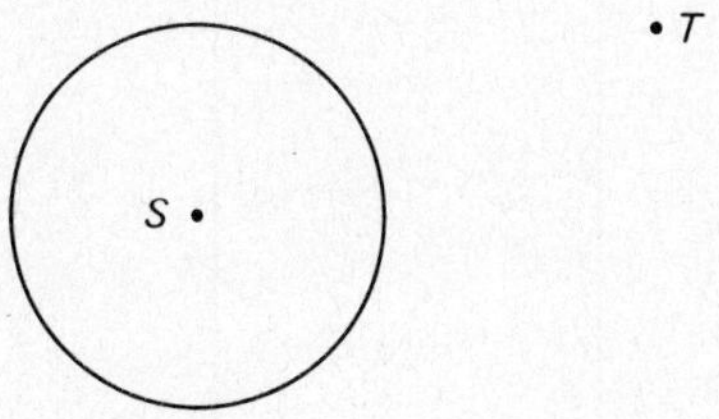

CONTENT

Often it is essential to construct simple geometric figures in order to solve complicated industrial problems. In this section, you will see how to construct triangles, squares, and hexagons.

Example 14-9

PROBLEM: Construct a triangle from three given line segments.

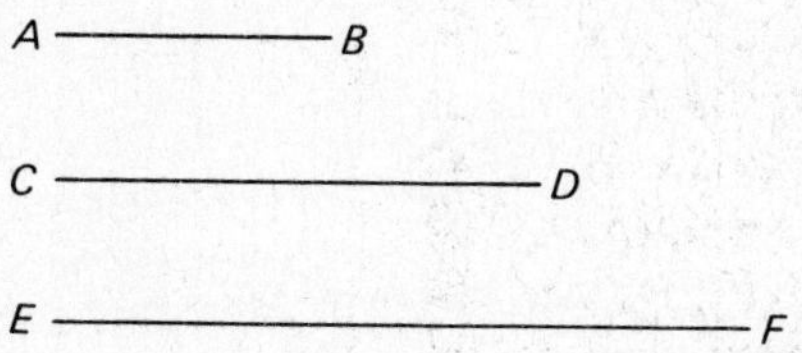

SOLUTION: First draw a working line and label it $\overline{XY}$.

Second, measure line EF with a compass and enter the points on the working line XY.

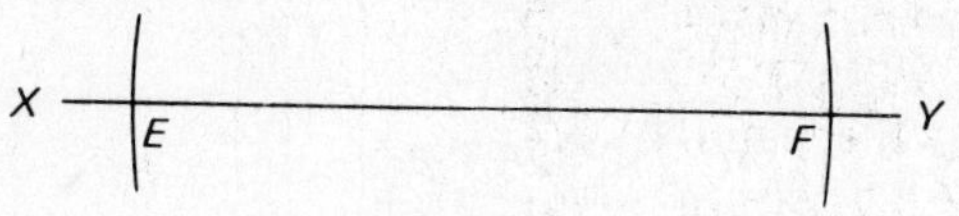

Third, measure line AB with the compass and swing an arc from point E.

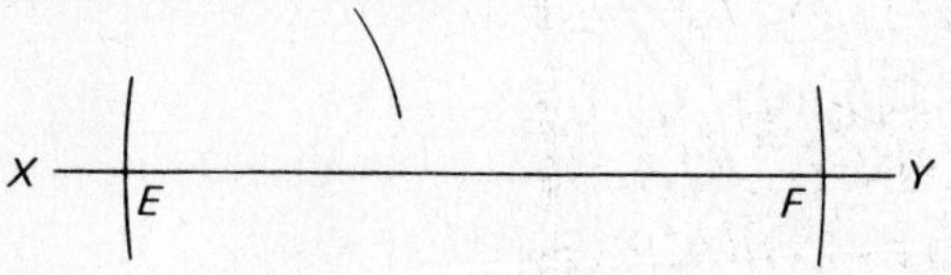

Fourth, measure line CD with the compass and swing an arc from point F so that it intersects the arc from point E. Designate the point of intersection as point G.

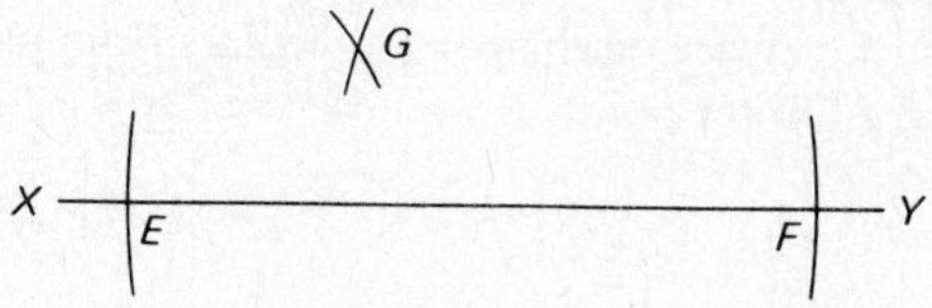

Fifth, draw in line segments EG and FG.

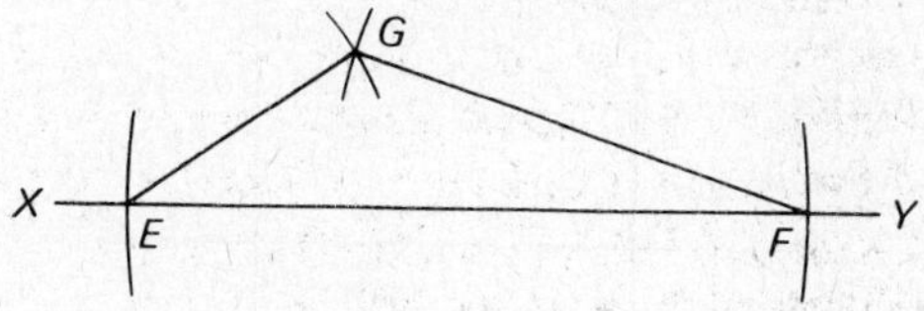

Sixth, measure with a compass to ensure that:

$$\overline{AB} = \overline{EG}$$
$$\overline{CD} = \overline{FG}$$
$$\overline{EF} = \overline{EF}$$

Example 14-10

PROBLEM: Construct a square, given the length of one side.

SOLUTION: First, draw a working line.

Second, measure line KJ and enter the points on the working line.

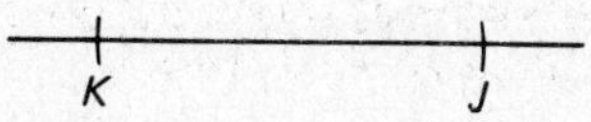

Third, construct perpendiculars at both points *K* and *J*.

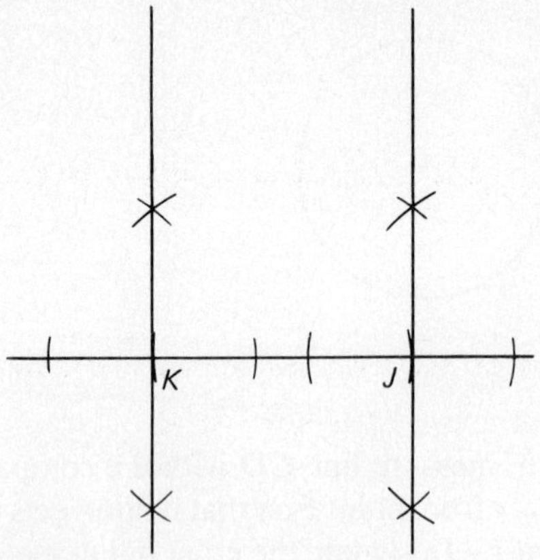

Fourth, measure the length of line segment *KJ* and strike an arc on the perpendiculars from points *K* and *J*. Enter points *L* and *M*.

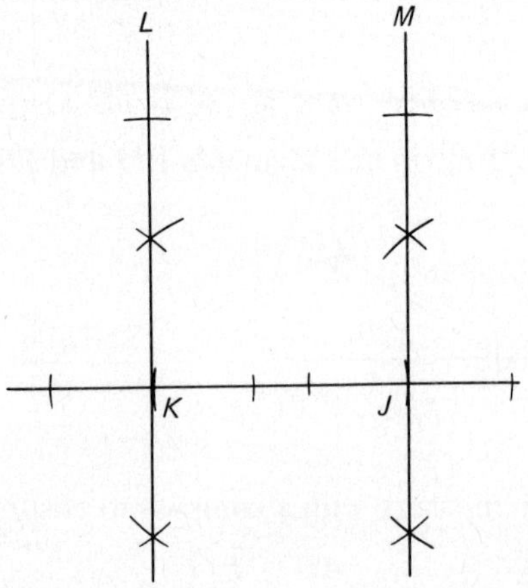

Fifth, draw in line segments *KL*, *LM*, and *JM*.

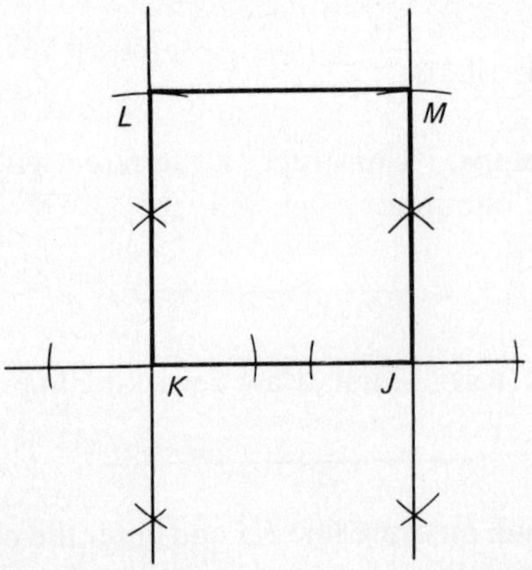

Sixth, measure with a compass to ensure that:

$$\overline{KJ} = \overline{KL} = \overline{LM} = \overline{JM}$$

Example 14-11

PROBLEM: Construct a regular hexagon, given the length of one side.

SOLUTION: Note that in a regular hexagon all sides are of equal length and all interior angles are equal. First, draw a circle using line segment *LM* as the radius of the circle.

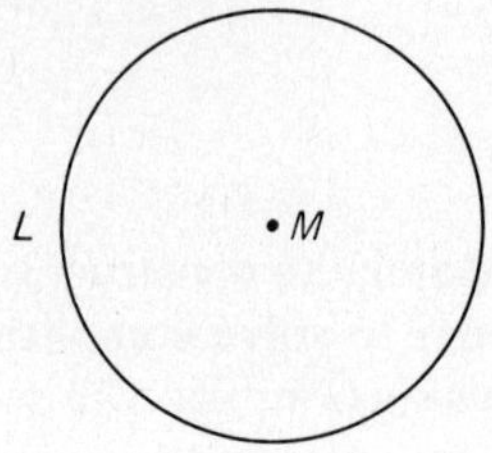

Second, using the same radius, swing an arc from point *L* to locate point *A*.

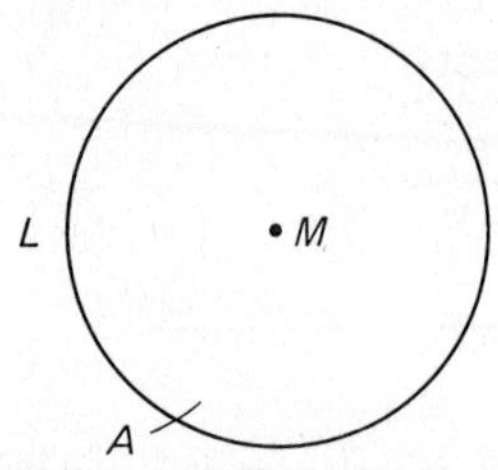

Third, using the same radius, swing an arc from point *A* to locate point *B*, from point *B* to locate point *C*, from point *C* to locate point *D*, and from point *D* to locate point *E*.

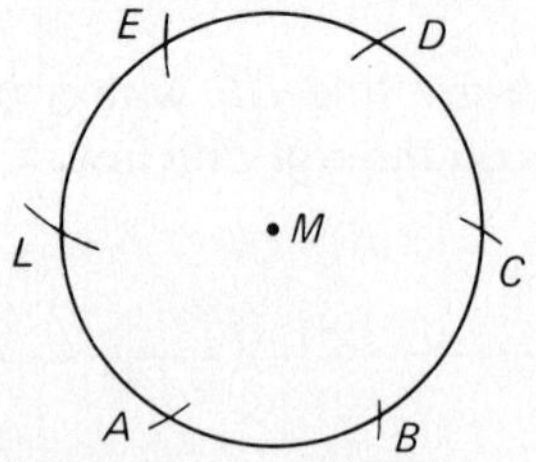

Fourth, draw in line segments *LA*, *AB*, *BC*, *CD*, *DE*, and *EL*.

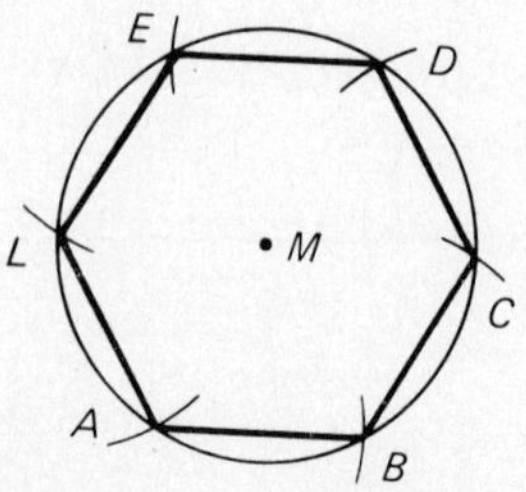

Fifth, measure with compass to ensure that:

$$LA = AB = BC = CD = DE = EL$$

Example 14-12

PROBLEM: Construct a 60° angle.

SOLUTION: The 60° angle will be constructed from an equilateral triangle. First, draw line segment *CK* on working line *MO*.

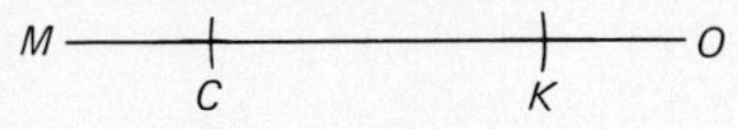

Second, from points *C* and *K*, swing arcs of equal length to line segment *CK*. Label the point of intersection as point *H*.

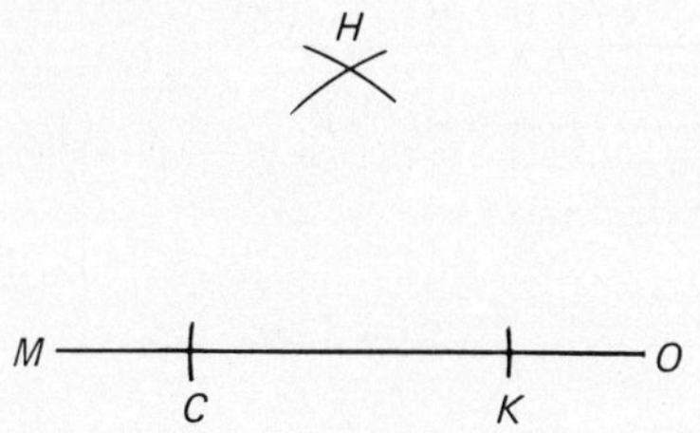

Third, draw line segments *CH* and *KH*.

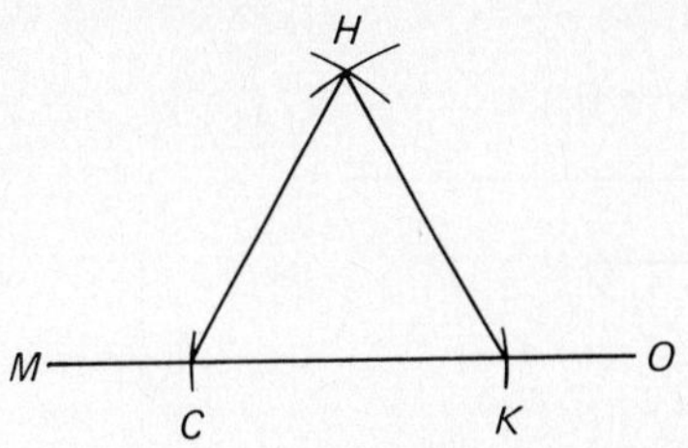

Fourth, measure with a compass to ensure that:

$$\overline{CH} = \overline{KH} = \overline{CK}$$

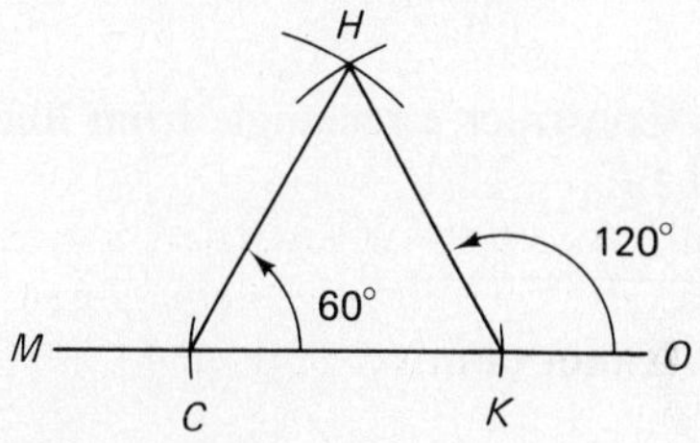

Fifth, since all sides are of equal length, each angle is 60° interior and 120° exterior.

EXERCISE 14-3

Construct triangles, squares, rectangles, angles, and other geometric figures as indicated in the following problems.

14-25. Construct a triangle from the given line segments.

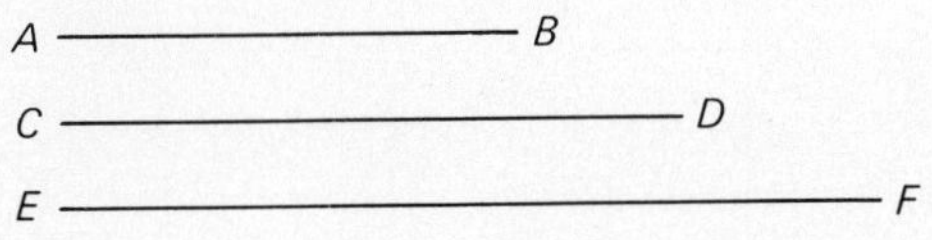

14-26. Construct a triangle from the given line segments.

14-27. Construct a triangle from the given line segments.

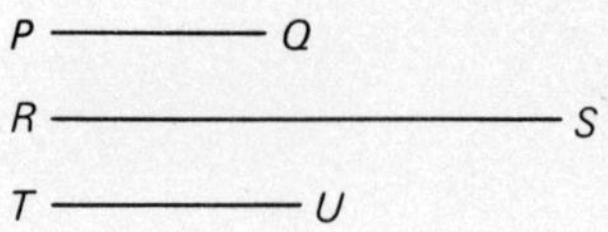

14-28. Construct a square from line segment *AB*.

14-29. Construct a rectangle from line segments *CD* and *EF*.

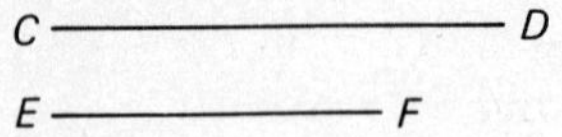

14-30. Construct a rectangle from line segments *GH* and *IJ*.

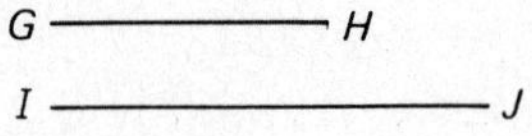

14-31. Construct a 120° angle.

14-32. Construct a 30° angle.

14-33. Construct a 45° angle.

14-34. Construct a 135° angle.

14-35. Construct a regular hexagon from line segment *MN*.

14-36. Construct a parallelogram from line segments *OP* and *QR*, with opposite angles of 60° and 120°, respectively.

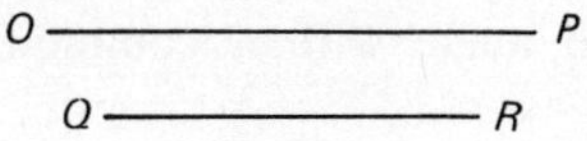

THINK TIME

Corporations often create special symbols, called logos, to represent their companies in the minds of consumers. Study the three logos which appear below, and then design some of your own using the compass and straight edge.

PROCEDURES TO REMEMBER

1. To copy a line segment:
   (a) Draw a working line with a straight edge.
   (b) Measure the given line segment with a compass.
   (c) Indicate a point on the working line and swing an arc with the compass.
   (d) The new line segment extends from the predetermined point on the working line to the point where the arc crossed the line.
2. To bisect a line segment:
   (a) Open the compass to a position greater than half the length of the line segment.
   (b) Swing an arc from each end point.
   (c) Connect the points of intersection above and below the given line segment with a straight edge.
   (d) This perpendicular line will bisect the given line segment.
3. To bisect an angle:
   (a) Swing an arc from the vertex of the angle so that it intersects both rays of the angle.
   (b) Swing an arc from each of the intersection points on the rays.
   (c) From the point of intersection of these arcs, draw a line to the vertex using a straight edge.
   (d) This line bisects the angle, thus creating two equal angles.
4. To copy an angle:
   (a) Draw a working line with a straight edge.
   (b) Indicate a point on the working line to serve as a vertex.
   (c) Swing an arc on the given angle and on the working line.
   (d) Set the compass on the point of intersection of one of the two rays of the given angle and open it so that it touches the points of intersection on both rays simultaneously.
   (e) Set the compass on the point of intersection on the working line. Swing the same arc as measured in step (d).
   (f) Draw a line from the vertex of the working line through the intersecting arcs above the working line.
   (g) This line should be considered as the second ray of the new angle.

5. To divide a line into a given number of equal segments:
   (a) Draw a working line 20° to 50° above the given line segment so that the two lines form an angle with a common vertex.
   (b) Divide the working line into the given number of segments by swinging preset arcs from the vertex to the first intersection of the working line; then from the first intersection, swing the second preset arc; from the second intersection, swing the third preset arc; etc.
   (c) Draw a line from the last intersection to the end point of the given line segment.
   (d) Copy each angle and draw lines to intersect the given line segment.
   (e) Each intersection will create equal line segments.
6. To construct a perpendicular line from a point on a line:
   (a) Place the compass on the given point on the line segment and swing an arc to cross the line segment at two different points.
   (b) From these two intersections, swing two arcs from each intersection, one below and one above the given line segment.
   (c) Draw a straight line from these points of intersection to intersect the given line in a perpendicular fashion.
7. To construct a perpendicular from a given point:
   (a) Open the compass and swing an arc from the given point so that two intersections are made on the given segment.
   (b) From these points of intersection, swing two additional arcs so that they intersect on the opposite side of the line from the given point.
   (c) Connect the given point with the point of intersection of two arcs. This line will be perpendicular to the given line segment.
8. To construct a line parallel to a given line:
   (a) Construct a working diagonal intersecting the given line segment at about 45°.
   (b) Swing an arc from the point of intersection of the given line and the working line.
   (c) Swing an arc from another point on the working line to intersect the first one.
   (d) Draw a line from the point on the working line through the point of intersection of the two arcs.
   (e) This line should be parallel to the given line.
9. To construct a tangent to a circle from a given point:
   (a) Draw a line from the given point to the center of the circle.
   (b) Construct a bisecting line to the line drawn in the previous step.
   (c) Swing an arc from the point of intersection to intersect the center of the circle and the edge of the circle at two points.
   (d) Draw lines from the given point to the two points on the edge of the circle.
   (e) These two lines form tangents to the circle.
10. To construct a triangle from three given line segments:
    (a) Draw a working line. Measure one of the given line segments with a compass, and enter this on the working line.
    (b) Measure a second line segment and swing an arc from one end point of the given line segment on the working line.
    (c) Measure the third line segment and swing an arc from the other end point of the given line segment so that the arc intersects the arc swing in step (b).
    (d) Draw lines from the point of intersection to each end of the given line segment.
    (e) The sides of the triangle should be equal to the given line segments.
11. To construct a square, given one side:
    (a) Draw a working line, measure the given side, and enter this on the working line.
    (b) Construct perpendiculars to the end points of the given side on the working line.
    (c) Measure the length of the given side, and from each of the end points of the given side, swing an arc on the perpendicular lines.
    (d) Draw a line between the perpendiculars to form a square.
    (e) Draw or darken in the four line segments.
    (f) Measure with a compass, if necessary, to ensure that all line segments or sides of the square are equal.
12. To construct a regular hexagon, given the length of one side:
    (a) Draw a circle using the line segment as the radius of the circle.
    (b) Using the same radius, swing an arc from the point where the radius touches

the circle to cut the edge of the circle on both sides of the point.
   (c) From the point where the previous arc intersected the circle, swing an additional arc. Continue to do this until the edge of the circle has been cut by arcs into six equal lengths.
   (d) Draw line segments to connect adjacent intersections.
   (e) Measure with a compass to insure that the line segments are of equal length.
13. To construct a 60° angle:
   (a) Draw a line segment.
   (b) From one end of the line segment, swing an arc of equal length to the line segment.
   (c) From the other end of the line segment, swing an arc of equal length to the line segment.
   (d) At the point of intersection, draw line segments to the end points of the original line segments.
   (e) Measure with a compass to insure that the line segments are of equal length.
   (f) The inside angles of the equilateral triangle are 60° angles.

CHAPTER SUMMARY

1. The following definitions are essential to the construction of simple geometric forms.
   (a) A *compass* is a tool used to draw circles or portions of circles and to measure line segments.
   (b) A *straight edge* is used to draw straight lines between points.
   (c) A *radius* is the opening between two points on a compass.
   (d) *Radii* is the plural of radius.
   (e) An *arc* is a portion of the circumference of a circle.
   (f) *Perpendicular* lines form angles of 90°.
   (g) Two lines are *parallel* when they run side by side and will never cross no matter how far they are extended.
   (h) A *tangent* line touches a circle at only one point.
2. Procedures to remember, included in this chapter, are:
   (a) Copy a line segment.
   (b) Bisect a line segment.
   (c) Bisect an angle.
   (d) Copy an angle.
   (e) Divide a line segment into a given number of equal segments.
   (f) Construct a perpendicular line from a point on a line.
   (g) Construct a perpendicular from a given point.
   (h) Construct a line parallel to a given line.
   (i) Construct a tangent to a circle from a given point.
   (j) Construct a triangle from three given line segments.
   (k) Construct a square, given one side.
   (l) Construct a regular hexagon, given the length of one side.
   (m) Construct a 60° angle.

CHAPTER TEST

T-14-1. Copy line segment $\overline{ST}$.

T-14-2. Copy angle *XYZ*.

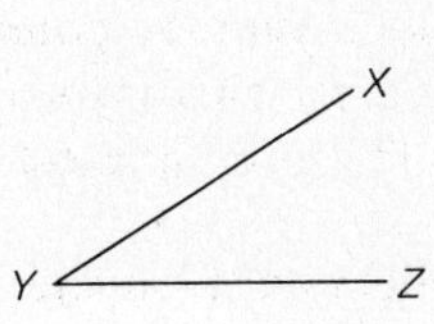

T-14-3. Bisect $\overline{PQ}$.

T-14-4. Bisect angle *KLM*.

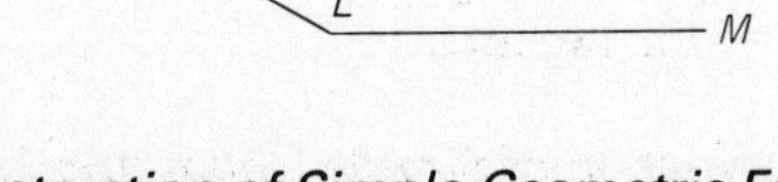

T-14-5. Divide $\overline{FG}$ into four equal parts.

T-14-6. Construct a perpendicular on $\overline{AB}$ at X.

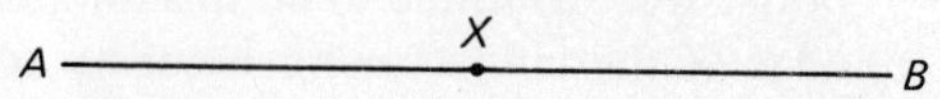

T-14-7. Construct a line parallel to $\overline{DE}$.

T-14-8. Construct a perpendicular to $\overline{AB}$ from point Z.

T-14-9. Using $\overline{JK}$, construct a hexagon with each side equal in length to $\overline{JK}$.

T-14-10. Construct a 60° angle.

T-14-11. Construct a tangent to circle O from point Y.

T-14-12. Construct a triangle from the line segments given below.

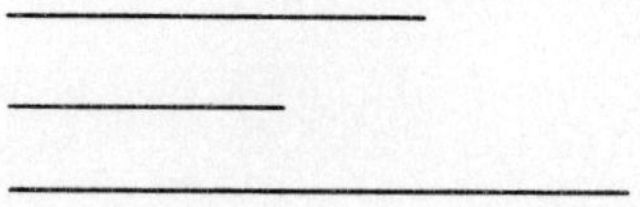

T-14-13. Construct a square, given side $\overline{LM}$.

T-14-14. Construct a perpendicular to $\overline{WX}$.

T-14-15. Divide $\overline{QR}$ into six equal parts.

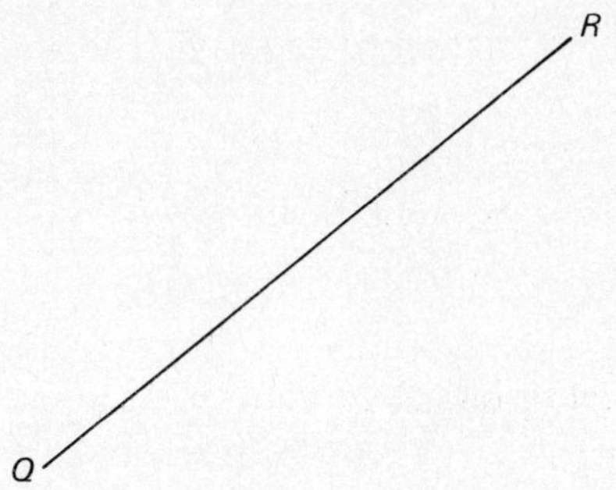

Section Four

TRIGONOMETRY

CHAPTER

15

FUNDAMENTALS OF TRIGONOMETRY

OBJECTIVES

1. To identify and explain the properties of triangles.
2. To identify and explain the properties of right triangles.
3. To identify and explain the trigonometric ratios of the sine, cosine, and tangent.
4. To solve similar triangles using ratio and proportion techniques.

SELF-TEST

The following test will evaluate your understanding and skill in working with basic trigonometric principles and functions. Satisfactory completion of these problems will indicate your ability to deal with the more complex trigonometric problems of Chapters 16 and 17.

S-15-1. Determine the degrees of $\angle A$ in triangle ABC.

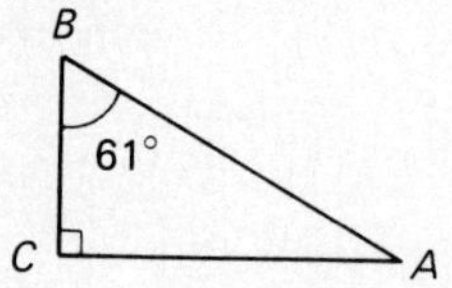

S-15-2. Determine the degrees of $\angle E$ in triangle DEF.

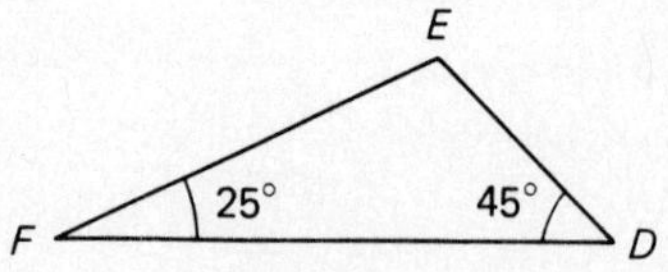

S-15-3. Letter the sides of triangle GHI.

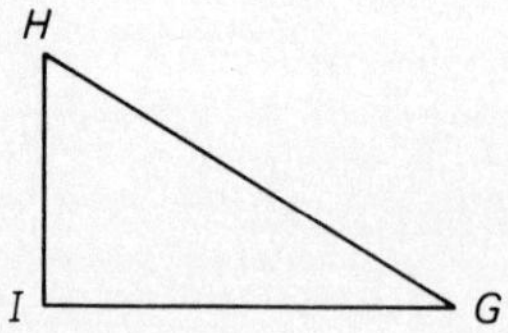

S-15-4. Letter the sides of triangle MLK.

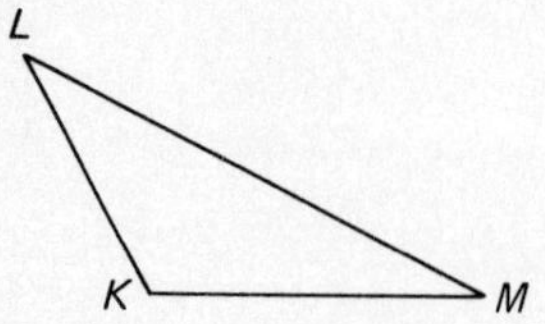

S-15-5. Letter the angles of the given triangle.

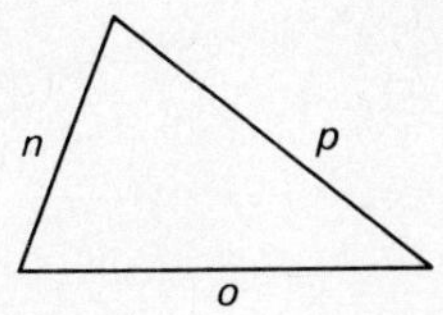

S-15-6. Given similar triangles QRS and TUV, find q.

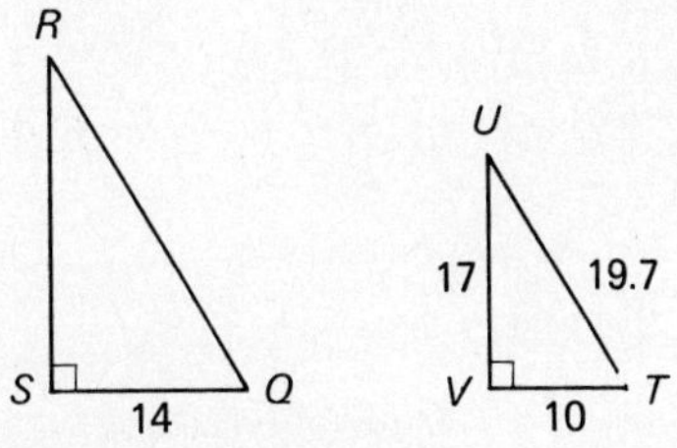

S-15-7. Find s in problem S-15-6.

S-15-8. Given similar triangles ABC and DEF, find $\angle D$.

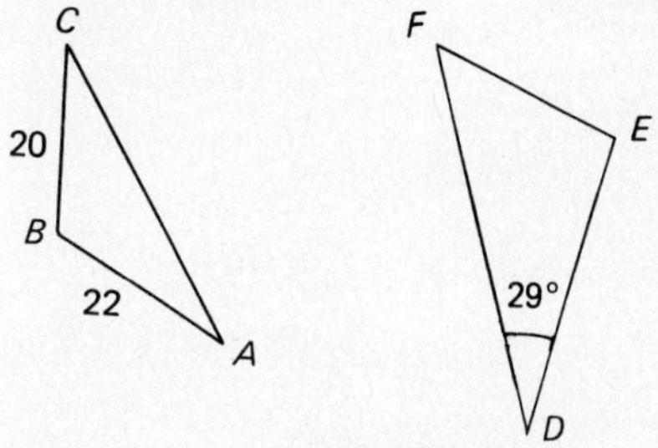

S-15-9. Find $\angle F$ in problem S-15-8.

S-15-10. Given similar triangles GHI and JKL, determine the length of i to the nearest tenth.

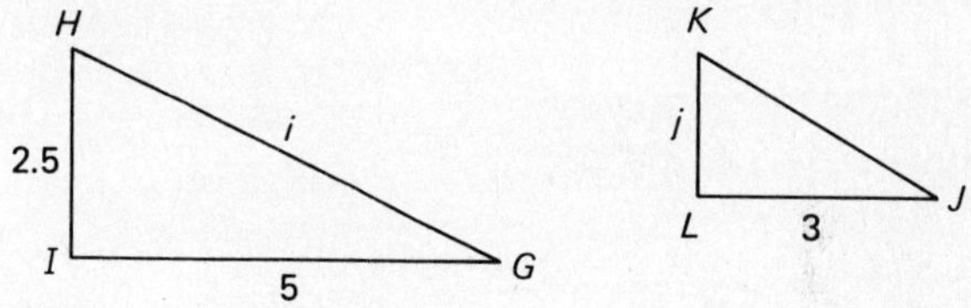

S-15-11. Determine the length of j in problem S-15-10 to the nearest tenth.

S-15-12. Determine the length of l in problem S-15-10 to the nearest tenth.

S-15-13. The sine of angle N is defined as the ratio of $\frac{n}{o}$. What is the sine of angle M?

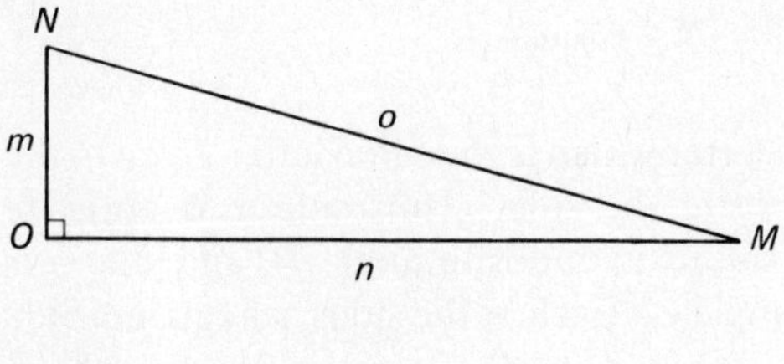

S-15-14. What is the tangent of angle N?

S-15-15. What is the cosine of angle M?

S-15-16. What trigonometric function equals the ratio $\frac{q}{r}$?

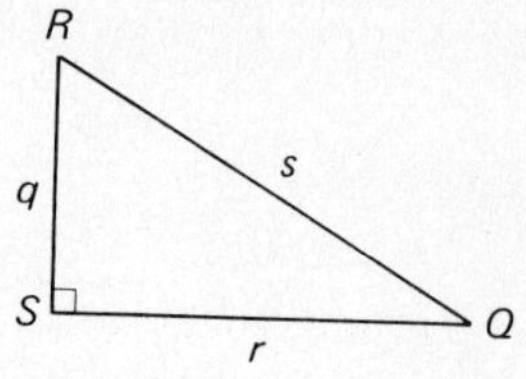

S-15-17. What trigonometric function equals the ratio $\frac{q}{r}$?

S-15-18. What two trigonometric functions equal the ratio $\frac{r}{s}$?

S-15-19. If the trigonometric function of the sine of angle A is .5000, what is the size of angle A?

S-15-20. If the trigonometric function of the tangent of angle R is .577, what is the size of angle R?

INTRODUCTION

Trigonometry is a combination of arithmetic, algebra, and geometry which is used by tradesmen and technicians to solve the complicated mathematical problems of industry. The word, trigonometry, is derived from the Greek words, *trigonon* and *metria*, which mean "triangle measure." However, trigonometry has outgrown its ancient definition and now includes the solution of problems dealing with squares, circles, rectangles, and other geometric figures. Trigonometry, or "trig" as it is commonly called, will be a distinct asset to you when you are confronted with industrial problems which call for the skills and techniques which only trigonometry can provide.

DEFINITIONS

All triangles have six parts: three sides and three angles. The determination of the six parts of a triangle is called the *solution of the triangle.* A *right triangle* is a triangle which has a right or 90° angle.

Figure A is a right triangle. Study the parts of the triangle and their appropriate identifications. Angle C is a right angle. Generally the letter C is assigned to the right angle. The symbol ⊿ is used to identify a right angle. Angles, like points, are identified by capital letters. Each side is identified by a

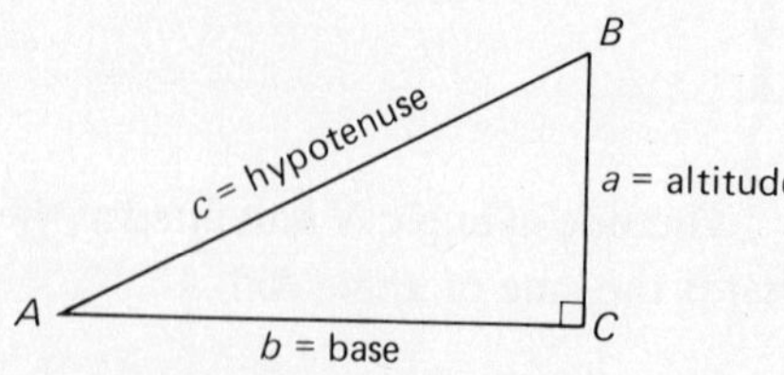

Figure A

lowercase letter which is the same letter given to the angle opposite the side. Thus side a is opposite angle A, side b is opposite angle B, and side c is opposite angle C. Each of the sides has a name. Side *a* is the altitude, side *b* is the base, and side *c* is the hypotenuse. Although the positioning and the shape of the right triangle may not always identify a base or altitude, the side opposite the right angle is always called the *hypotenuse.*

CONTENT

Through the problems and solutions that follow, other distinctive properties of the right triangle will be illustrated.

Example 15-1

PROBLEM: Determine the sum of the degrees of the angles of a right triangle as illustrated by Figure B.

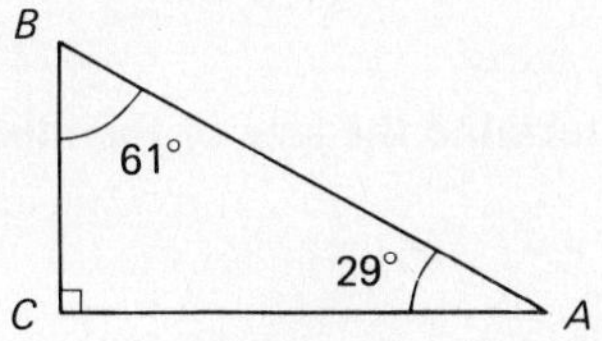

Figure B

SOLUTION: The sum of the degrees of the angles of any triangle will always be 180°. First, measure the two unknown angles with a protractor if necessary. If only one angle is unknown, subtract the sum of the two known angles from 180°.

Second, check by adding the degrees of the angles to total 180°. Thus 90° + 61° + 29° = 180°.

EXERCISE 15-1

Identify the triangles and their parts as directed in the following problems.

Another way to verify that the total degrees of the angles is 180° is to tear the angles from a paper triangle and then glue them together to form a straight angle or 180°. Note the illustration below.

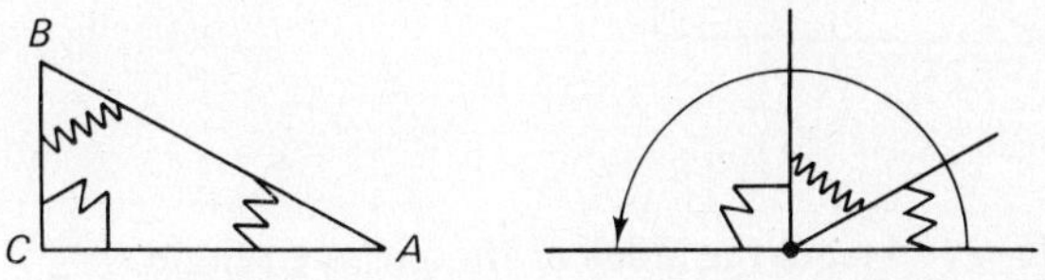

Example 15-2

PROBLEM: Determine the total degrees of the angles of *any* triangle as illustrated by Figure C.

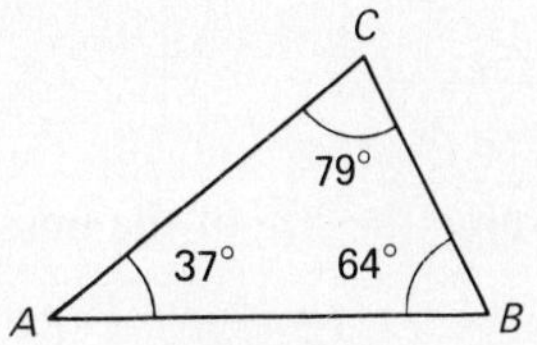

Figure C

SOLUTION: First, measure two unknown angles with a protractor if necessary. If one angle is known, measure only one angle.

Second, subtract the sum of the degrees of the two known angles from 180°.

Third, check by adding the degrees of the angles to total 180°. Thus 37° + 64° + 79° = 180°.

15-1. What kind of a triangle is *ABC*?

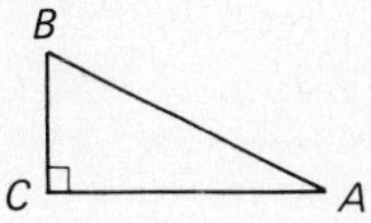

15-2. What type of an angle is $\angle C$?

15-3. Letter the sides of the given triangle.

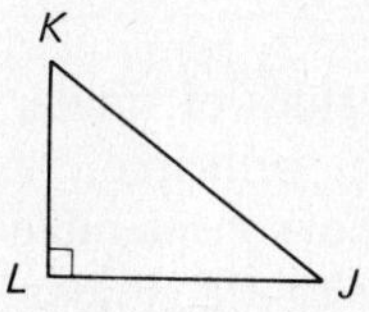

15-4. Letter the sides of the given triangle.

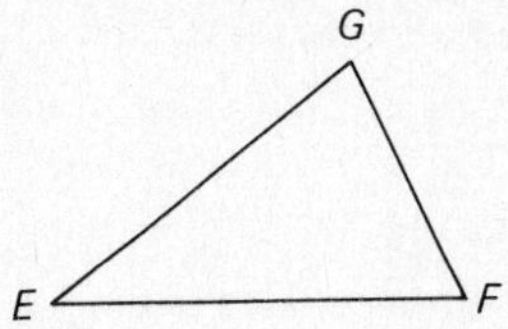

15-5. Letter the angles of the given triangle.

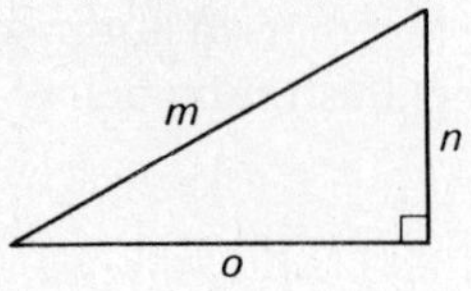

15-6. Label the sides of the given triangle.

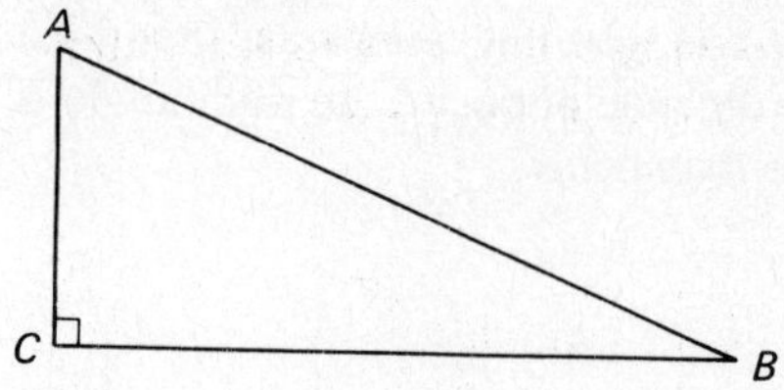

15-7. Label the sides of the given triangle.

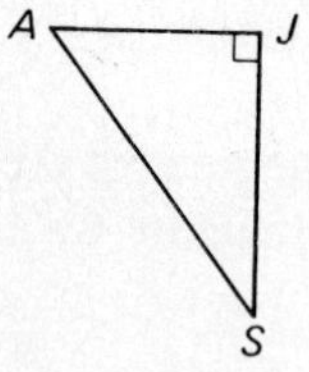

15-8. Determine the size of the missing angle.

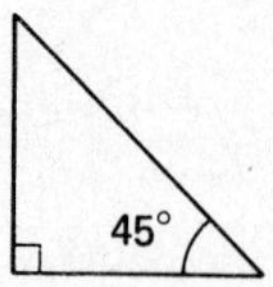

15-9. Determine the size of the missing angle.

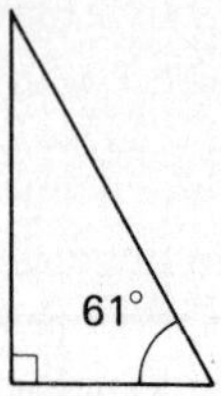

15-10. Determine the size of the missing angle.

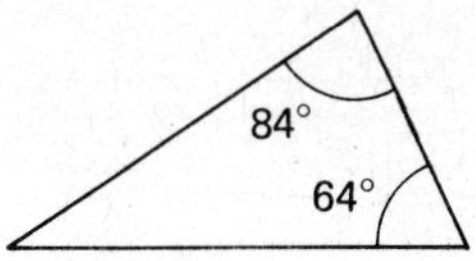

CONTENT

Triangles are *similar* if their corresponding angles are equal. The triangles below are similar right triangles. The equality of angles can be determined by comparing the ratios of known sides. Study the triangles and the ratios given below.

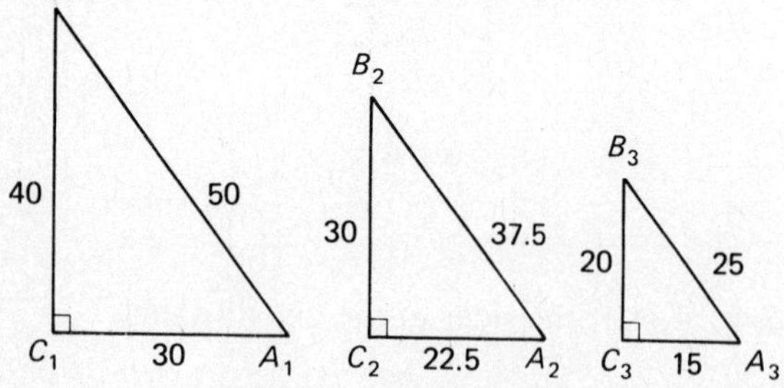

$$\frac{C_1A_1}{A_1B_1} = \frac{C_2A_2}{A_2B_2} = \frac{C_3A_3}{A_3B_3}$$

$$\frac{30}{50} = \frac{22.5}{37.5} = \frac{15}{25}$$

$$.6 = .6 = .6$$

Identical ratios of the lengths of the sides indicate equality of angles. Study the triangles and ratios given below and verify the equality of the angles.

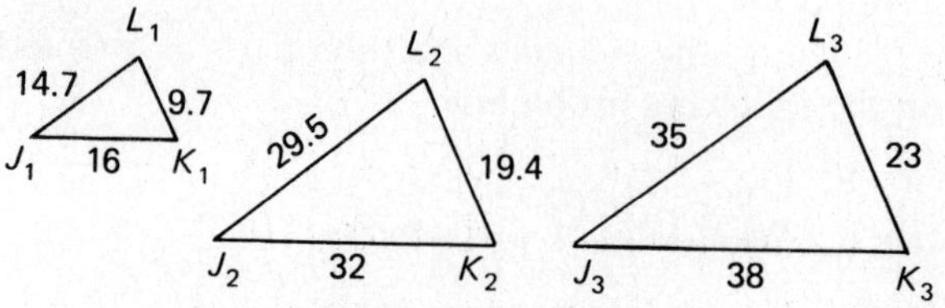

$$\frac{J_1L_1}{J_1K_1} = \frac{J_2L_2}{J_2K_2} = \frac{J_3L_3}{J_3K_3}$$

$$\frac{14.7}{16} = \frac{29.5}{32} = \frac{35}{38}$$

$$.92 = .92 = .92$$

The fact that the ratios of the sides of similar triangles are equal is a basic principle of trigonometry. This principle will be of considerable importance in determining solutions to trigonometrical problems.

Example 15-3

PROBLEM: Given two similar triangles, determine the lengths of the remaining sides.

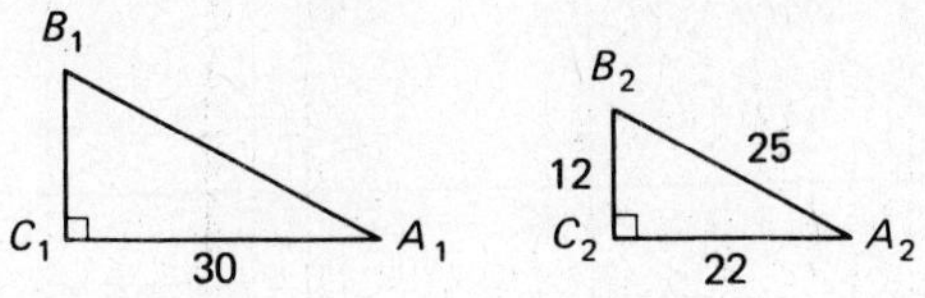

SOLUTION: Since corresponding sides of similar triangles have similar ratios, ratios can be set up and unknowns can be determined using simple algebraic procedures.

First, set up the ratios.

$$\frac{C_1A_1}{C_2A_2} = \frac{A_1B_1}{A_2B_2} = \frac{B_1C_1}{B_2C_2}$$

Second, substitute known values.

$$C_1A_1 = 30; \qquad C_2A_2 = 22;$$

$$A_2B_2 = 25; \qquad B_2C_2 = 12$$

$$\frac{C_1A_1}{C_2A_2} = \frac{A_1B_1}{A_1B_2}$$

$$\frac{30}{22} = \frac{A_1B_1}{25}$$

Third, solve for the unknown A_1B_1.

$$\frac{30 \times 25}{22} = A_1B_1$$

$$34.1 = A_1B_1$$

Fourth, substitute known values.

$$C_1A_1 = 30; \qquad C_2A_2 = 22; \qquad B_2C_2 = 12$$

$$\frac{C_1A_1}{C_2A_2} = \frac{B_1C_1}{B_2C_2}$$

$$\frac{30}{22} = \frac{B_1C_1}{12}$$

Fifth, solve for the unknown B_1C_1.

$$\frac{30 \times 12}{22} = B_1C_1$$

$$16.4 = B_1C_1$$

EXERCISE 15-2

Solve for the unknown sides of the similar triangles in the following problems.

15-11. Given the similar triangles $A_1B_1C_1$ and $A_2B_2C_2$, solve for A_2B_2.

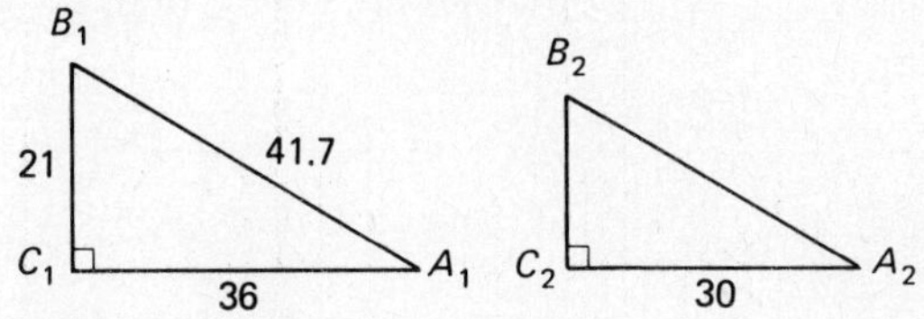

15-12. Solve for B_2C_2.

15-13. Given the similar triangles $J_1K_1L_1$ and $J_2K_2L_2$, solve for J_1K_1.

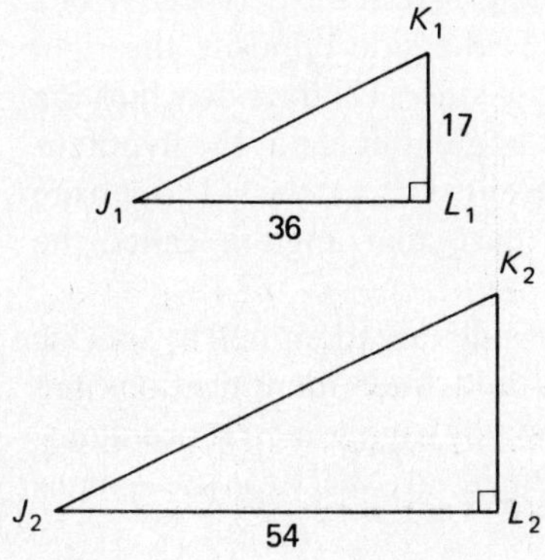

15-14. Solve for L_2K_2.

15-15. Solve for K_2J_2.

15-16. Given the similar triangles *DEF* and *XYZ*, solve for *y*.

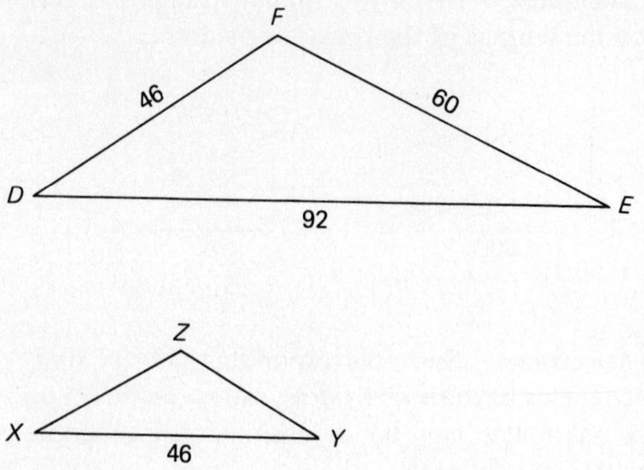

15-17. Solve for *x*.

15-18. Given the similar triangles *MNO* and *EFG*, solve for *e*.

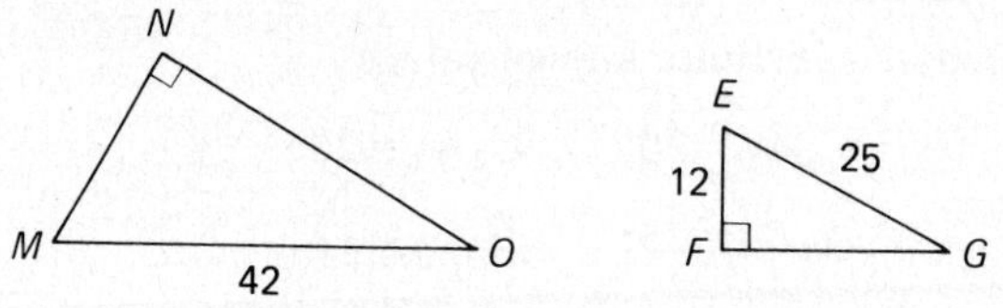

15-19. Solve for *o*.

15-20. Solve for *m*.

CONTENT

For the purpose of identification, the sides of a right triangle are named in relation to a particular angle. In triangle *ABC*, angle *A* dictates that side *b* is its adjacent side, side *a* is its opposite side, and side *c* is the hypotenuse.

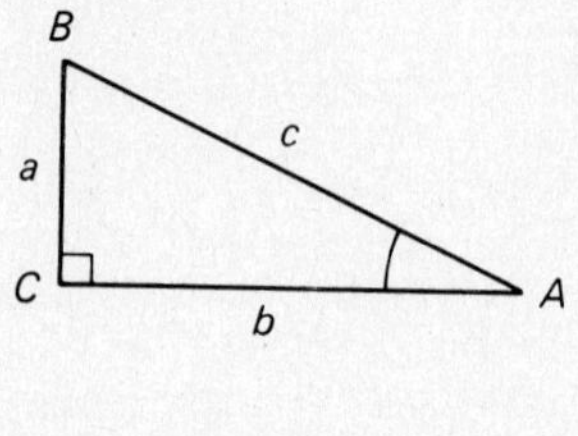

From the perspective of angle *B*, side *b* is its opposite side, side *a* is its adjacent side, and side *c* is the hypotenuse. A simple rule to follow for identifying the sides of a right triangle from the perspective of a particular angle is: (1) the side opposite the right angle is called the *hypotenuse*; (2) the side which is a part of the particular angle, but is not the hypotenuse, is called the *adjacent* side; (3) the side opposite or farthest from the particular angle is called the *opposite side.*

The ratios of these sides are identified by specific names. These ratios and their identifications are essential to the solution of trigonometric problems. In triangle *ABC*, angle *A* dictates the names assigned to the sides. The names of the various ratios created from angle A are as follows:

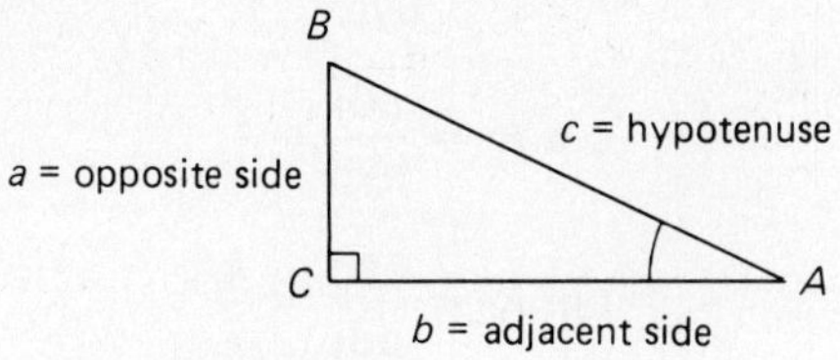

$$\text{sine of angle } A = \frac{\text{opposite side}}{\text{hypotenuse}}$$

or

$$\sin A = \frac{\text{opp}}{\text{hyp}} = \frac{a}{c}$$

$$\text{cosine of angle } A = \frac{\text{adjacent side}}{\text{hypotenuse}}$$

or

$$\cos A = \frac{\text{adj}}{\text{hyp}} = \frac{b}{c}$$

$$\text{tangent of angle } A = \frac{\text{opposite side}}{\text{adjacent}}$$

or

$$\tan A = \frac{\text{opp}}{\text{adj}} = \frac{a}{b}$$

It is essential that these ratios and their identifications become a part of your trigonometry vocabulary. The ratios of the sides of right triangles are called the *trigonometric functions* of the right triangles. Since the angles will not always be designated as A, B, and C, it is important to identify the trigonometric functions with the names of the sides and not with given angles. The two examples below illustrate the identification of the three basic trigonometric functions.

Example 15-4

PROBLEM: Identify the three basic trigonometric functions of angle D and angle E in triangle DEF.

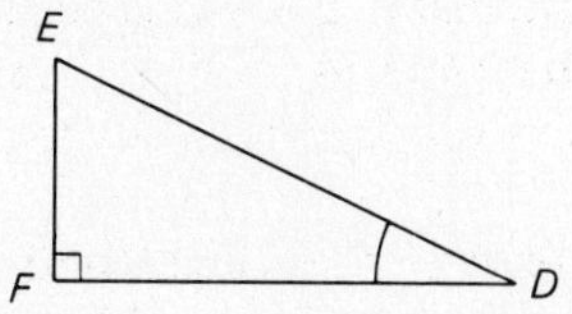

SOLUTION: First, identify the sides in relation to angle D as shown.

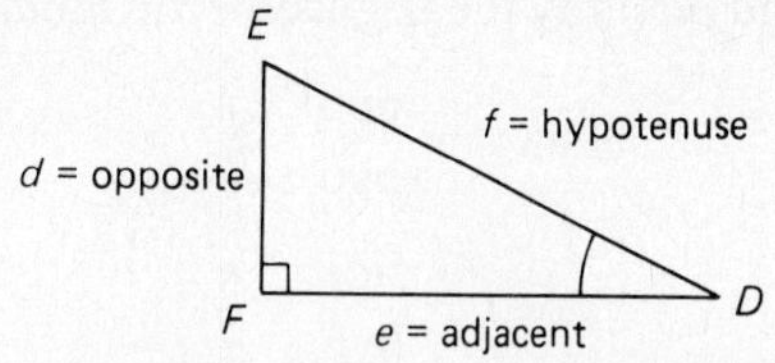

Second, identify the trigonometric functions.

$$\sin D = \frac{\text{opp}}{\text{hyp}} = \frac{d}{f}$$

$$\cos D = \frac{\text{adj}}{\text{hyp}} = \frac{e}{f}$$

$$\tan D = \frac{\text{opp}}{\text{adj}} = \frac{d}{e}$$

Third, identify the sides in relation to angle E as shown.

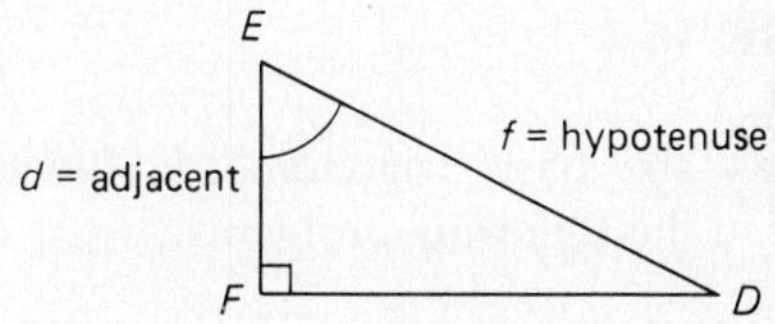

Fourth, identify the trigonometric functions.

$$\sin E = \frac{\text{opp}}{\text{hyp}} = \frac{e}{f}$$

$$\cos E = \frac{\text{adj}}{\text{hyp}} = \frac{d}{f}$$

$$\tan E = \frac{\text{opp}}{\text{adj}} = \frac{e}{d}$$

Example 15-5

PROBLEM: Identify the basic trigonometric functions of angle J and angle K in triangle JKL.

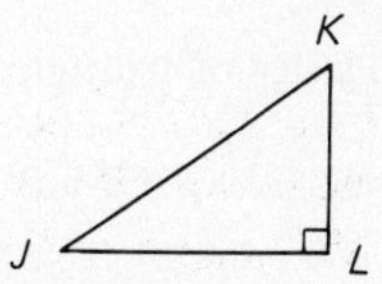

SOLUTION: First, identify the sides in relation to angle J as shown.

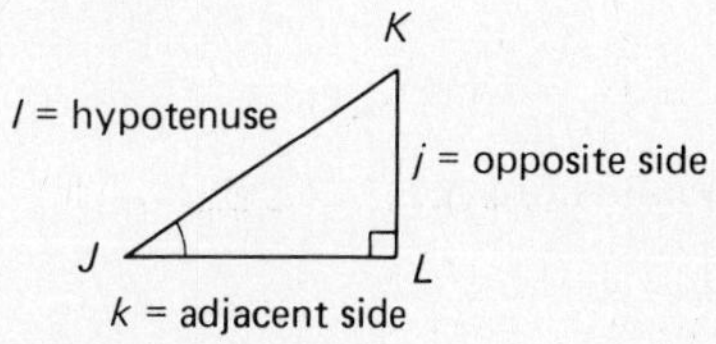

Second, identify the trigonometric functions.

$$\sin J = \frac{\text{opp}}{\text{hyp}} = \frac{j}{l}$$

$$\cos J = \frac{\text{adj}}{\text{hyp}} = \frac{k}{l}$$

$$\tan J = \frac{\text{opp}}{\text{adj}} = \frac{j}{k}$$

Third, identify the sides in relation to angle K as shown.

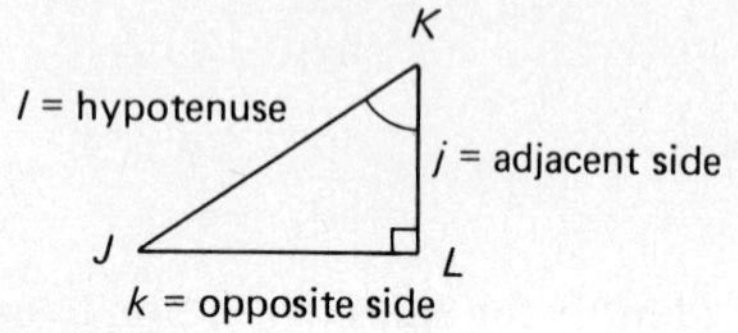

EXERCISE 15-3

Identify the basic trigonometric functions as indicated in the following problems.

15-21. Given right triangle HIJ, what is the sine of $\angle H$?

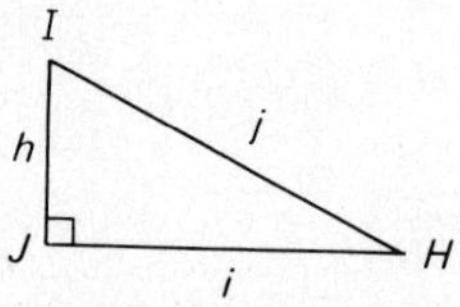

15-23. What is the cosine of $\angle H$?

15-25. What is the cosine of $\angle I$?

15-27. Letter the sides of triangle XYZ.

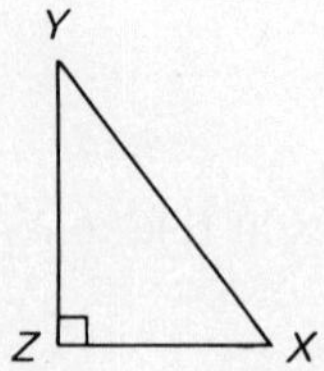

15-29. What is tan X?

Fourth, identify the trigonometric functions.

$$\sin K = \frac{\text{opp}}{\text{hyp}} = \frac{k}{l}$$

$$\cos K = \frac{\text{adj}}{\text{hyp}} = \frac{j}{l}$$

$$\tan K = \frac{\text{opp}}{\text{adj}} = \frac{k}{j}$$

15-22. What is the tangent of $\angle I$?

15-24. What is the sine of $\angle I$?

15-26. What is the tangent of $\angle H$?

15-28. What is sin X?

15-30. What is cos Y?

15-31. What is tan Y?

15-32. What is sin Y?

15-33. What is cos X?

15-34. Letter the angles of the following triangle.

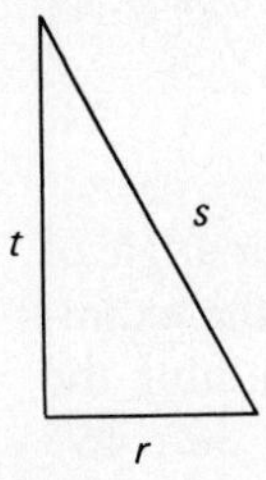

15-35. What is sin R?

15-36. What is cos T?

15-37. What is tan T?

15-38. Complete the lettering of the following triangle.

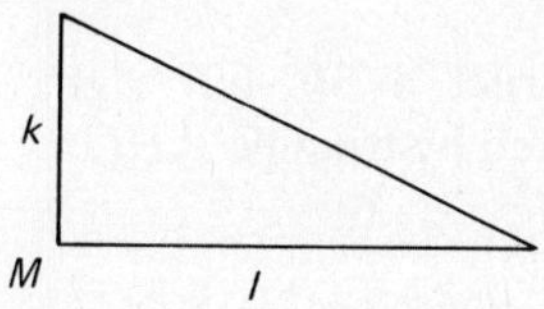

15-39. What is sin K?

15-40. What is tan K?

15-41. Given the ratio $\frac{k}{l}$ of $\angle L$, what is the trigonometric function?

15-42. Given the ratio $\frac{k}{m}$ of $\angle K$, what is the trigonometric function?

15-43. Given the ratio $\frac{l}{m}$ of $\angle L$, what is the trigonometric function?

15-44. Complete the lettering of the following triangle.

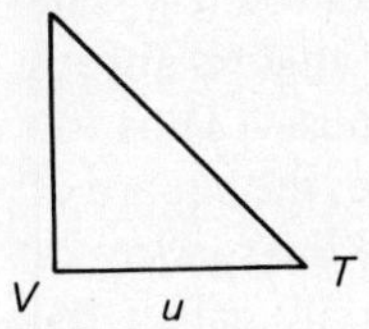

15-45. Given the ratio $\frac{t}{u}$ of $\angle T$, what is the trigonometric function?

15-46. Given the ratio $\frac{u}{v}$ of $\angle T$, what is the trigonometric function?

15-47. Given the ratio $\frac{u}{t}$ of $\angle T$, what is the trigonometric function?

CONTENT

A trigonometric function of a particular angle of a right triangle has a constant value. In the examples that follow, the processes for determining the values of the trigonometric functions of 30°, 45°, and 60° angles will be illustrated.

Example 15-6

PROBLEM: Determine the values of the three basic trigonometric functions (sin, cos, and tan) of a 30° angle.

SOLUTION: First, construct a 30°–60° right triangle and letter as illustrated by triangle *ABC*.

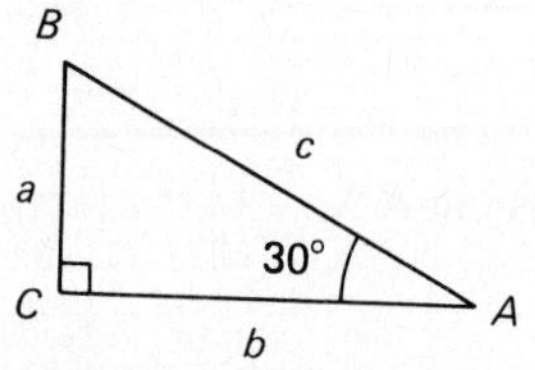

Second, enter values as indicated in triangle *ABC*.

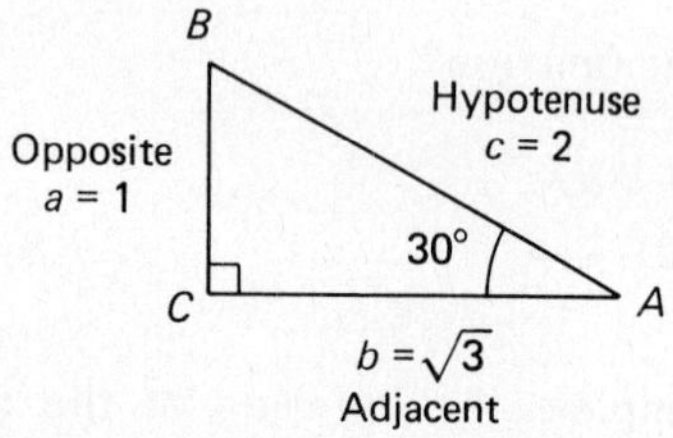

The value of 2 is given to the hypotenuse (c). The value of the side (a) opposite the 30° angle is always one half of the value of the hypotenuse. Thus the value of side a is 1. In a right triangle, the square of the hypotenuse is equal to the sum of the squares of the other two sides. This is commonly called the Pythagorean theorem. When necessary, follow the procedures below to determine the value of side b.

$$\begin{aligned} b &= \sqrt{c^2 - a^2} \\ &= \sqrt{(2)^2 - (1)^2} \\ &= \sqrt{4 - 1} \\ &= \sqrt{3} \end{aligned}$$

Enter the value of b as illustrated.

Third, determine the sine of 30°

$$\sin 30° = \frac{\text{opp}}{\text{hyp}} = \frac{1}{2} = .500$$

Fourth, determine the cosine of 30°.

$$\cos 30° = \frac{\text{adj}}{\text{hyp}} = \frac{\sqrt{3}}{2} = \frac{1.732}{2} = .866$$

Fifth, determine the tangent of 30°.

$$\begin{aligned} \tan 30° &= \frac{\text{opp}}{\text{adj}} = \frac{1}{\sqrt{3}} = \frac{1}{\sqrt{3}}\left(\frac{\sqrt{3}}{\sqrt{3}}\right) = \frac{\sqrt{3}}{3} \\ &= \frac{1.732}{3} = .577 \end{aligned}$$

Thus, the values of the basic trigonometric functions of a 30° angle are:

$$\sin 30° = .500$$

$$\cos 30° = .866$$

$$\tan 30° = .577$$

Example 15-7

PROBLEM: Determine the values of the three basic trigonometric functions (sin, cos, and tan) of a 45° angle.

SOLUTION: First, construct a 45°–45° right triangle as illustrated by triangle *DEF*.

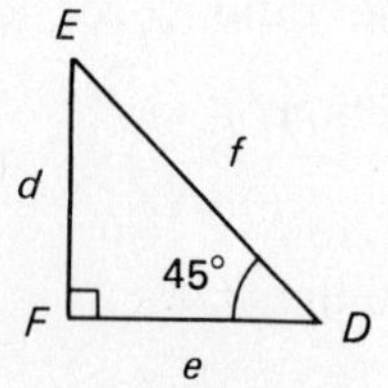

Second, enter the values as indicated in triangle *DEF*.

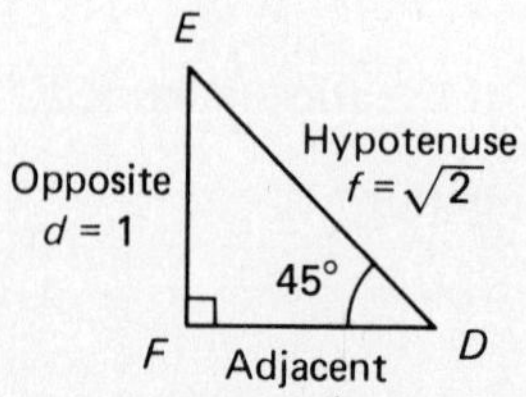

The value of 1 is assigned to the adjacent (e) and opposite (d) sides. The values of the two sides of a right triangle with 45° angles which are not the hypotenuse are equal. When necessary, follow the procedure below to determine the value of side f.

$$f = \sqrt{d^2 + e^2}$$
$$= \sqrt{(1)^2 + (1)^2}$$
$$= \sqrt{1 + 1}$$
$$= \sqrt{2}$$

Third, determine the sine of 45°.

$$\sin 45° = \frac{\text{opp}}{\text{hyp}} = \frac{1}{\sqrt{2}} = \frac{1}{\sqrt{2}}\left(\frac{\sqrt{2}}{\sqrt{2}}\right) = \frac{\sqrt{2}}{2}$$
$$= \frac{1.414}{2} = .707$$

Fourth, determine the cosine of 45°.

$$\cos 45° = \frac{\text{adj}}{\text{hyp}} = \frac{1}{\sqrt{2}} = \frac{1}{\sqrt{2}}\left(\frac{\sqrt{2}}{\sqrt{2}}\right) = \frac{\sqrt{2}}{2} = \frac{1.414}{2} = .707$$

Fifth, determine the tangent of 45°.

$$\tan 45° = \frac{\text{opp}}{\text{adj}} = \frac{1}{1} = 1 = 1.000$$

Thus, the values of the basic trigonometric functions of a 45° angle are:

$\sin 45° = .707$

$\cos 45° = .707$

$\tan 45° = 1.000$

EXERCISE 15-4

Determine the values of the basic trigonometric functions (sin, cos, and tan) as indicated in the following problems.

15-48. Given the triangle ABC, find b.

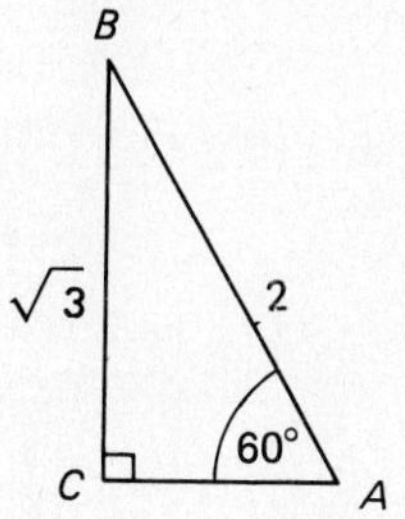

15-49. Find B.

15-50. Find the sine of 60°.

15-51. Find the cosine of 60°.

15-52. Find the tangent of 60°.

15-53. Given right triangle JKL, determine the value of l.

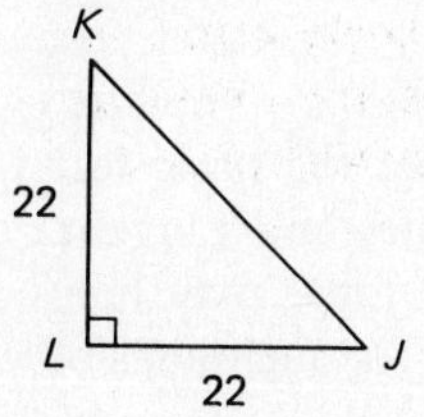

15-54. Determine the tangent of $\angle J$.

15-55. Determine the sine of $\angle K$.

15-56. Determine the cosine of $\angle J$.

15-57. How many degrees are there in angle J?

15-58. How many degrees are there in angle K?

15-59. Given right triangle PQR, determine the value of side p.

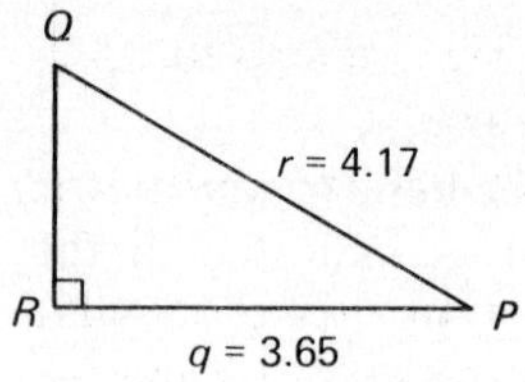

15-60. Determine the value of the tangent of $\angle P$.

15-61. Determine the value of the sine of $\angle Q$.

15-62. Determine the value of the tangent of $\angle Q$.

15-63. Determine the value of the cosine of $\angle Q$.

15-64. How many degrees are there in $\angle Q$?

15-65. How many degrees are there in $\angle P$?

THINK TIME

Often it is difficult to remember the basic formulas for determining sine, cosine, and tangent. Three acronyms, soph, cash, and topa, may help you to remember these formulas. The meanings of the acronyms are as follows:

soph (*s*ine = ***op***posite over ***h***ypotenuse)

cash (*c*osine = ***a***djacent ***s***ide over ***h***ypotenuse)

topa (*t*angent = ***op***posite over ***a***djacent)

PROCEDURES TO REMEMBER

1. To determine the unknown lengths of sides in similar triangles:

(a) Set up ratios between similar sides.
(b) Substitute known values for letter identification of the sides.
(c) Solve for the unknown values.

2. The formulas for determining the three basic trigonometric functions are:

$$\text{sine} \angle = \frac{\text{opposite side}}{\text{hypotenuse}}$$

$$\text{cosine} \angle = \frac{\text{adjacent side}}{\text{hypotenuse}}$$

$$\text{tangent} \angle = \frac{\text{opposite side}}{\text{adjacent side}}$$

CHAPTER SUMMARY

The following definitions are essential to working with trigonometric functions.

1. All triangles have six parts: three sides and three angles.
2. The determination of the six parts of a triangle is called the solution of the triangle.
3. A right triangle is a triangle which has a right or 90° angle.
4. Capital letters are used to identify angles.
5. Lowercase letters are used to identify the sides of a triangle.
6. In a right triangle: (a) the side opposite the right angle is the hypotenuse, (b) the side opposite or farthest from a given angle is the opposite side, and (c) the side next to a given angle is the adjacent side.
7. The sum of the degrees of the angles of a triangle is 180°.
8. Triangles are similar if their corresponding angles are equal.
9. Identical ratios of the lengths of the sides of triangles indicate equality of the angles.
10. The ratios of the sides of right triangles are called the trigonometric functions.
11. The value of the side opposite a 30° angle of a right triangle is always one half of the value of the hypotenuse.
12. The square of the hypotenuse is equal to the sum of the squares of the other two sides.

CHAPTER TEST

T-15-1. Determine the degrees of angle *B* in triangle *ABC*.

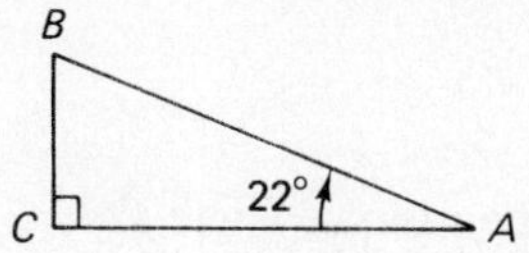

T-15-2. Determine the degrees of angle *E* in triangle *DEF*.

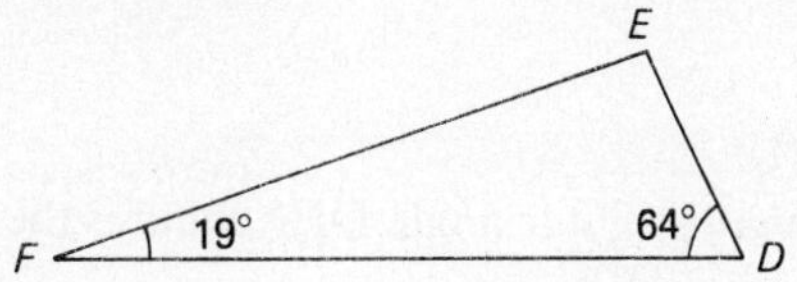

T-15-3. Letter the sides of triangle *GHI*.

T-15-4. Letter the sides of triangle *XYZ*.

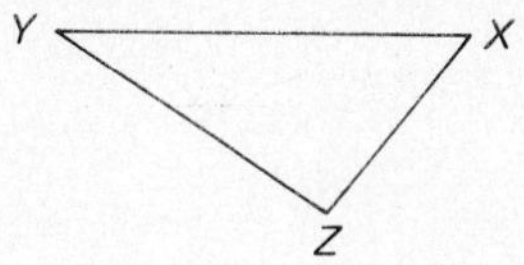

T-15-5. Letter the angles of the given triangle.

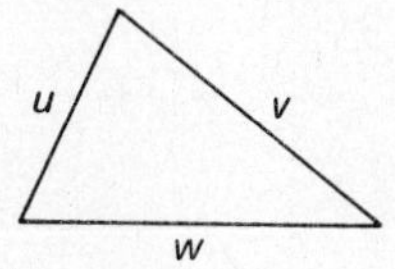

T-15-6. Given similar triangles *JKL* and *MNO*, determine the length of *m* in triangle *MNO*.

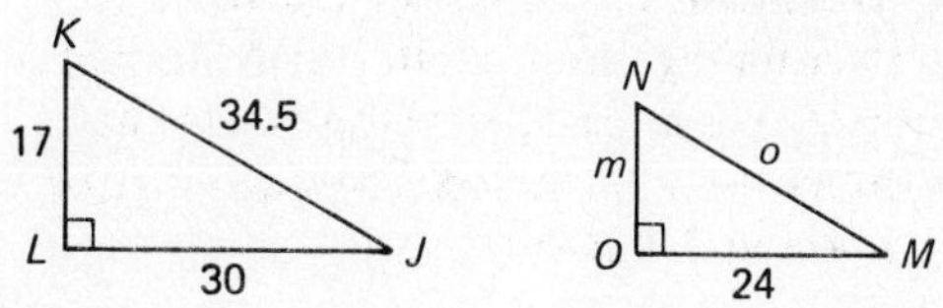

T-15-7. Determine the length of *o* in triangle *MNO*.

T-15-8. Given similar triangles *PQR* and *STU*, determine the length of *r* in triangle *PQR*.

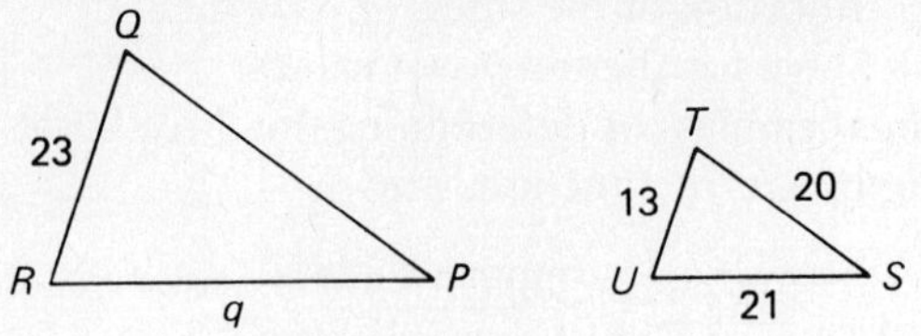

T-15-9. Determine the length of *q* in triangle *PQR*.

T-15-10. Given similar triangles *XYZ* and *ABC*, determine the length of *z* in triangle *XYZ* to the nearest tenth.

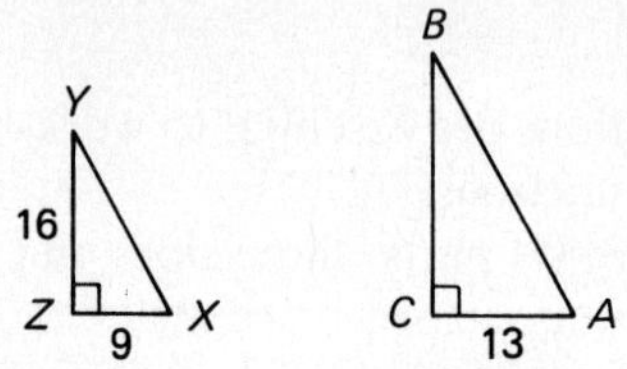

T-15-11. Determine the length of *a* in triangle *ABC* to the nearest tenth.

T-15-12. Determine the length of *c* in triangle *ABC* to the nearest tenth.

T-15-13. Given the right triangle *DEF*, what is the tangent ratio of angle *E*?

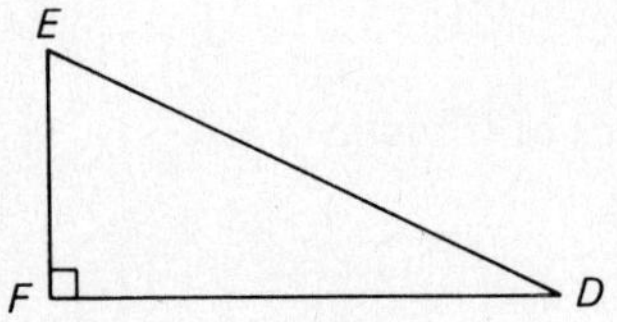

T-15-14. What is the cosine ratio of angle *D*?

T-15-15. What is the sine ratio of angle *E*?

T-15-16. Given the right triangle *GHI*, what trigonometric functions equal the ratio $\frac{h}{g}$?

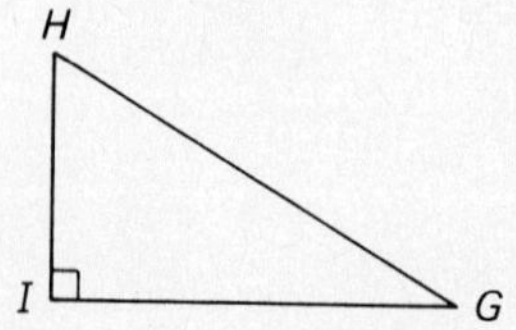

T-15-17. What trigonometric functions equal the ratio $\frac{h}{i}$?

T-15-18. What trigonometric functions equal the ratio $\frac{g}{i}$?

T-15-19. If the tangent of angle D is 1.000, what is the size of angle D?

T-15-20. If the sine of angle K is .866, what is the size of angle K?

CHAPTER

16

SOLUTION OF RIGHT TRIANGLES

OBJECTIVES

1. To identify and read the trigonometric functions table.
2. To interpolate for the values of trigonometric functions not included in the trigonometric functions table.
3. To solve problems involving right triangles through the use of trigonometric functions.
4. To solve industrial problems using trigonometric functions.

SELF-TEST

The following test will measure your ability to work with trigonometric functions to solve measurement problems involving right triangles. You will need a table of trigonometric functions to solve these problems. Such a table is located on the inside back cover. Satisfactory completion of this test will indicate your mastery of the skills taught in this chapter and your readiness to develop trigonometric skills in working with oblique triangles. Points of failure will indicate need for development of skills taught within this chapter.

S-16-1. Find the sine of 43°.

S-16-2. Find the cosine of 5° 40′.

S-16-3. Find the tangent of 48° 20′.

S-16-4. Find the angle whose tangent is .2186.

S-16-5. Find the angle whose tangent is 4.1653.

S-16-6. Find the tangent of 11° 37′.

S-16-7. Find the angle whose cosine is .5476.

S-16-8. Determine the degrees in angle B of triangle ABC.

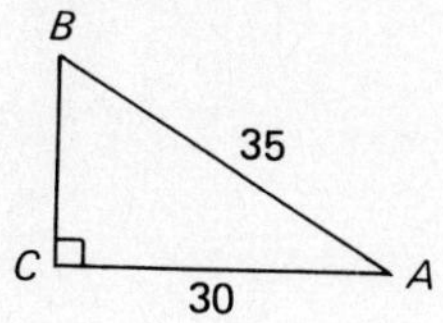

S-16-9. Given triangle PQR, determine the length of q.

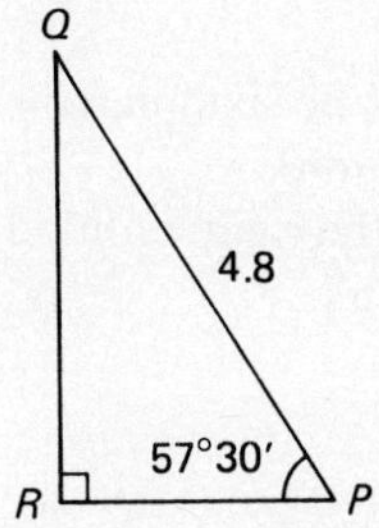

S-16-10. Find the length of p in triangle PQR.

S-16-11. Given triangle DEF, determine the size of angle D.

S-16-12. Determine the length of e in triangle DEF.

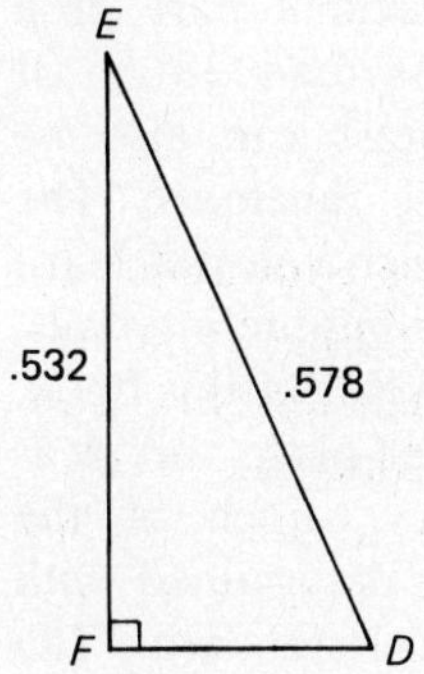

S-16-13. An airplane flies 1250 miles northeast. How far east is the plane from its original position?

S-16-14. Determine the included taper angle on a crankshaft having a taper of .75 inches per foot.

S-16-15. Six bolt holes are to be drilled on the circumference of a 6-inch hub of a wheel. Determine the straight line distance between the holes.

INTRODUCTION

Few trigonometric functions are as easily determined as those for the more common angles of 30°, 45°, and 60°. In order to aid people who work with trigonometric functions, special tables have been developed. These tables give the values of the basic functions of all angles. The preciseness of the function and angles is dependent upon the "completeness" or "exactness" of the table. For example, some tables are set up with 10° intervals, others in 1° intervals, and still others to the nearest minute. It should also be noted that some tables contain values to the nearest tenth or hundredth of a degree. This kind of table is read the same way a degree-and-minute table is read. Minutes are converted to a decimal part of a degree. For example, 38° 42′ is read $38°\frac{42'}{60}$ or 38.7°. Some electronic calculators can also be used to determine trigonometric functions. The trigonometric functions table which you will learn to read and use is set up with 10-minute intervals. It is located on the inside back cover of this book. The table is set up for four-place or ten-thousandths degree of accuracy. Each of the numbers in the table has been determined with preciseness and is a mathematical standard. The procedure used to read this table is basically the same procedure used to read any trigonometric table. Each number represents the ratio of one side of a triangle to another side. The trigonometric functions table saves the mathematician a great deal of time. Without the table, triangles would have to be precisely constructed, the angles and sides would have to be accurately measured, and the trigonometric ratios would have to be calculated according to the constructed angles.

Example 16-1

PROBLEM: Find the sine value of 38° in Table 2 of Trigonometric Functions.

SOLUTION: First, locate Table 2 of Trigonometric Functions on the inside back cover of this book.

Second, locate 38° in the table (the sixth narrow column from the right, at mid-column).

Third, read the next column to the right, entitled at the top "sin," and the value .6157.

Example 16-2

PROBLEM: Find the tangent value of 14° 31′ in Table 2 of Trigonometric Functions.

SOLUTION: First, locate Table 2 of Trigonometric Functions.

Second, locate 14° 31′ in the table (first narrow column on the left near the bottom).

Third, read the fourth column from the left, entitled at the top "tan," and the value .2586.

Example 16-3

PROBLEM: Find the cosine value of 76° 20′ in Table 2 of Trigonometric Functions.

SOLUTION: First, locate Table 2 of Trigonometric Functions.

Second, locate 76° 20′ in the table (the sixth narrow column from the left, near the bottom).

Third, since the angle is *greater* than 45°, read the titles of the columns at the bottom of the page. Thus, you will read the second narrow column from the left, entitled "cos," and the value .2532.

Remember, when an angle is *less* than 45°, read the table from *left to right* and use the titles at the *top* of the table; when an angle is *greater* than 45°, read the table from *right to left* and use the titles at the *bottom* of the table.

EXERCISE 16-1

Using Table 2 of Trigonometric Functions, find the value of the trigonometric functions as indicated below.

16-1. Find the tangent value of 31°.

16-2. Find the sine value of 21°.

16-3. Find the cosine value of 30° 30′.

16-4. Find the cosine value of 86° 30′.

16-5. Find the tangent value of 63° 50′.

16-6. Find the cosine value of 37° 20′.

16-7. Tan 17° 40′ =

16-8. Sin 38° 50′ =

16-9. Tan 78° 10′ =

16-10. Sin 45° 20′ =

CONTENT

Using the locational skills just acquired, but in reverse order, you can determine the degrees of the angle of a given trigonometric function. The examples which follow will help you to locate the degrees of angles in Table 2 of Trigonometric Functions, given the trigonometric function.

Example 16-4

PROBLEM: Find the angle whose sine is .3907.

SOLUTION: First, locate Table 2 of Trigonometric Functions.

Second, locate the sine value of .3907 (the eighth narrow column from the left, at mid-column).

Third, read the next column to the left, entitled "angle," and 23°. Note that the following abbreviations may be used:

$$\sin A = .3907$$

$$A = 23°$$

Example 16-5

SOLUTION: Find the angle whose tangent is 2.2113.

SOLUTION: First, locate Table 2 of Trigonometric Functions.

Second, locate the tangent value of 2.2113 (the eighth narrow column from the right, at mid-column).

Third, read the next column to the right, entitled "angle." Since the tangent is greater than 45°, read from the bottom up to locate 65° 40′. Thus,

$$\tan B = 2.2113$$

$$B = 65° 40'$$

Remember, when the tangent value is greater than 1, then the angle of the tangent is greater than 45°.

EXERCISE 16-2

Using Table 2 of Trigonometric Functions, find the angles indicated by the trigonometric functions.

16-11. Find angle A, given $\sin A = .1219$.

16-12. Find angle X, given $\tan X = 1.2954$.

16-13. Determine angle R when $\cos R = .4566$.

16-14. How large is angle W when $\tan W = 20.206$?

16-15. Given $\sin J = .6225$, find angle J.

16-16. $\cos P = .9528$; $P =$

16-17. $\tan K = 2.9319$; $K =$

16-18. $\sin M = .0640$; $M =$

16-19. $\cos N = .4488$; $N =$

16-20. $\sin Y = .9644$; $Y =$

CONTENT

Not all function values and angles can be located in Table 2 of Trigonometric Functions. When a function value or angle cannot be located on the table, the value or angle can be determined through a process called *interpolation*. This process involves locating a value between given values in Table 2 of Trigonometric Functions. The following examples illustrate one way you may interpolate to determine an unknown value.

Example 16-6

PROBLEM: Determine the tangent value of 36° 24′.

SOLUTION: First, locate the angle greater than (36° 30′) and the angle lesser than (36° 20′) the given angle (36° 24′) in Table 2 of Trigonometric Functions.

Second, locate the tangents of 36° 30′ and 36° 20′ in Table 2 of Trigonometric Functions.

Third, arrange the angles and tangents as illustrated below:

$$10'\left[\begin{array}{l}\tan 36°\,30' = .7400 \\ 4'\left[\begin{array}{l}\tan 36°\,24' = \;\;? \\ \tan 36°\,20' = .7355\end{array}\right]?\end{array}\right].0045$$

Fourth, subtract the tangent of the lesser angle (.7355) from the tangent of the greater angle (.7400) to determine the difference in value between the tangents (.0045).

$$\begin{array}{rl} .7400 & (\tan \text{ of } 36°\,30') \\ -\,.7355 & (\tan \text{ of } 36°\,20') \\ \hline .0045 & \end{array}$$

Fifth, set up an equation and solve for the unknown value of the tangent of 36° 24′.

$$\frac{4'}{10} = \frac{?}{.0045}$$

$$\frac{4'(.0045)}{10} = ?$$

$$.0018 = \tan 4'$$

Sixth, add the value of 4′ (.0018) to the value of the tangent of 36° 20′ (.7355) to determine the value of the tangent of 30° 24′ (.7373).

$$\begin{array}{rl} .7355 & = \tan 36°\,20' \\ +.0018 & = \tan \qquad 4' \\ \hline .7373 & = \tan 36°\,24' \end{array}$$

Example 16-7

PROBLEM: Determine the cosine of 18° 38′.

SOLUTION: First, locate the angle greater than (18° 40′) and the angle lesser than (18° 30′) the given angle (18° 38′) in Table 2 of Trigonometric Functions.

Second, locate the cosines of 18° 40′ and 18° 30′ in the table.

Third, arrange the angles and cosines as illustrated below. Remember that the cosine values decrease as the size of the angle increases.

$$10'\left[\begin{array}{l} 8'\left[\begin{array}{l}\cos 18^\circ 30' = .9483 \\ \cos 18^\circ 38' = \quad ?\end{array}\right] ? \\ \\ \cos 18^\circ 40' = .9474 \end{array}\right] .0009$$

Fourth, subtract the cosine of the greater angle (.9474) from that of the lesser angle (.9483) to determine the difference in value between cosines (.0009).

$$\begin{array}{rl} .9483 &= \cos 18^\circ 30' \\ -\ .9474 &= \cos 18^\circ 40' \\ \hline .009 & \end{array}$$

Fifth, set up an equation and solve for the unknown value of the cosine of 18° 38′.

$$\frac{8}{10} = \frac{?}{.009}$$

$$\frac{8'(.0009)}{10'} = ?$$

$$.00072 = \cos 8'$$

Sixth, subtract the value of 8′ (.00072) from the value of the cosine of 18° 30′ (.9483) to determine the value of the cosine of 18° 38′ (.9476).

$$\cos 18^\circ 38' = .9483 - .00072$$

$$= .94758$$

$$= .9476$$

Example 16-8

PROBLEM: Determine the angle whose sine is .6894.

SOLUTION: First, locate the angle greater than (18° 40′) and the angle lesser than (18° 30′) the given angle (18° 38′) in Table 2 of Trigonometric Functions.

Second, locate the angles of sines .6884 and .6905 in the table.

Third, arrange the angles and sines as illustrated below. Remember that sine values increase as the size of the angle increases.

$$10'\left[\begin{array}{l} ?\left[\begin{array}{l}\sin 43^\circ 30' = .6884 \\ \sin \quad ? \quad = .6894\end{array}\right] .0010 \\ \\ \sin 43^\circ 40' = .6905 \end{array}\right] .0021$$

Fourth, subtract the sine of 43° 30′ (.6884) from the sine of the unknown angle (.6894) to determine the difference in value between sines (.0010).

$$\begin{array}{rl} .6894 &= \sin ? \\ -\ .6884 &= \sin 43^\circ 30' \\ \hline .0010 & \end{array}$$

Fifth, set up an equation and solve for the unknown angle whose sin is .6844.

$$\frac{?}{10'} = \frac{.0010}{.0021}$$

$$? = \frac{10'(.0010)}{.0021}$$

$$= 4.76' \quad \text{or, rounded to nearest whole, } 5'$$

Sixth, add 5′ to 43° 30′ to determine the angle whose sin is .6894.

$$\begin{array}{rl} 43^\circ 30' &= \text{angle whose sin is } .6884 \\ +\qquad 5' &= \text{angle whose sin is } .0010 \\ \hline 43^\circ 35' &= \text{angle whose sin is } .6894 \end{array}$$

Example 16-9

PROBLEM: Determine the size of angle S whose cosine is .1270.

SOLUTION: First, locate the cosine greater than (.1276) and cosine lesser than (.1248) the given cosine (.1270) in Table 2 of Trigonometric Functions and make note of the angles.

Second, arrange the angles and cosines as illustrated below and determine the difference of value between cosines. Remember that cosine values decrease as the size of the angle increases.

$$10'\left[\begin{array}{l} ?\left[\begin{array}{l}\cos 82^\circ 40' = .1276 \\ \cos 82^\circ \ ?' = .1270\end{array}\right] .0006 \\ \\ \cos 82^\circ 50' = .1248 \end{array}\right] .0028$$

Third, set up an equation and solve for angle S of cosine .1270.

$$\frac{?}{10'} = \frac{.0006}{.0028}$$

$$? = \frac{10'(.0006)}{.0028}$$

$$= 2.14' \quad \text{or, rounded to nearest whole, } 2'$$

Fourth, add 2′ to 82° 40′ to determine the degrees in angle S.

$$\text{Angle } S = 82^\circ 42'$$

$$\cos 82^\circ 42' = .1270$$

EXERCISE 16-3

Using Table 2 of Trigonometric Functions as a guide, determine the value of the unknown trigonometric functions and angles as indicated.

16-21. Determine the sine of 2° 15′.

16-22. Determine the tangent of 19° 33′.

16-23. Determine the cosine of 54° 12′.

16-24. Determine the tangent of 38° 5′.

16-25. Determine the sine of 87° 41′.

16-26. Determine the cosine of 45° 8′.

16-27. Determine the tangent of 87° 44′.

16-28. Determine the tangent of 8° 8′.

16-29. Determine angle A whose sine is .9850.

16-30. Determine angle D whose cosine is .4530.

16-31. Determine angle X whose tangent is 2.2800.

16-32. Determine angle P whose cosine is .8580.

16-33. Determine angle Q whose sine is .8809.

16-34. Determine angle R whose cosine is .7777.

16-35. Determine angle V whose tangent is .0330.

CONTENT

Using the skills you have developed to determine unknown trigonometric functions and angles from Table 2 of Trigonometric Functions, you will be able to solve for unknown angles and sides of right triangles. The examples which follow illustrate ways to solve for unknown angles and sides of right triangles.

Example 16-10

PROBLEM: Given right triangle ABC, determine the length of side a.

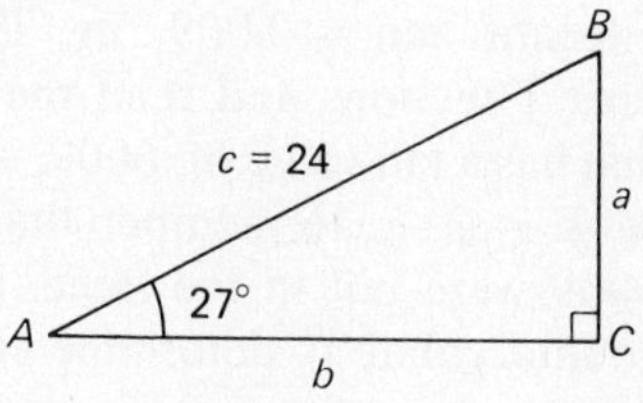

SOLUTION: First, set up a ratio involving one of the given values. This ratio should relate the opposite side (a) to the hypotenuse (c) of the given angle A. Use the formula,

$$\sin A = \frac{\text{opp}}{\text{hyp}}$$

Thus,

$$\sin 27° = \frac{a}{24}$$

Second, locate sin 27° in Table 2 of Trigonometric Functions (.4540) and set this value into the ratio.

Third, solve the equation for a. Remember, the "loop" terms multiplied together will equal side a.

$$.4540 = \frac{a}{24}$$

$$24(.4540) = a$$

$$10.896 = a$$

Example 16-11

PROBLEM: Given right triangle ABC, determine the length of side b.

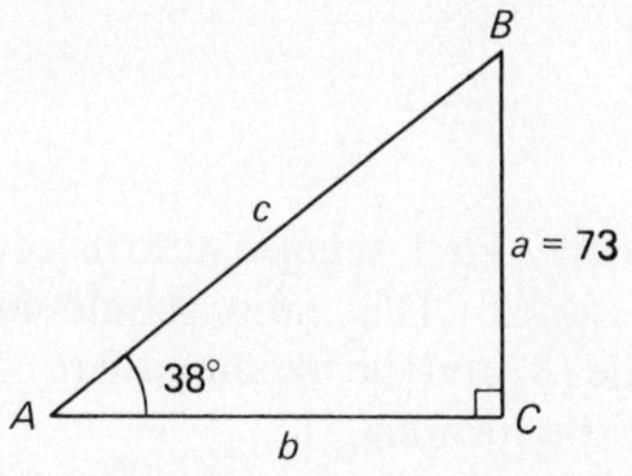

SOLUTION: First, set up a ratio involving one of the given values. This ratio should involve the opposite side (a) and the adjacent side (b) of angle A. Remember the formula,

$$\tan A = \frac{\text{opp}}{\text{adj}}$$

Thus,

$$\tan 38° = \frac{73}{b}$$

Second, locate tan 38° in Table 2 of Trigonometric Functions (.7813) and set this value into the ratio.

Third, multiply both sides of the equation by b.

$$.7813 = \frac{73}{b}$$

$$b(.7813) = 73$$

Fourth, divide both sides of the equation by .7813.

$$\frac{b(\cancel{.7813})}{(\cancel{.7813})} = \frac{73}{.7813}$$

$$b = 93.434$$

Fifth, round to the nearest tenth (93.4). Remember, accuracy cannot be created through calculation.

Example 16-12

PROBLEM: Given right triangle *ABC*, determine the length of side *c*.

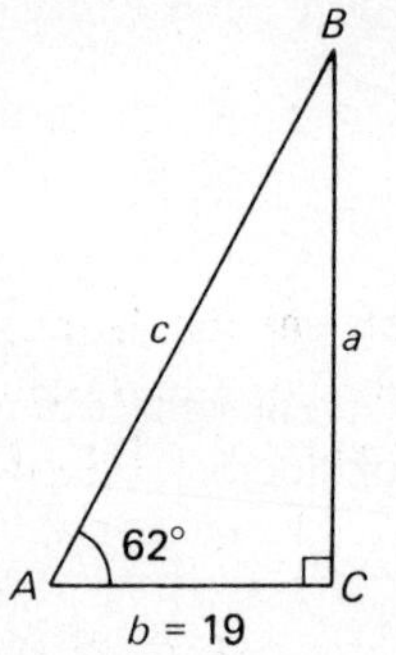

SOLUTION: First, set up a ratio involving one of the given values. This ratio should involve the adjacent side (b) and the hypotenuse (c) of angle A. Remember the formula,

$$\cos A = \frac{\text{opp}}{\text{hyp}}$$

Thus,

$$\cos 62^\circ = \frac{19}{c}$$

Second, locate cos 62° in Table 2 of Trigonometric Functions (.4695) and set this value into the ratio.

Third, multiply both sides of the equation by c.

$$.4695 = \frac{19}{c}$$

$$c(.4695) = 19$$

Fourth, divide both sides of the equation by .4695.

$$\frac{c(\cancel{.4695})}{(\cancel{.4695})} = \frac{19}{(.4695)}$$

$$= 40.2556$$

$$= 40.3$$

Fifth, round to the nearest tenth (40.3).

Example 16-13

PROBLEM: Given the triangle *JKL*, determine the degrees of angle *K*.

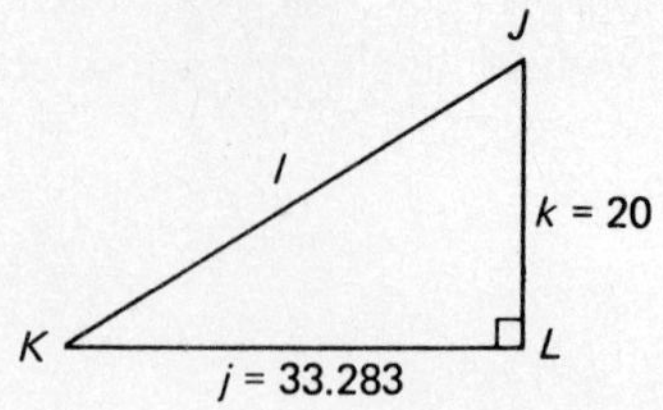

SOLUTION: First, set up a ratio involving the given values. This ratio should involve the opposite side ($k = 20$) and the adjacent side ($j = 33.283$) of angle D. Remember the formula,

$$\tan K = \frac{\text{opp}}{\text{adj}}$$

Thus,

$$\tan K = \frac{20}{33.283}$$

Second, complete the division indicated:

$$\tan K = \frac{20}{33.283}$$

$$= .6009$$

Third, locate tan = .6009 in Table 2 of Trigonometric Functions and read the degrees of the angle that has a tan value of .6009, which is 31°. Thus, angle $K = 31°$. Remember that if the tan value of .6009 were not in the table, it would be necessary to interpolate to determine angle K.

Example 16-14

PROBLEM: Solve the right triangle *ABC*.

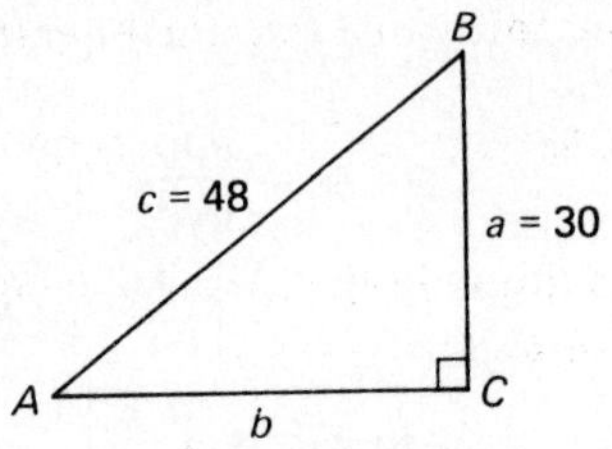

SOLUTION: The phrase "solve the right triangle" means to determine all of the unknown parts of the right triangle. Therefore, angles A and B and side b must be determined.

First, determine the degrees in angle A. Set up a ratio involving the given values, the opposite side ($a = 30$) and the hypotenuse ($c = 48$). Remember the formula,

$$\sin A = \frac{\text{opp}}{\text{hyp}}$$

Thus,

$$\sin A = \frac{30}{48}$$

$$= .625$$

Second, locate the sin value of .625 in Table 2 of Trigonometric Functions. Note that $\sin 38° 40' = .6248$. Since this is very close to the sin value of .625 and interpolation would only result in the determination of fractions of a second for 38° 40′, for our purposes 38° 40′ will be acceptable for the size of angle A.

Third, since the sum of angles A and B equals 90°, set up a simple equation and solve for $\angle B$.

$$\angle A + \angle B = 90°$$

$$38° 40' + \angle B = 90°$$

$$\angle B = 90° - 38° 40'$$

$$= 51° 20'$$

Fourth, determine the length of side b. Set up a ratio involving the given values, angle A (38° 40′) and the hypotenuse (48). Recall the formula,

$$\cos A = \frac{\text{adj}}{\text{hyp}}$$

Thus,

$$\cos 38° 40' = \frac{a}{48}$$

Fifth, locate cos 38° 40′ in Table 2 of Trigonometric Functions and note the value, .7808. Substitute this value into the equation.

$$\cos 38° 40' = \frac{b}{48} \quad \text{or} \quad .7808 = \frac{b}{48}$$

Sixth, multiply both sides of the equation by 48 and solve for b.

$$48\,(.7808) = \frac{b}{48} \times 48$$

$$37.4784 = b$$

$$b = 37.5$$

Example 16-15

PROBLEM: Given the following parts of a right triangle, $P = 23°$, $R = 90°$, and side $r = 55$, solve the right triangle.

SOLUTION: First, sketch triangle PQR as illustrated.

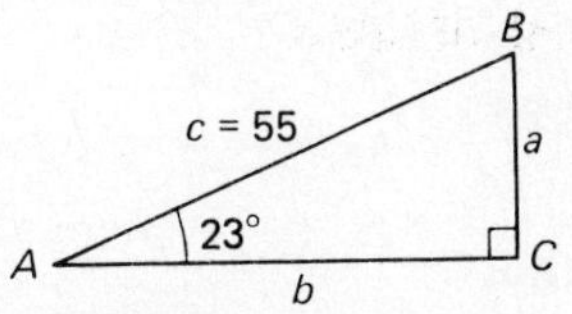

Second, determine the degrees in angle B. Since the sum of angles A and B equals 90°, set up a simple equation and solve for $\angle B$.

$$\angle A + \angle B = 90°$$

$$23° + \angle B = 90°$$

$$\angle B = 90° - 23°$$

$$= 67°$$

Third, determine the length of side a. Set up a ratio involving the given values, angle A (23°) and the hypotenuse c (55). Then,

$$\sin A = \frac{\text{opp}}{\text{hyp}}$$

$$\sin 23° = \frac{a}{55}$$

Fourth, locate sin 23° in Table 2 of Trigonometric Functions and note the value, .3907. Substitute the value into the equation.

$$\sin 23° = \frac{a}{55}$$

$$.3907 = \frac{a}{55}$$

Fifth, multiply both sides of the equation by 55 and solve for p.

$$55(.3907) = a$$

$$21.4885 = a$$

$$a = 21.5$$

Sixth, determine the length of side b. Set up a ratio involving the given values, angle A (23°) and the hypotenuse c (55). Then,

$$A = \frac{\text{opp}}{\text{hyp}}$$

$$\cos 23° = \frac{b}{55}$$

Seventh, locate cos 23° in Table 2 of Trigonometric Functions and note the value, .9205. Substitute this value into the equation.

$$\cos 23° = \frac{b}{55} \quad \text{or} \quad .9205 = \frac{b}{55}$$

Eighth, multiply both sides of the equation by 55 and solve for b.

$$55(.9205) = b$$

$$50.6275 = b$$

$$b = 50.6$$

EXERCISE 16-4

Determine the value of the unknown sides and angles as indicated.

16-36. Given the triangle ABC, find c.

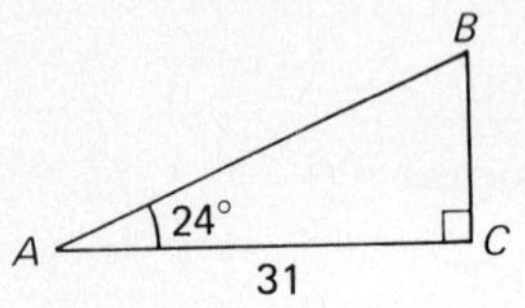

16-37. Given the triangle DEF, find f.

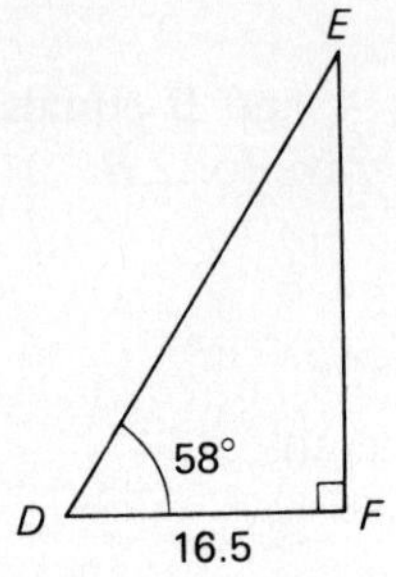

16-38. Determine the degrees of $\angle G$ in triangle GHI.

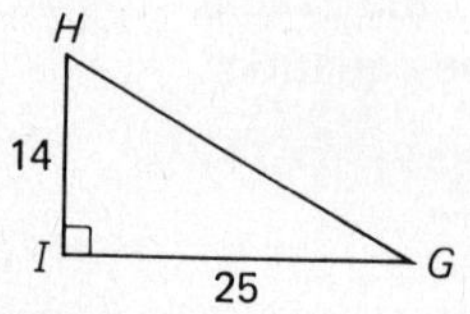

16-39. Determine the length of side l in triangle JKL.

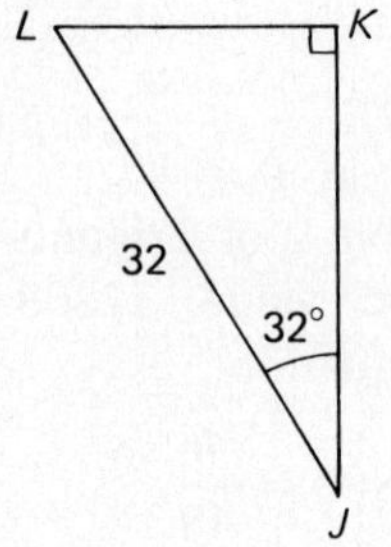

16-40. Determine the length of side n in triangle MNO.

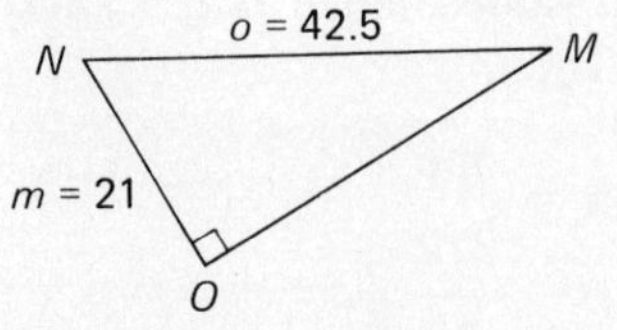

16-41. Determine the degrees of $\angle P$ in triangle PQR.

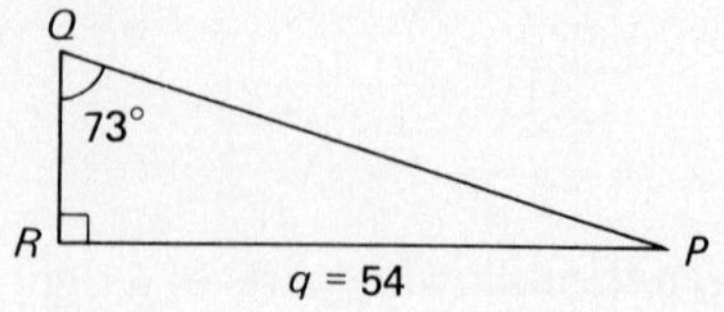

16-42. Determine the length of side r in triangle PQR.

16-43. Determine the length of side p in triangle PQR.

16-44. Determine the degrees of $\angle S$ in triangle STU.

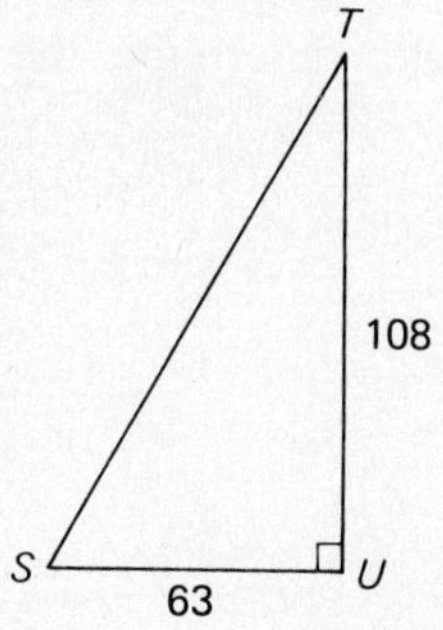

16-45. Determine the degrees of $\angle T$ in triangle STU.

16-46. Determine the length of side u in triangle STU.

16-47. Determine the degrees of $\angle X$ in triangle XYZ.

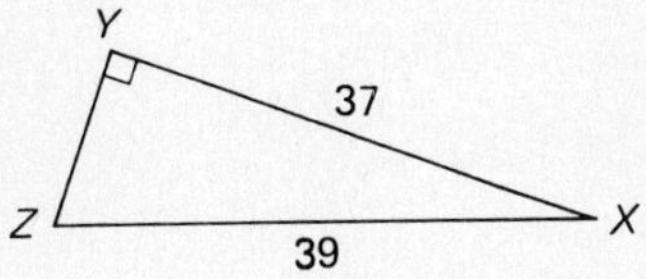

16.48. Determine the degrees of $\angle Y$ in triangle XYZ.

16-49. Determine the length of side x in triangle XYZ.

16-50. Determine the degrees of $\angle B$ in triangle ABC.

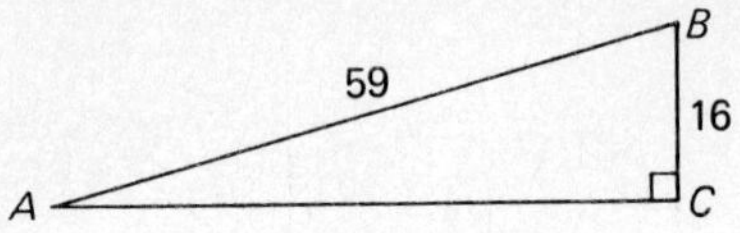

16-51. Determine the degrees of $\angle A$ in triangle ABC.

16-52. Determine the length of side b in triangle ABC.

16-53. Given the right triangle, determine the degrees of $\angle D$ in triangle DEF.

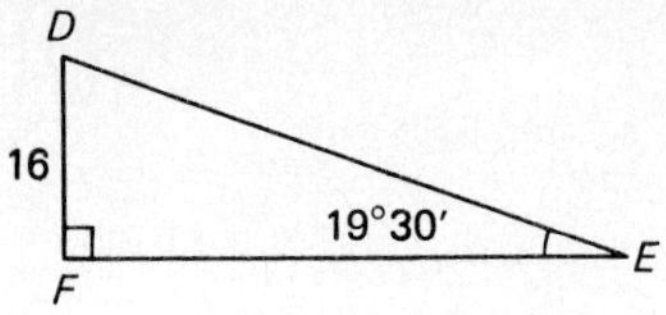

16-54. Determine the length of side d in triangle DEF.

16-55. Determine the length of side f in triangle DEF.

16-56. Sketch triangle GHI, with $\angle I = 90°$, $H = 21° 50'$, and side $g = 175$.

16-57. Determine the length of side h in triangle GHI.

16-58. Determine the degrees of $\angle G$ in triangle *GHI*.

16-59. Determine the length of side i in triangle *GHI*.

16-60. Determine the length of side j in triangle *JKL* when side $l = 12.48$, $\angle L = 90°$, and $\angle J = 83° 45'$.

CONTENT

The ability to solve right triangles can be most helpful in solving problems in industry. In fact, some industrial problems can only be solved through the use of trigonometry. In many other cases, the use of trigonometry can help you solve problems more easily and efficiently.

The examples which follow illustrate some of the applications of the solution of right triangles to industry. Your imagination will help you find other uses for your ability to solve right triangles.

Example 16-16

PROBLEM: A contractor is asked to build an A-frame cabin 24 feet high with a peak angle of 55°. Determine the length of the rafters.

SOLUTION: First, draw a sketch of the cabin and insert the known values as illustrated.

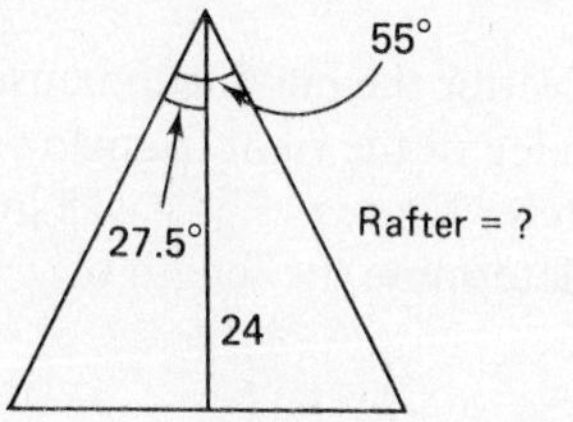

Second, set up a ratio involving the given values, 27.5° and the height (24 feet).

$$\cos A = \frac{\text{adj}}{\text{hyp}} \quad \text{or} \quad \cos 27.5° = \frac{24 \text{ feet}}{\text{rafter length}}$$

Third, insert the value of cos 27.5° and solve for the rafter length.

$$\cos 27.5° = \frac{24 \text{ feet}}{\text{rafter length}}$$

$$.8870 = \frac{24 \text{ feet}}{\text{rafter length}}$$

$$\text{Rafter length } (.8870) = 24 \text{ feet}$$

$$\frac{\text{Rafter length } \cancel{(.8870)}}{\cancel{.8870}} = \frac{24 \text{ feet}}{.8870}$$

$$\text{Rafter length} = \frac{24 \text{ feet}}{.8870}$$

$$= 27.057 \text{ feet}$$

$$= 27.1 \text{ feet}$$

Example 16-17

PROBLEM: The captain of an iron-ore freighter spots a lighthouse at an angle of 2° 30′ from his position (corrected for the height of the control room from the surface of the water). His map indicates the light is 158 feet above the surface of the water. How far is the ship from the lighthouse?

SOLUTION: First, draw a sketch of the situation and insert the known values.

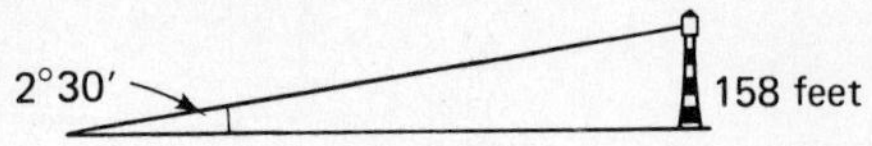

Second, set up a ratio involving the given values, 2° 30′ and the height of the light, 158 feet. Recall,

$$\tan A = \frac{\text{opp}}{\text{adj}}$$

Thus,

$$\tan 2^\circ 30' = \frac{\text{opp (158 feet)}}{\text{adj (distance from light)}}$$

Third, insert the value of tan 2° 30′ and solve for the unknown distance.

$$\tan 2^\circ 30' = \frac{158 \text{ feet}}{\text{distance}}$$

$$.0437 = \frac{158 \text{ feet}}{\text{distance}}$$

$$\text{distance} = \frac{158 \text{ feet}}{.0437}$$

$$= 3615.5606 \text{ feet}$$

$$= 3615.6 \text{ feet}$$

Example 16-18

PROBLEM: Determine the missing dimension (x) on the drawing.

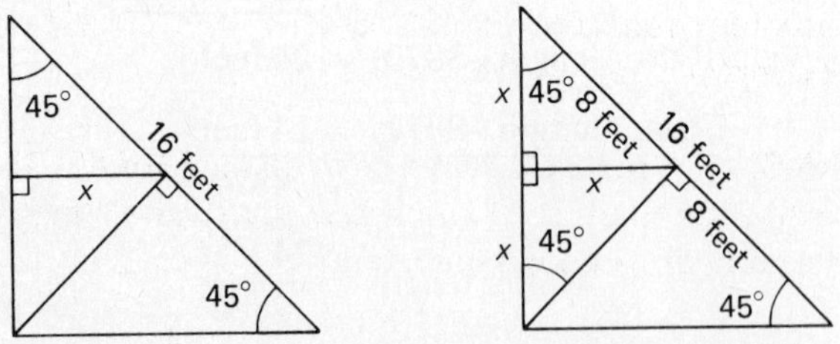

SOLUTION: First, study the sketch at the left, recalling the properties of a 45° right triangle, and then enter obvious facts on the sketch at the right. The indicated values of x are all equal.

Second, set up a ratio involving the given values, 45° and the length of one side (8 feet). Recall,

$$\sin A = \frac{\text{opp}}{\text{hyp}}$$

Thus,

$$\sin 45^\circ = \frac{\text{opp } (x)}{\text{hyp}} = \frac{x}{8 \text{ feet}}$$

Third, insert the value of sin 45° and solve for x.

$$\sin 45^\circ = \frac{x}{8 \text{ feet}}$$

$$.7071 = \frac{x}{8 \text{ feet}}$$

Fourth, multiply both sides of the equation by 8 feet.

$$(8 \text{ feet})(\sin 45^\circ) = x$$

$$(8 \text{ feet})(.7071) = x$$

$$x = 5.6568 \text{ feet}$$

Fifth, if desirable, convert .6568 feet to inches.

$$.6568 \times 12 \text{ inches} = 7.8816 \text{ inches}$$

Depending on the need for accuracy, the inches could be rounded off to the nearest eighth, which would be $7\frac{7}{8}$ inches, or to the nearest inch, which would be 8 inches. Thus the length of the missing dimension is 5 feet $7\frac{7}{8}$ inches.

Example 16-19

PROBLEM: Determine the depth of cut required to mill a square on the end of a $1\frac{1}{2}$ inch shaft.

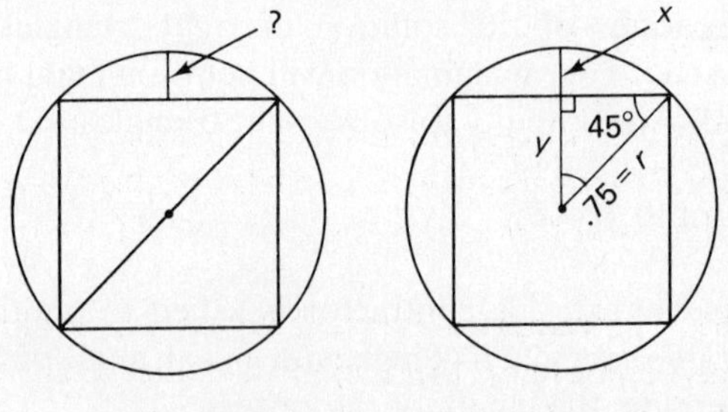

SOLUTION: First, study the sketch and then enter the facts as shown on the second sketch. Recall the facts of plane geometry to determine needed dimensions.

Second, letter the missing dimension x. Use the letter y as a leg of the right triangle as shown in the second sketch. Thus, $x + y = .75$ inch.

Third, determine the length of y. Set up a ratio

involving the given values, 45° and the length of the hypotenuse (.75 inch). Then,

$$\sin A = \frac{\text{opp}}{\text{hyp}}$$

$$\sin 45° = \frac{y}{.75}$$

Fourth, insert the value of sin 45° and solve for y.

$$.7071 = \frac{y}{.75 \text{ inch}}$$

Fifth, multiply both sides of the equation by .75 inch.

$$(.75 \text{ inch})(.7071) = y$$

$$.530325 = y$$

$$y = .53$$

Sixth, solve for x.

$$x + y = .75$$

$$x + .53 = .75$$

$$x = .75 - .53$$

$$= .22$$

Thus the depth of cut is .22 inch.

Example 16-20

PROBLEM: If a skycycle is "shot" from a ramp at 57° at 220 miles per hour, how fast is it traveling horizontally?

SOLUTION: First, draw a sketch of the problem. Insert the given values (220 miles per hour and 57°) and decide what is to be determined.

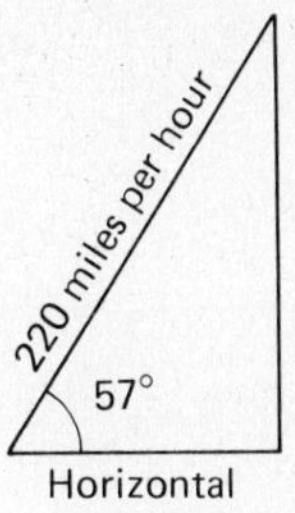

Second, recall the trigonometric function which involves the adjacent side and hypotenuse of the given angle.

$$\cos A = \frac{\text{adj}}{\text{hyp}}$$

Third, substitute the given values and solve for the horizontal.

$$\cos 57° = \frac{\text{horizontal}}{220 \text{ miles per hour}}$$

$$(220 \text{ miles per hour})(\cos 57°) = \text{horizontal}$$

$$(220 \text{ miles per hour})(.5446) = \text{horizontal}$$

$$119.812 \text{ miles per hour} = \text{horizontal}$$

$$\text{Horizontal} = 119.8 \text{ miles per hour}$$

EXERCISE 16-5

The following problems will help you to develop your skills in working with everyday problems which can be solved through the use of trigonometry. Draw sketches as necessary to better understand the problem you are solving. Refer to the preceding examples if necessary.

16-61. A triangle is the most rigid of all geometric forms. Determine the length of a diagonal for a cattle gate 10 feet long and 4 feet high.

16-62. A rocket is launched at an angle of 82° 30′ with a speed of 2800 kilometers per hour. What is its vertical speed in kilometers?

16-63. A roof is to rise 20 feet in a horizontal distance of 32 feet. What is the slope of the roof?

16-64. How long will a ladder need to be if it is used to paint the top of the gable of a house if the house is 18 feet 6 inches high and the base of the ladder is 10 feet from the base of the house?

16-65. Determine the missing dimension, x.

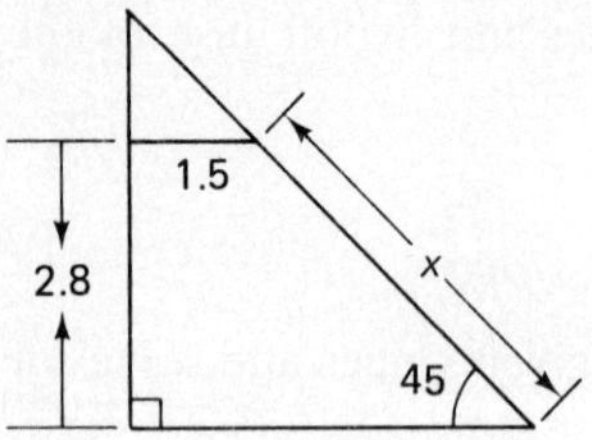

16-66. A utility pole is supported by a guy wire 35 feet from the ground. If the wire is inclined at an angle of 49° 30′ from the ground, how many feet of wire will be needed?

16-67. Point B is 80 air miles north of point A. Point C is 125 air miles east of point B. What is the shortest air mileage between points A and C?

16-68. Determine the length of the air ventilation shaft of the tunnel shown.

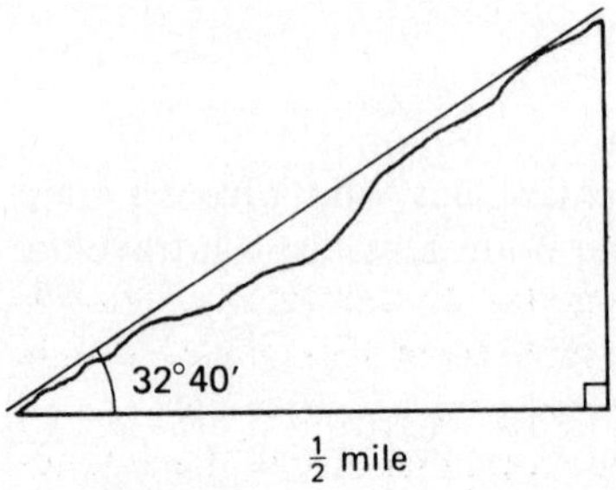

16-69. A land surveyor measures 50 feet along a river bank. Then with a transit point directly across the river from the original point, he reads an angle of 76°. How wide is the river?

16-70. Determine the missing dimension in the following illustration of a flat-head screw.

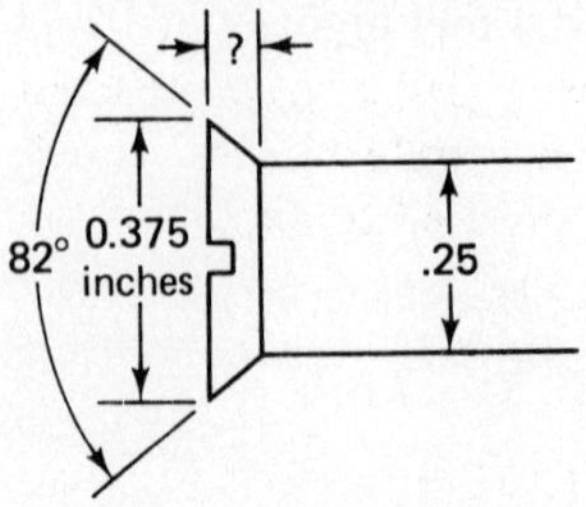

16-71. Determine the included taper angle of a $2\frac{1}{2}$-inch shaft having a taper of $\frac{1}{2}$ inch per foot.

16-72. The Washington Monument is 555 feet high. What is the angle of elevation to the top from a distance of 1 mile?

16-73. A new freeway is constructed through a city at an angle of 68° to the local streets. If each street is 380 feet apart, how long will the freeway be between intersections of north/south streets. Observe the illustration below. Remember to determine distances from median points of the highways.

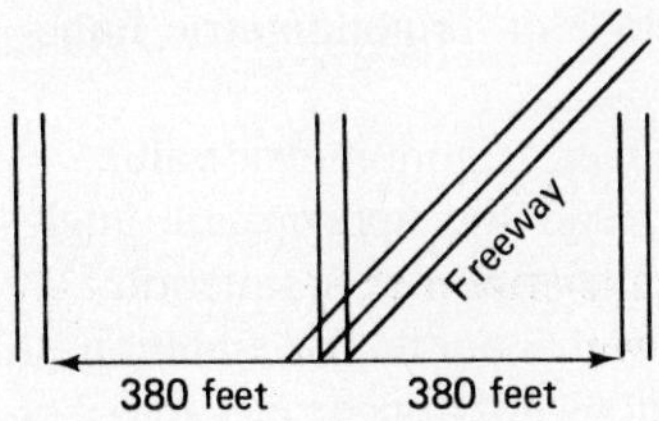

16-74. Determine the volume of the prism in the drawing below.

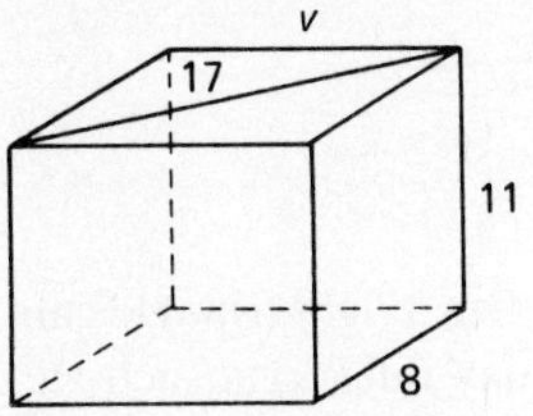

16-75. A sheet of metal 25 centimeters wide is bent along its centerline to form a V-shaped trough. Will the trough have a greater capacity when it is 12 centimeters wide at the top or when it is 10 centimeters deep?

16-76. A helicopter flying at an altitude of 1 mile observes a heliport landing point at an angle of depression of 24° 20′. How far is the helicopter from the landing point?

16-77. A pentagonal flower garden is constructed inside a 16-foot-diameter circle. How much decorative fence must be ordered to surround the garden?

16-78. If the slope of the land is 4° 10′, how far will the water be backed up behind a 45-foot dam?

16-79. The end of an 8-centimeter shaft is milled as shown. Determine the radius of the smaller circle.

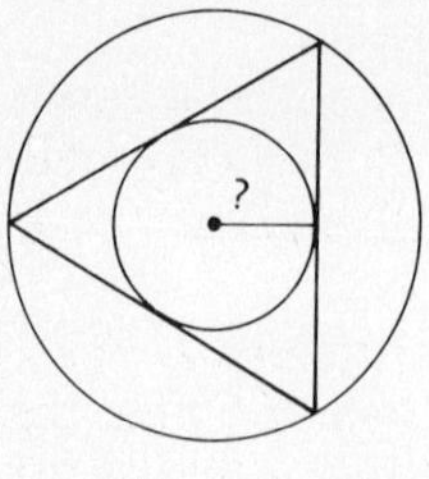

16-80. The 30° arc of a circular curve of a mountain road has a chord of 75 feet. Determine the radius of the curve.

THINK TIME

The procedures used to solve a right triangle can also be used to solve everyday questions such as: How high is a particular building or tree? How far has one sailed, flown, or snowmobiled in a particular direction? The use of these procedures to solve everyday problems is limited only to the creative thinking of the person. For example, managers of golf courses occasionally wish to determine distances across natural barriers as an aid to golfers. See if you can determine the shortest distance, in yards, from the tee to the hole in the illustration.

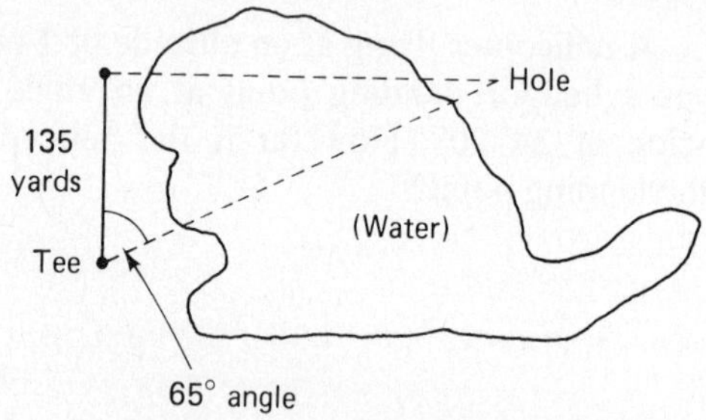

Set up other problems of your own and determine the unknown values.

PROCEDURES TO REMEMBER

1. To determine the trigonometric functions of angles:
   (a) Locate Table 2 of Trigonometric Functions.
   (b) Locate the degrees of the given angle.
   (c) Locate and read the appropriate trigonometric function of the angle.
2. To determine the angle of a trigonometric function:
   (a) Locate Table 2 of Trigonometric Functions.
   (b) Locate the given trigonometric value.
   (c) Locate and read the appropriate angle for the given trigonometric function.
3. To determine the trigonometric function of an angle which does not appear in Table 2 of Trigonometric Functions:
   (a) Locate the angles that are greater and lesser than the given angle in Table 2.
   (b) Locate the trigonometric functions of the greater and lesser angles.
   (c) Subtract the lesser trigonometric function from the greater.
   (d) Set up an equation and solve for the unknown trigonometric value.
   (e) Combine the determined value with the known value.
4. To determine the angle of a trigonometric function which does not appear in Table 2 of Trigonometric Functions.
   (a) Locate the trigonometric functions that are greater and less than the given function in Table 2.
   (b) Locate the angles of the greater and lesser trigonometric functions.
   (c) Subtract the given trigonometric function from the greater trigonometric function.
   (d) Set up an equation to determine the minutes or parts of a minute represented by the portion of the trigonometric function determined in step (c).
   (e) Combine the determined angle with the known angle.

5. To determine the unknown side of a right triangle:
   (a) Set up a ratio involving the given values.
   (b) Solve the equation for the unknown side.
   (c) Round off to the nearest tenth if desired.
6. To determine an unknown angle of a right triangle:
   (a) Set up a ratio involving the given lengths of the sides to determine the value of the trigonometric function.
   (b) Solve the equation for the unknown trigonometric function.
   (c) Locate the angle of the determined trigonometric function in Table 2.
   (d) Interpolate to determine the angle if necessary.

CHAPTER SUMMARY

The following definitions and principles are essential to working with trigonometric functions.

1. *Interpolation* is the process of determining the value of an unknown trigonometric function or angle using known trigonometric functions or angles from Table 2 of Trigonometric Functions.
2. The determination of the *arc function* is based on the value of a trigonometric function and the angle it represents.
3. When using Table 2 of Trigonometric Functions, for angles *less* than 45°, read the table from *left* to *right* and use the titles at the *top* of the table.
4. When using Table 2 of Trigonometric Functions, for angles *greater* than 45°, read the table from *right* to *left* and use the titles at the *bottom* of the table.
5. When the tangent value is greater than 1, the angle of the tangent is greater than 45°.
6. Sine values *increase* as the size of the angle increases.
7. Cosine values *decrease* as the size of the angle increases.
8. Tangent values *increase* as the size of the angle increases.
9. To solve a right triangle means to determine all of the unknown parts of the right triangle.

CHAPTER TEST

T-16-1. Determine the tangent of 76°.

T-16-2. Determine the cosine of 32° 40′.

T-16-3. Sin 81° 20′ =

T-16-4. Determine the angle whose cosine is .4823.

T-16-5. Cos A = .7679

A =

T-16-6. Determine the cosine of 29° 33′.

T-16-7. Tan A = .6796

A =

T-16-8. Determine the length of side b in triangle ABC.

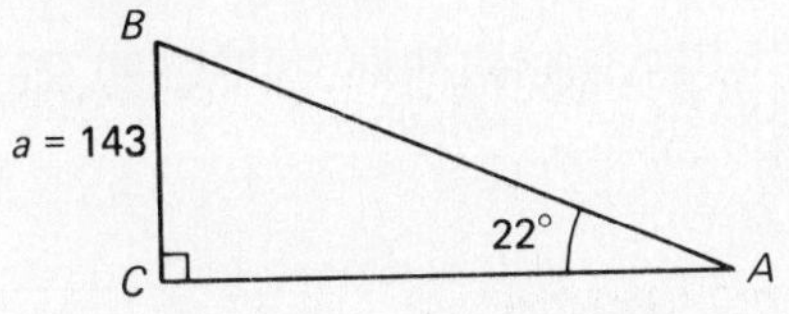

T-16-9. Given triangle JKL, determine the degrees in angle K.

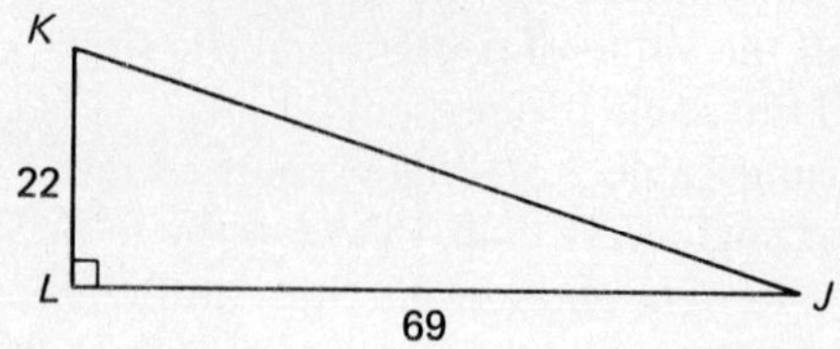

T-16-10. Determine the length of side a if $A = 37°$, $C = 90°$, and side $c = 303$ centimeters.

T-16-11. Determine the length of side s in triangle RST.

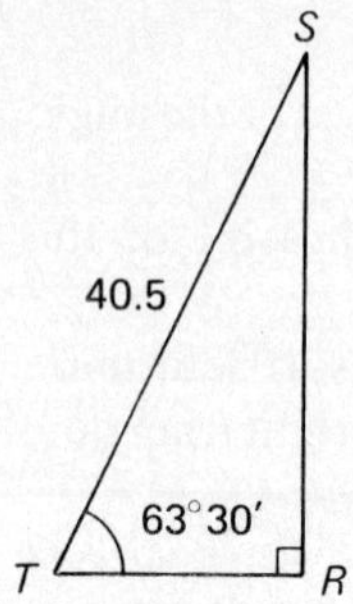

T-16-12. Determine the length of side t in triangle RST.

T-16-13. A 24-foot ladder leans against a wall. Find the angle the ladder makes with the wall if the ladder is $9\frac{1}{2}$ feet from the base of the wall.

T-16-14. From point A, a sailboat sails $1\frac{1}{2}$ miles due west to point B and then 4 miles south to point C. What is the distance from point A to point C?

T-16-15. What is the largest square that can be milled from the end of a $1\frac{1}{2}$-inch shaft?

CHAPTER

17

SOLUTION OF OBLIQUE TRIANGLES

OBJECTIVES

1. To read the trigonometric functions of Table 2 for angles greater than 90°.
2. To solve oblique triangles using the Law of Sines and the Law of Cosines.
3. To solve industrial problems involving the use of oblique triangles.

SELF-TEST

This test will evaluate your skills and ability to solve oblique triangles. You will need to use Table 2 of Trigonometric Functions to solve some of these problems. It is located on the inside back cover of this book. If you are able to solve these problems, no further development of these skills is necessary. Be sure to show your work so that possible areas of trouble can be easily identified.

S-17-1. Determine the cos 119°.

S-17-2. Sin 169° 33′ = ?

S-17-3. Determine side a in oblique triangle ABC.

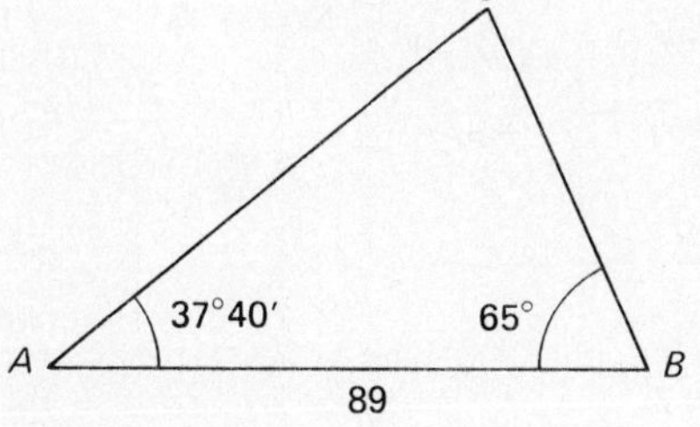

S-17-4. Determine the length of side d if $\angle D = 118°$, $\angle E = 27°$, and side $f = 59.4$.

S-17-5. Determine angle I of oblique triangle GHI.

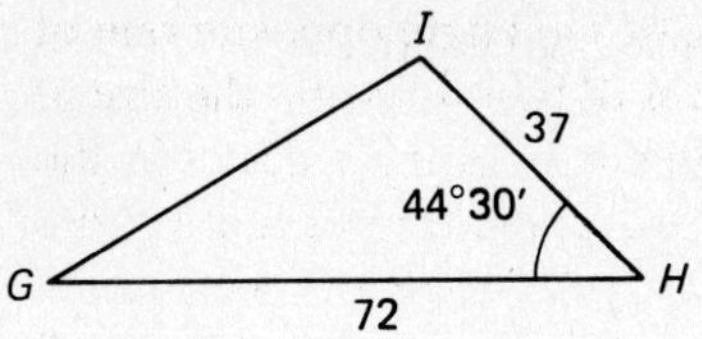

S-17-6. Determine the length of side k if side $j = 426$, side $l = 304$, and $\angle K = 32° 40'$.

S-17-7. Determine the length of side o in triangle MNO.

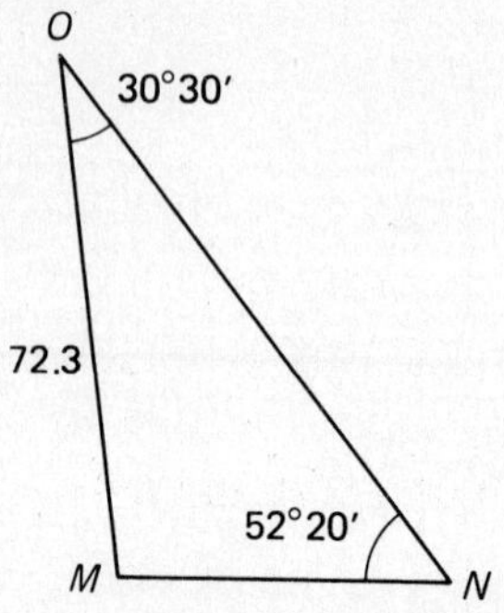

S-17-8. Determine $\angle M$ in triangle MNO.

S-17-9. Determine the length of side m in triangle MNO.

S-17-10. Determine the shortest distance across the ravine from point A to point B. Use the illustration below as a guide.

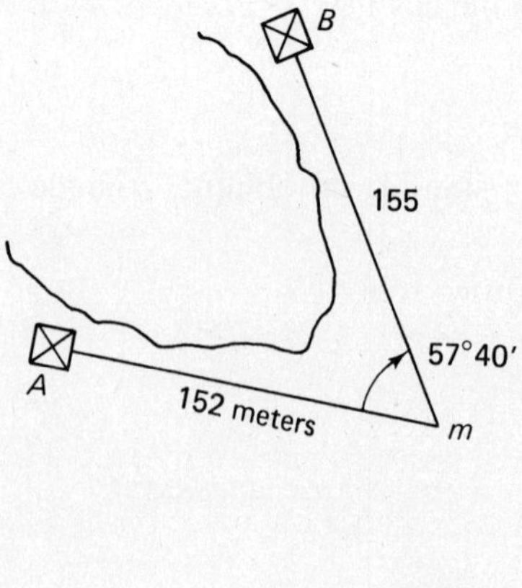

INTRODUCTION

Recall that oblique triangles are triangles which do not contain a right angle. The solution of oblique triangles is somewhat more difficult than right triangles. The Laws of Sines and Cosines are used to solve oblique triangles. At times, an oblique triangle can be divided into two or more right triangles for simple solution. All of the angles of an oblique triangle may be acute, or one of them may be obtuse. Like the right triangle, the sum of the interior angles is a straight angle of 180° and there are six parts: three sides and three angles.

DEFINITIONS

Two facts regarding sines and cosines greater than 90° must be known before the Laws of Sines and Cosines can be used. First, if an angle is greater than 90° and less than 180°, then the sine of that angle is equivalent to the sine of 180° minus the given angle. Stated another way: If $\angle A$ is greater than 90°, then

$$\sin A = \sin(180° - \angle A)$$

So if $\angle A = 127°$,

$$\sin 127° = \sin(180° - 127°)$$

$$\sin 127° = \sin 53°$$

From Table 2, $\sin 53° = .7986$. Thus $\sin 127° = \sin 53° = .7986$. So if $\angle A = 90°$, then $\sin A = \sin(180° - A)$.

Second, if an angle is greater than 90° and less than 180°, then the cosine of that angle is equivalent to a negative cosine value of 180° minus the given angle. Stated another way: If $\angle A$ is greater than 90°, then

$$\cos A = -\cos(180° - \angle A)$$

Thus if $\angle A = 164°$,

$$\cos 164° = -\cos(180° - 164°)$$

$$\cos 164° = -\cos 16°$$

From Table 2 we know $-\cos 16° = -.9613$. Thus $\cos 164° = -\cos 16° = -.9613$. So if $\angle A = 90°$, then $\cos A = -\cos(180° - \angle A)$.

Remember that the cosine value of an angle greater than 90° is always negative. Further in-depth proof of these concepts may be obtained in most texts in trigonometry.

CONTENT

In order to solve oblique triangles using the Laws of Sines and Cosines, one of four combinations of values must be given: (1) the sizes of two angles and the length of one side; (2) the lengths of two sides and the size of the angle opposite one of the sides; (3) the lengths of two sides and the size of the angle between the two given sides; (4) the lengths of all three sides. Thus, at least three of the six parts of the triangle must be given. The specific parts which are given will determine which law will be selected as a tool to solve the oblique triangle.

Although it is not essential to know how the Law of Sines is developed, a simple study of the process can have some value. Observe the illustration and the process as given below using triangle *ABC*.

First, draw a perpendicular from point *C* to point *D* or side *c* to form right triangles *ACD* and *BCD*.

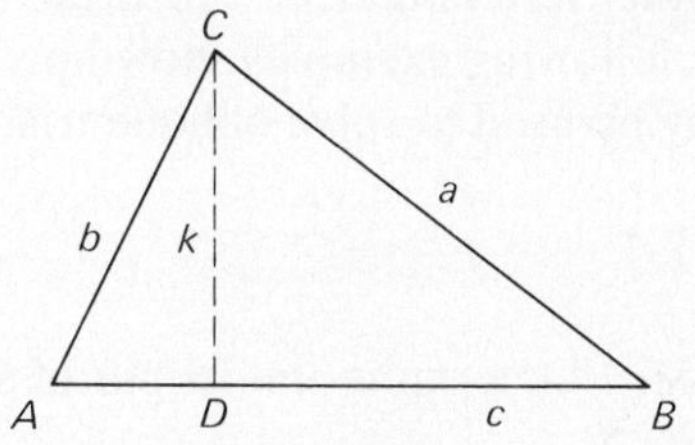

Second, state the sine ratios of $\angle A$ and $\angle B$.

$$\sin A = \frac{\text{opposite}}{\text{hypotenuse}} = \frac{k}{b}$$

$$\sin B = \frac{\text{opposite}}{\text{hypotenuse}} = \frac{k}{a}$$

Third, solve each ratio for *k*.

$$\sin A = \frac{k}{b} = b \sin A = k$$

$$\sin B = \frac{k}{a} = a \sin B = k$$

Since $k = k$,

$$b \sin A = a \sin B$$

Fourth, divide both sides of the equation by $\sin A \sin B$.

$$\frac{b \cancel{\sin A}}{\cancel{\sin A} \sin B} = \frac{a \cancel{\sin B}}{\sin A \cancel{\sin B}}$$

$$\frac{b}{\sin B} = \frac{a}{\sin A}$$

In like manner, a perpendicular could be dropped from point *A* which would result in similar ratios, such as:

$$\frac{b}{\sin B} = \frac{c}{\sin C}$$

Thus, we could develop the Law of Sines:

$$\frac{a}{\sin A} = \frac{b}{\sin B} = \frac{c}{\sin C}$$

Note that the Law of Sines states that the ratio of a side to the sine of its angle is equal to the ratios of the other sides and their angles.

The Law of Sines could also be stated in the inverse formula below:

$$\frac{\sin A}{a} = \frac{\sin B}{b} = \frac{\sin C}{c}$$

To solve triangles with the Law of Sines, two or more of these ratios are used and known quantities are substituted into the formula. Naturally, letter identification of triangles may be different and consequently other letters may be substituted into the ratios. The following examples show how the Law of Sines may be used to solve oblique triangles.

Example 17-1

PROBLEM: Determine the length of side a in triangle ABC.

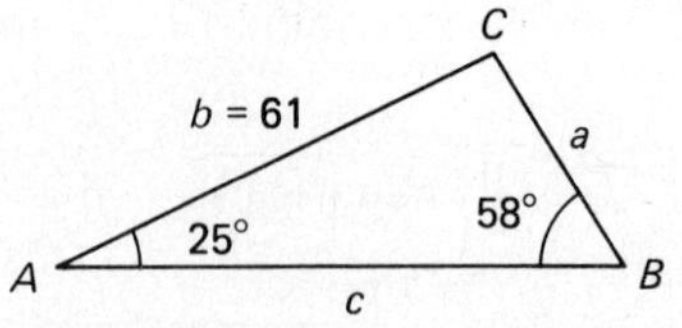

SOLUTION: First, note the given values, $\angle A = 25°$, $\angle B = 58°$, $\angle C = 97°$, and side $b = 61$. We are given two angles and one side; therefore, the Law of Sines should be used to solve the problem.

$$\frac{a}{\sin A} = \frac{b}{\sin B} = \frac{c}{\sin C}$$

Second, select two of the three ratios which will make use of the given values to determine the length of side a. The ratios are:

$$\frac{a}{\sin A} = \frac{b}{\sin B}$$

Third, substitute the letters with known values and the sine values from Table 2.

$$\frac{a}{\sin 25°} = \frac{61}{\sin 58°}$$

$$\frac{a}{.4226} = \frac{61}{.8480}$$

Fourth, multiply both sides of the equation by .4226 and solve the equation.

$$(\cancel{.4226})\frac{a}{\cancel{.4226}} = \frac{61}{.8480}(.4226)$$

$$a = \frac{25.7786}{.8480}$$

$$= 30.399 \text{ or } 30.4$$

Example 17-2

PROBLEM: Determine the degrees of C in oblique triangle ABC.

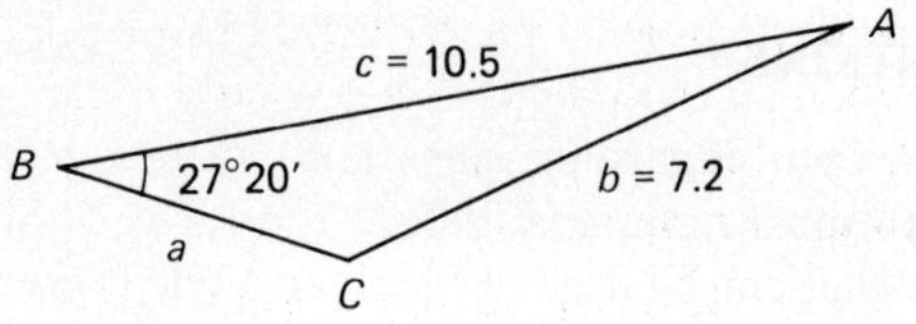

SOLUTION: First, note the given values, $\angle B = 27° 20'$. side $b = 7.2$, and side $c = 10.5$. We are given two angles and one side; therefore, the Law of Sines should be used to solve the problem.

Second, set up two ratios which will make use of the given values to determine the size of $\angle C$. The ratios are:

$$\frac{b}{\sin B} = \frac{c}{\sin C}$$

Third, substitute the letters with known values and the sine values from Table 2 and solve for angle C.

$$\frac{b}{\sin B} = \frac{c}{\sin C}$$

$$\frac{7.2}{\sin 27° 20'} = \frac{10.5}{\sin C}$$

$$\frac{\cancel{7.2} \sin C}{\cancel{7.2}} = \frac{(10.5)(.4592)}{7.2}$$

$$\sin C = .66966$$

$$\angle C = 42°$$

Fourth, from the sketch, $\angle C$ is seen to be greater than 90°. Using the definition $\sin A = \sin(180° - \angle A)$, substitute the facts of this problem.

$$\sin 42° = \sin(180° - 42°)$$

$$= \sin 138°$$

$$\angle C = 138°$$

Example 17-3

PROBLEM: Solve oblique triangle ABC.

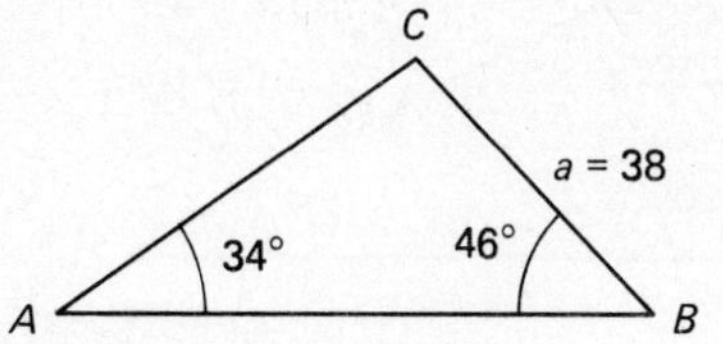

SOLUTION: First, note that $\angle A = 34°$, $\angle B = 46°$, and side $a = 38$, and that $\angle C$ and sides b and c are to be determined. Since we are given two sides and one angle, the Law of Sines should be used to solve the problem.

Second, determine $\angle C$.

$$\angle A + \angle B + \angle C = 180°$$

$$34° + 46° + \angle C = 180°$$

$$\angle C = 180° - 34° - 46°$$

$$= 100°$$

Third, set up two ratios which will make use of the given values to determine the length of side b. The ratios are:

$$\frac{a}{\sin A} = \frac{b}{\sin B}$$

Fourth, substitute letters with known values and the sine values from Table 2.

$$\frac{a}{\sin A} = \frac{b}{\sin B}$$

$$\frac{38}{\sin 34°} = \frac{b}{\sin 46°}$$

$$\frac{38}{.5592} = \frac{b}{.7193}$$

$$\frac{38 \times .7193}{.5592} = b$$

$$48.8794 = b$$

$$48.9 = b$$

Fifth, set up two ratios which will make use of the given values to determine the length of side c. The ratios are:

$$\frac{a}{\sin A} = \frac{c}{\sin C}$$

Sixth, substitute letters with known values and the sine values from Table 2 and solve for c.

$$\frac{a}{\sin B} = \frac{c}{\sin C}$$

$$\frac{38}{\sin 34°} = \frac{c}{\sin 100°}$$

Since $\sin (180° - 100°) = \sin 80°$,

$$\frac{38}{\sin 34°} = \frac{c}{\sin 80°}$$

$$\frac{38}{.5592} = \frac{c}{.9848}$$

$$\frac{38 \times .9848}{.5592} = c$$

$$66.9213 = c$$

or

$$c = 66.9$$

EXERCISE 17-1

Using the Law of Sines, determine the degrees of angles and lengths of sides as indicated in each problem.

17-1. Determine the length of side b in triangle ABC.

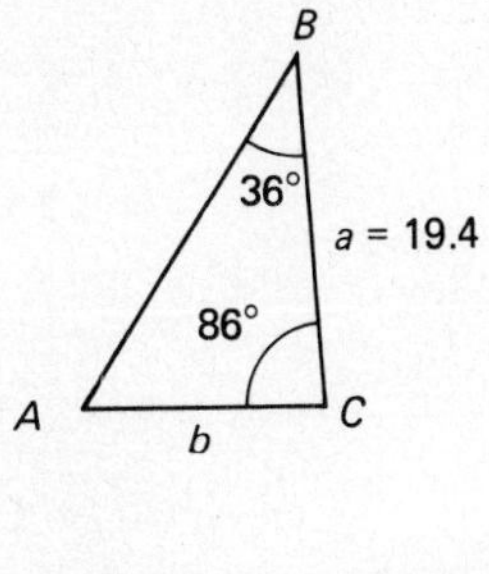

17-2. Given $\angle A = 24°$, $\angle C = 65°$, and side $b = 24.3$, determine the length of side c.

17-3. Determine the degrees in $\angle B$ of triangle *ABC*.

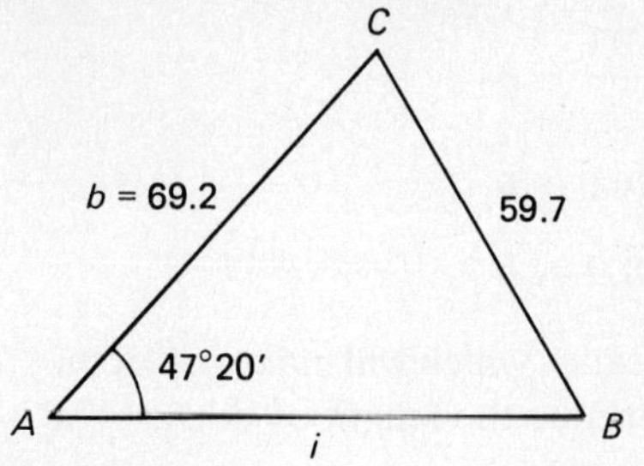

17-4. Given the oblique triangle *JKL*, find the length of side *k*.

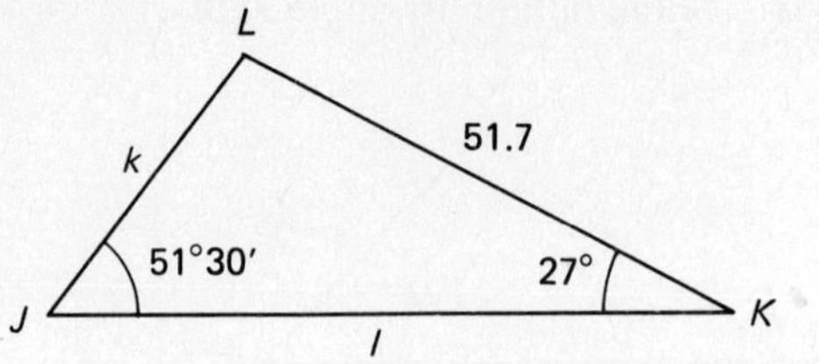

17-5. Given $\angle A = 98°\,6'$, $\angle B = 43°\,42'$, and side $b = 78.23$, determine the length of side *a*.

17-6. Determine the length of side *c* in triangle *ABC*.

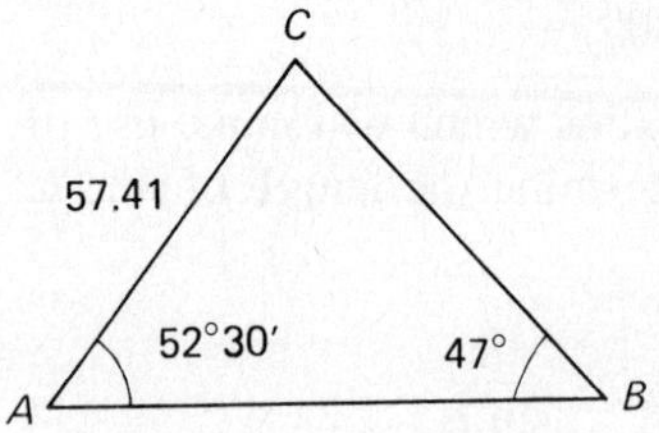

17-7. Given the triangle *STU*, determine the degrees in angle *U*.

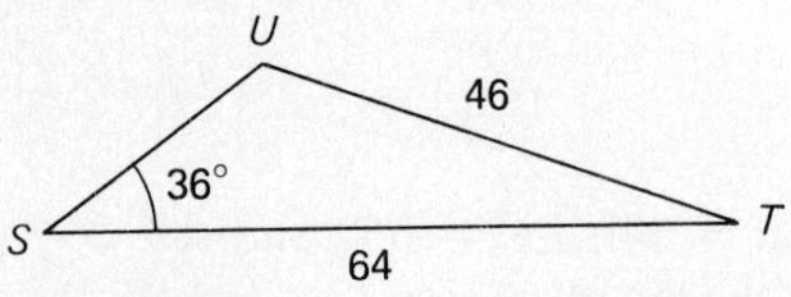

17-8. Determine the degrees in $\angle T$ of triangle *STU*.

17-9. Determine the length of side t in triangle STU.

17-10. Given the triangle XYZ, determine the degrees in $\angle Z$ of triangle XYZ.

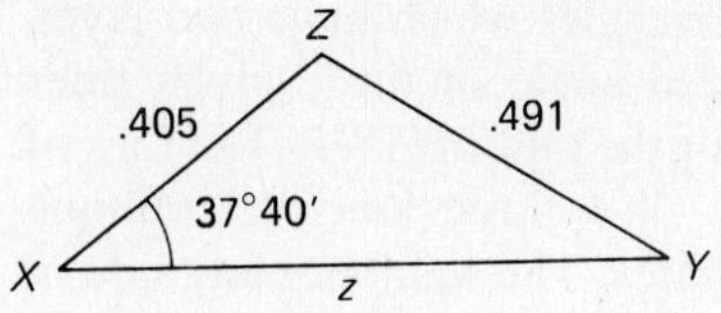

17-11. Determine the degrees in $\angle Y$ of triangle XYZ.

17-12. Determine the length of side z in triangle XYZ.

17-13. Solve the oblique triangle ABC given $\angle A = 63°$, $\angle B = 42°$, and side $c = 670$.

17-14. Side a =

17-15. Side b =

CONTENT

The Law of Sines cannot be used to solve all oblique triangles. Triangles which have two given sides and the included angle or three given sides cannot be solved using the Law of Sines. The Law of Cosines is used to solve three kinds of oblique triangles. This law states: The square of any side of a triangle is equal to the sum of the squares of the other two sides minus two times the product of those sides and the cosine of the included angle.

The Law of Cosines is stated by the following formulas:

$$a^2 = b^2 + c^2 - 2bc \cos A \quad \text{or} \quad \cos A = \frac{b^2 + c^2 - a^2}{2bc}$$

$$b^2 = a^2 + c^2 - 2ac \cos B \quad \text{or} \quad \cos B = \frac{a^2 + c^2 - b^2}{2ac}$$

$$c^2 = a^2 + b^2 - 2ab \cos C \quad \text{or} \quad \cos C = \frac{a^2 + b^2 - c^2}{2ab}$$

Because the Law of Cosines is much more difficult to prove than the Law of Sines, we will not go through the process. However, should you desire to demonstrate the Law of Cosines, most advanced trigonometry textbooks will contain the proof.

The Law of Cosines is much more difficult to work with than the Law of Sines; it should only be used when necessary. The following examples illustrate the use of the Law of Cosines to solve oblique triangles.

Example 17-4

PROBLEM: Given oblique triangle ABC, find the length of side c.

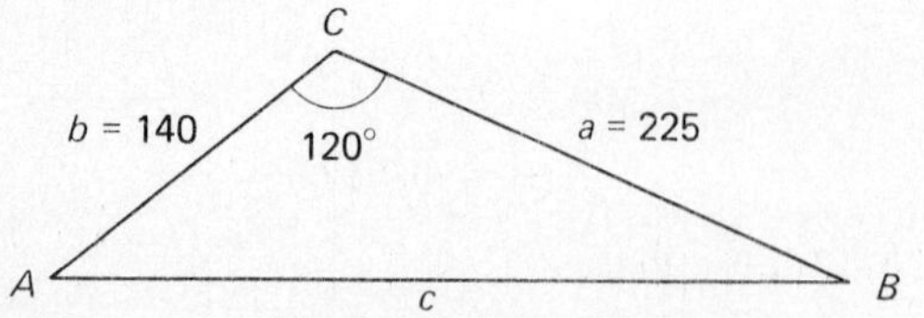

SOLUTION: First, note the given values: side $a = 225$, side $b = 140$, and $\angle C = 120°$. We are given two sides and the included angle; therefore the Law of Cosines should be used to solve the problem.

Second, select the Law of Cosines formula to solve for side c.

$$c^2 = a^2 + b^2 - 2ab \cos C$$

or

$$\sqrt{c^2} = \sqrt{a^2 + b^2 - 2ab \cos C}$$

$$c = \sqrt{a^2 + b^2 - 2ab \cos C}$$

Third, determine the cosine value of $\angle C$ (120°). Remember that $\angle C$ is greater than 90°. Therefore the formula $\cos A = \cos(180° - \angle A)$ should be used.

$$\cos 120° = -\cos(180° - 120°)$$
$$= -\cos 60°$$
$$= -.5000$$

Fourth, substitute the letters with known values and solve for side c.

$$c = \sqrt{a^2 + b^2 - 2ab \cos C}$$
$$= \sqrt{225^2 + 140^2 - 2(225)(140)(-.5000)}$$
$$= \sqrt{50{,}625 + 19{,}600 + 31{,}500}$$
$$= \sqrt{101{,}725}$$
$$= 318.9$$

Note that the multiplication of two negatives resulted in a positive.

Example 17-5

PROBLEM: Given the oblique triangle JKL, determine the degrees in $\angle J$.

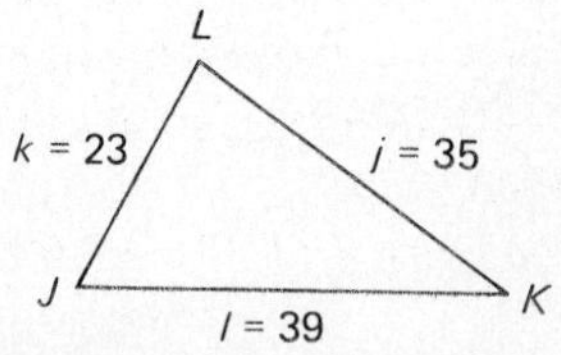

SOLUTION: First, note the given values: side $j = 34$, side $k = 23$, and side $l = 39$. We are given three sides; therefore the Law of Cosines should be used to solve the problem.

Second, select the Law of Cosines formula to solve for $\angle J$. Recall that the square of a side of any triangle is equal to the sum of the squares of the other two sides minus two times the product of those two sides and the cosine of the included angle. Then solve the formula for the angle.

$$\cos J = \frac{k^2 + l^2 - j^2}{2kl}$$

Third, substitute the letters with known values and solve for cos J.

$$\cos J = \frac{(23)^2 + (39)^2 - (35)^2}{2(23)(39)}$$

$$= \frac{529 + 1521 - 1225}{1794}$$

$$= \frac{825}{1794}$$

$$= .4599$$

Fourth, using Table 2 and interpolation, determine $\angle J$.

$$\angle J = 62° \, 37'$$

Example 17-6

PROBLEM: Solve oblique triangle ABC; $\angle A = 110°$, side $b = 24$, and side $c = 15$.

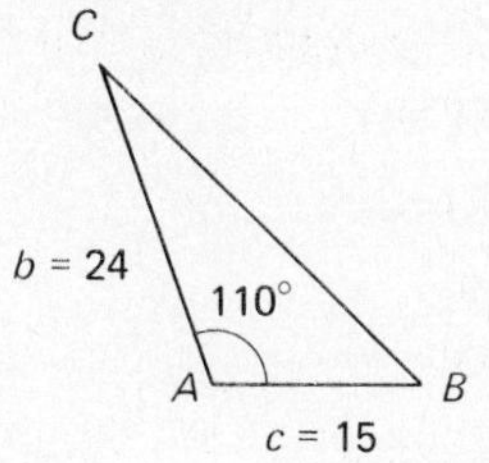

SOLUTION: First, note the given values ($\angle A$, sides b and c) and that $\angle B$, $\angle C$, and side a are to be determined. Since we are given two sides and the included angle, the Law of Cosines should be used to solve the triangle.

Second, select the Law of Cosines formula to solve for side a.

$$a^2 = b^2 + c^2 - 2bc \cos A$$

$$\sqrt{a^2} = \sqrt{b^2 + c^2 - 2bc \cos A}$$

$$a = \sqrt{b^2 + c^2 - 2bc \cos A}$$

Third, determine the cosine value of $\angle A$ (110°). Realize that $\angle A$ is greater than 90° and use the formula, $\cos A = -\cos(180° - \angle A)$:

$$\cos 110° = -\cos(180° - 110°)$$

$$\cos 110° = -\cos 70°$$

$$-\cos 70° = -.3420$$

Fourth, substitute the letters with known values and solve for side a.

$$a = \sqrt{b^2 + c^2 - 2bc \cos A}$$

$$= \sqrt{(24)^2 + (15)^2 - 2(24)(15)(-.3420)}$$

$$= \sqrt{576 + 225 + 246.24}$$

$$= \sqrt{1047.24}$$

$$= 32.36$$

Note that the multiplication of two negatives $-2(24)(15)$ and $-.3420$ resulted in a positive 246.24. The process of determining $\sqrt{1047.24}$ was not shown. (See Chapter 5, Square Root, if necessary.)

Fifth, select the Law of Sines formula to determine the degrees in $\angle B$.

$$\frac{a}{\sin A} = \frac{b}{\sin B}$$

Sixth, determine the sine value of $\angle A$ (110°) to use in the Law of Sines formula. Realize that $\angle A > 90°$ and use the formula, $\cos A = -\cos(180° - \angle A)$:

$$\sin 110° = \sin(180° - 110°)$$

$$= \sin 70° = .9397$$

Seventh, substitute the letters with known values and solve for sin B.

$$\frac{a}{\sin A} = \frac{b}{\sin B}$$

$$\frac{32.36}{.9397} = \frac{24}{\sin B}$$

$$(\sin B)(32.36) = (24)(.9397)$$

$$\frac{\sin B\,(32.36)}{32.36} = \frac{(24)(.9397)}{32.36}$$

$$\sin B = .6969$$

Eighth, using Table 2 and interpolation, determine B.

$$\sin B = .6969$$

$$\angle B = 44° \, 11'$$

Ninth, determine $\angle C$.

$$\angle A + \angle B + \angle C = 180°$$

$$\angle C = 180° - 110° - 44° \, 11'$$

$$= 25° \, 11'$$

EXERCISE 17-2

Using the Laws of Cosines and Sines, determine the degrees of angles and lengths of sides as indicated in each problem.

17-16. Determine the length of side b in oblique triangle ABC.

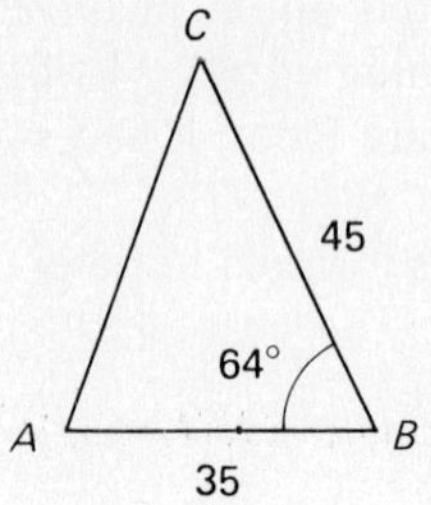

17-17. Determine the degrees in angle F in oblique triangle DEF.

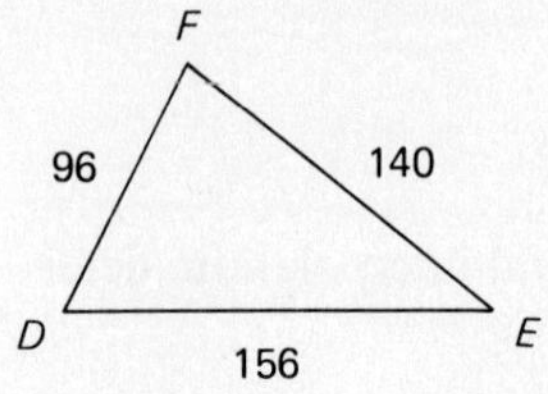

17-18. Given oblique triangle GHI, find side i.

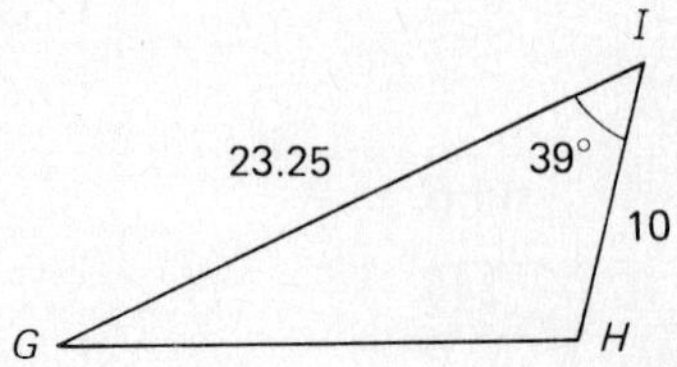

17-19. Determine the length of side k in oblique triangle JKL.

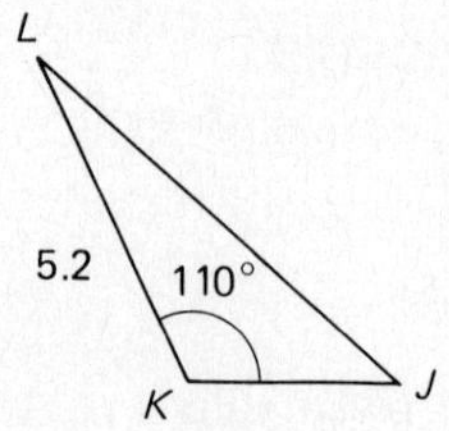

17-20. Determine the length of side m in oblique triangle MNO when side $n = 48$, side $o = 28.6$, and $\angle M = 128°$.

17-21. Determine the degrees of angle R in oblique triangle PQR when side $p = 61$, side $q = 65$, and side $r = 76$.

17-22. Given triangle TUV, find angle U.

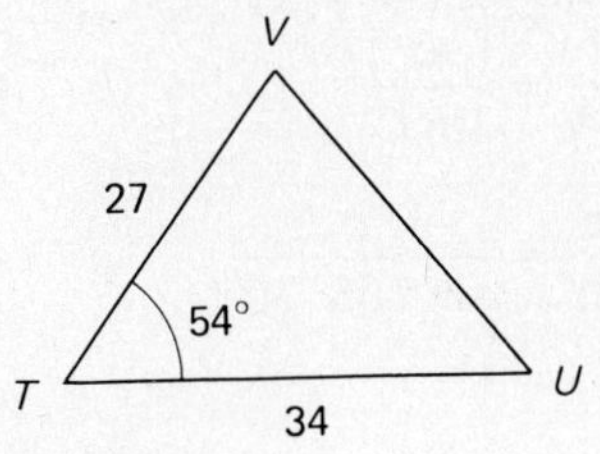

17-23. Find angle V.

17-24. Side t =

17-25. Given the oblique triangle GHI, angle H =

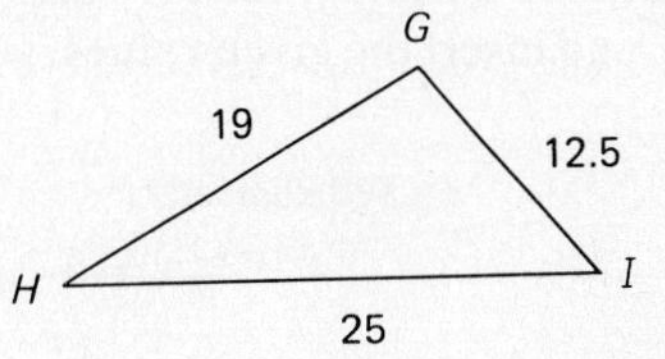

17-26. Angle I =

17-27. Angle G =

17-28. Solve the oblique triangle ABC given sides $a = 4.3$, $b = 8.4$, and $c = 5.2$. Angle A =

17-29. Angle B =

17-30. Angle C =

CONTENT

The Laws of Sines and Cosines have innumerable uses in industry. Problems which deal with navigation, alternating current, and vectors are just a few of the situations in which the skills of determining unknowns in oblique triangles can be effective. The examples which follow illustrate some of these industrial usages.

Example 17-7

PROBLEM: A light plane flew for 85 kilometers and then discovered a course error of 8° 30′. The destination was reached in 120 kilometers. What would have been the total distance if the course error had not been made?

SOLUTION: First, make a sketch, identify angles and sides by letters, and insert the given values.

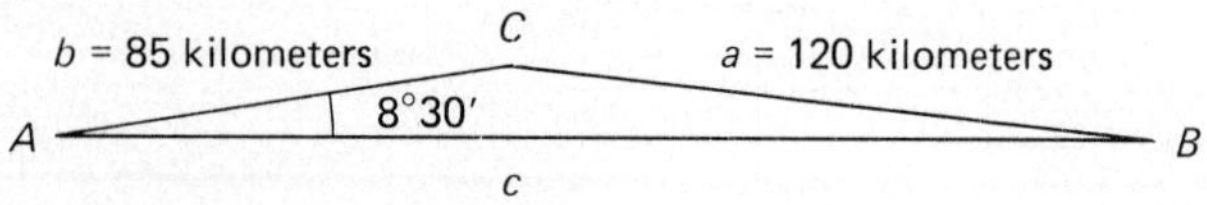

Second, select the Law of Sines formula to solve for $\angle B$, and then solve for $\angle B$.

$$\frac{a}{\sin A} = \frac{b}{\sin B}$$

$$\frac{120 \text{ km}}{\sin 8° 30'} = \frac{85 \text{ km}}{\sin B}$$

$$\frac{120 \text{ km}}{.1478} = \frac{85 \text{ km}}{\sin B}$$

$$\sin B\,(120 \text{ km}) = (85 \text{ km})(.1478)$$

$$\frac{\sin B\,(\cancel{120} \text{ km})}{\cancel{120} \text{ km}} = \frac{(85 \text{ km})(.1478)}{120 \text{ km}}$$

$$\sin B = .1047$$

$$\angle B = 6° \text{ (approximately)}$$

Third, determine the degrees in $\angle C$.

$$\angle A + \angle B + \angle C = 180°$$

$$8° 30' + 6° + \angle C = 180°$$

$$\angle C = 180° - 8° 30' - 6°$$

$$= 165° 31'$$

Fourth, determine the sine value of C.

$$\sin 165° 30' = \sin (180° - 165° 30')$$

$$= \sin 14° 30' = .2504$$

Fifth, select the Law of Sines formula to solve for side c, the unknown distance.

$$\frac{a}{\sin A} = \frac{c}{\sin C}$$

$$\frac{120 \text{ km}}{\sin 8° 30'} = \frac{c}{\sin 165° 30'}$$

Remember,

$$\sin 165° 30' = \sin (180° - 165° 30')$$

$$= \sin 14° 5' = .2504$$

Thus,

$$\frac{120 \text{ km}}{.1478} = \frac{c}{.2504}$$

$$\frac{(120 \text{ km})(.2504)}{.1478} = c$$

$$203.3 \text{ km} = c$$

Thus, the course distance without the course directional error would have been 203.3 kilometers.

Example 17-8

PROBLEM: Given the force diagram representing two concurrent forces acting on a pin connection of the truss on a shear machine, determine the resultant.

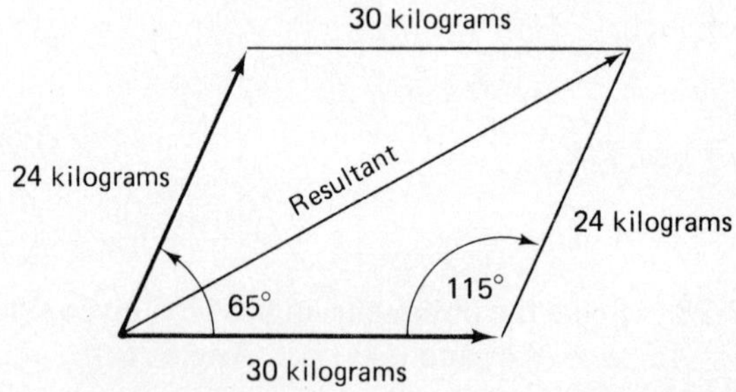

SOLUTION: First, note the given values and the nature of the problem. The two forces acting concurrently form a parallelogram, and the resultant cuts the parallelogram into two oblique triangles.

Second, select the Laws of Cosines formula which will enable you to determine the resultant.

$$c^2 = a^2 + b^2 - 2ab \cos C$$

Resultant

$$= \sqrt{(\text{force } 1)^2 + (\text{force } 2)^2 - 2(\text{force } 1)(\text{force } 2) \cos 115°}$$

$$= \sqrt{(24)^2 + (30)^2 - 2(24)(30)(\cos 115°)}$$

Third, determine the cosine value of 115°.

$$\cos 115° = -\cos (180° - 115°)$$

$$\cos 115° = -\cos 65°$$

$$\cos 65° = -.4226$$

Fourth, determine the resultant.

$$\text{Resultant} = \sqrt{(24)^2 + (30)^2 - 2(24)(30)(-.4226)}$$

$$= \sqrt{576 + 900 + 608.544}$$

$$= \sqrt{2084.544}$$

$$= 45.6568$$

$$= 45.7 \text{ km}$$

EXERCISE 17-3

Using the Laws of Cosines and Sines, determine the required data in the following industrial and daily living problems. Make a sketch where necessary, designating given values and unknown quantities.

17-31. From the sketch of a triangular-shaped lot, determine the length of the missing side.

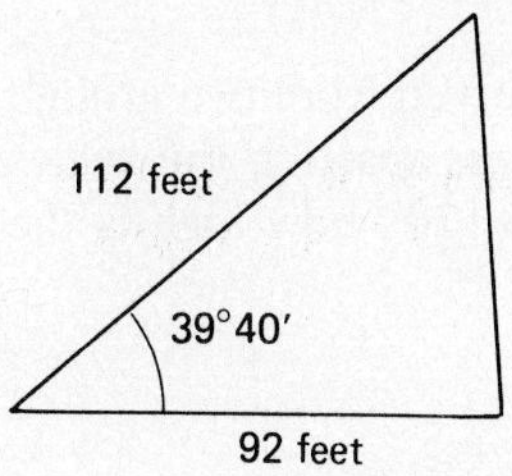

17-32. If 18-foot rafters are used for the building shown, how wide would the building be?

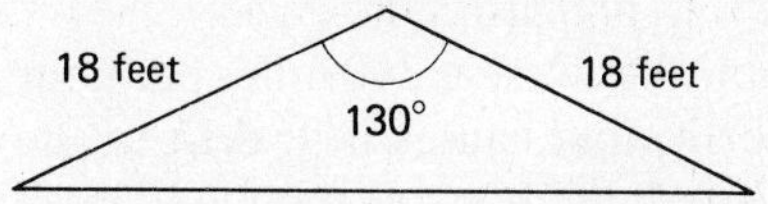

17-33. What is the angle of slope of the rafters in Problem 17-32?

17-34. What is the resultant force in the sketch below?

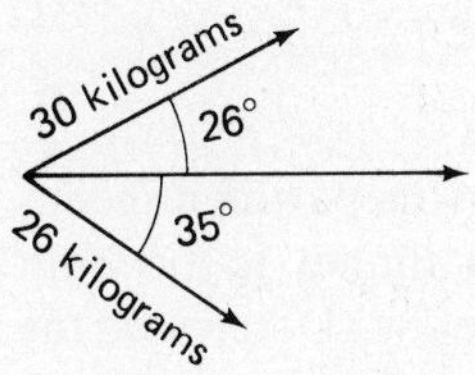

17-35. Determine the resultant force in the sketch below.

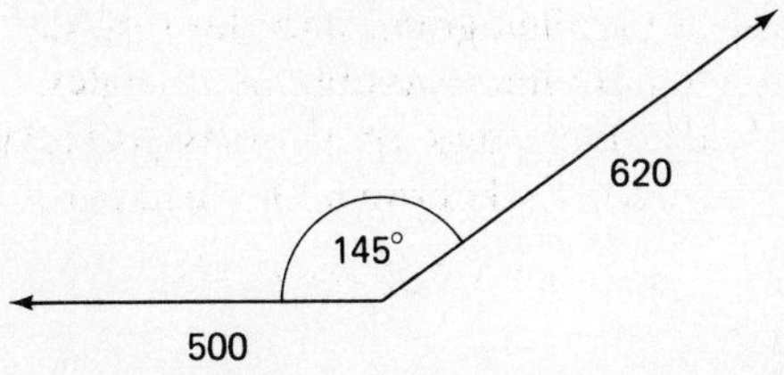

17-36. Determine the lengths of the guy wires on the utility pole illustrated below.

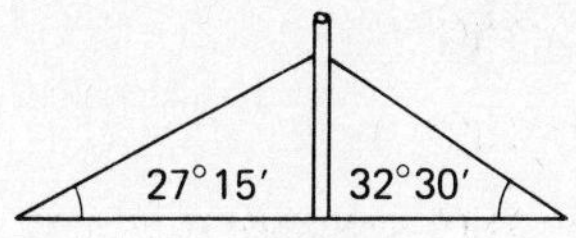

17-37. Determine the length of a private duck-hunting pond, using the information provided in the sketch.

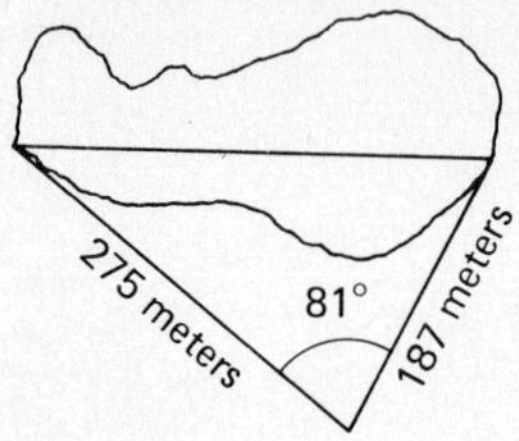

17-38. A surveyor determines the length of one side of a tract of land to be 520 rods. From each end of this side, he sights a pole and finds the angles to be 74° 10′ and 65°. Determine the length of the other sides.

17-39. An airplane flies west at 240 miles per hour for 4 hours. A north wind causes it to drift to the south at 6° 20′. Determine how far the plane must fly to make the correction as indicated and reach the intended destination.

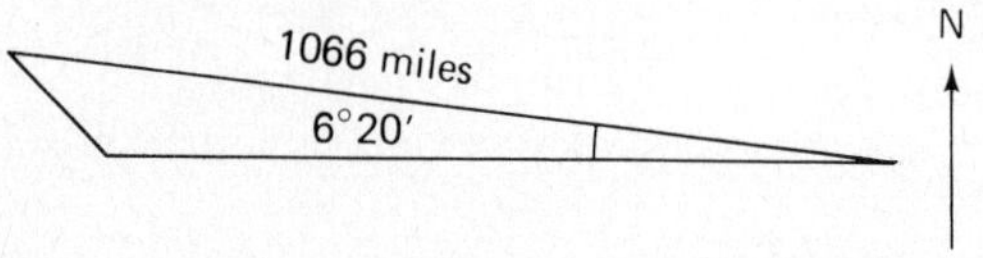

17-40. A helicopter is observed from two ground stations which are 3560 meters apart. If the angles of elevation are 54° and 28° 20′, how high is the helicopter?

17-41. The Montana Highway Department needs to determine the length of a tunnel to be built according to the sketch given below. Determine the length of the projected tunnel.

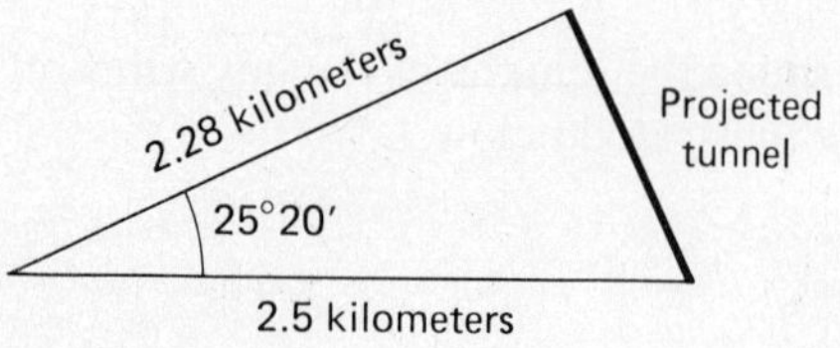

17-42. Two freeways intersect at an angle of 132°. If two cars start from the intersection and travel on the rays of the angle, how far apart will they be in 4 hours if one car is traveling at 55 miles per hour and the other car is traveling at 50 miles per hour?

17-43. A snowmobile is stuck in a snowbank. A force of 100 pounds is pulling at 25° from the rear of the snowmobile on the right side. Another force of 185 pounds is pulling at 25° from the rear of the snowmobile on the left side. Determine the magnitude of the resultant.

17-44. A hunter walks in straight lines for 281 meters, 385 meters, and 582 meters. Determine the smallest angle he makes in his triangular search for game.

17-45. Determine the height of the given building from the sketch below.

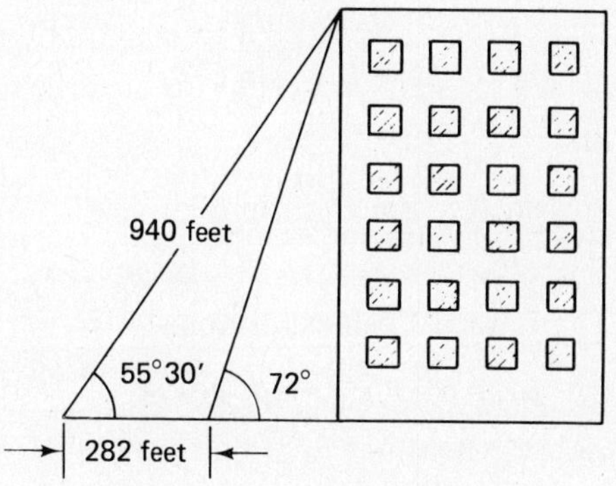

THINK TIME

There are numerous uses of the Laws of Sines and Cosines. Perhaps you can think of ways in which you can use the laws to assist you in your occupation or around your home. It may take a little imagination to think of potential uses, but you will find that the law can save you time and needless work.

The example which follows show how the law can be used to develop recreational and natural-resource facilities. A park and recreation director decides to set up a marsh trail. Because of the low ground, a boardwalk is to be built. He is able to place the walk around the water areas as shown in the sketch. After he has completed the boardwalk around the marsh, he determines to place bridges across the water areas from point D to G and point B to G. How long will these bridges have to be? Note that $\angle x = \angle y$.

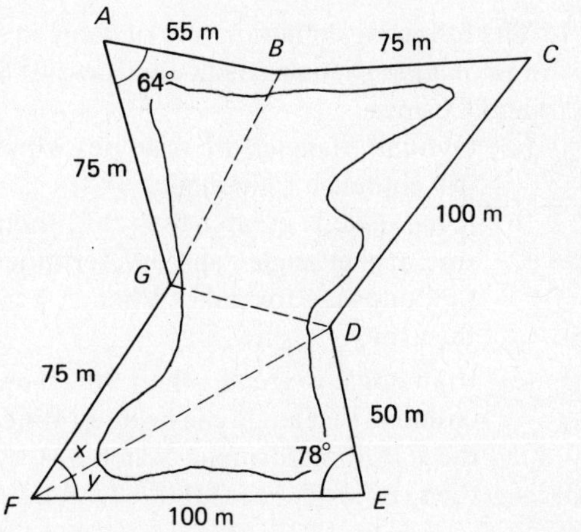

PROCEDURES TO REMEMBER

1. The procedure for determining the length of a side or the degrees of an angle of an oblique triangle using the Law of Sines is as follows:
   (a) Make a sketch of the given triangle.
   (b) Note the given values.
   (c) Select two of the three ratios which make use of the given values to determine the length of the side.
   (d) Substitute known values and sine values for letters.
   (e) Solve the ratios for the unknowns.
2. The procedure for determining the length of a side of an oblique triangle using the Law of Cosines is as follows:
   (a) Make a sketch of the triangle to be solved.
   (b) Note the given values.
   (c) Select the Law of Cosines formula.
   (d) Determine the cosine value of the given angle.
   (e) Substitute known values and cosine values for letters.
   (f) Solve for the unknown side.
3. The procedure for determining the degrees in an unknown angle of an oblique triangle, given the length of three sides, using the Law of Cosines is as follows:
   (a) Make a sketch of the triangle to be solved.
   (b) Note the given values.
   (c) Select the Law of Cosines formula to solve for the unknown angle.
   (d) Substitute known values for letters.
   (e) Use Table 2 and interpolate if necessary to determine the unknown angle.

CHAPTER SUMMARY

1. The following definitions are of value in solving oblique triangles using the Laws of Sines and Cosines.
   (a) Oblique triangles are triangles which do not contain a right angle.
   (b) If an angle is greater than 90°, then the sine of that angle can be determined by the following formula: When $\angle A > 90°$, then $\sin A = \sin(180° - \angle A)$.
   (c) If an angle is greater than 90°, then the cosine of that angle can be determined by the following formula: When $\angle A > 90°$, then $\cos A = -\cos(180° - \angle A)$.
   (d) One of four combinations of values must be given if the Laws of Sines and Cosines are to be used: (1) two angles and the length of one side; (2) the length of two sides and the angle opposite one of the sides; (3) the lengths of two sides and the angle between the two sides; (4) the lengths of all three sides.
   (e) The Law of Sines states that the ratio of a side to the sine of its angle is equal to the ratios of the other sides and their angles.
   (f) The Law of Cosines states that the square of any side of a triangle is equal to the sum of the squares of the other two sides minus two times the product of those sides and the cosine of the included angle.
2. The following formulas are of value in solving oblique triangles:
   (a) The Law of Sines formulas:

$$\frac{a}{\sin A} = \frac{b}{\sin B} = \frac{c}{\sin C}$$

or

$$\frac{\sin A}{a} = \frac{\sin B}{b} = \frac{\sin C}{c}$$

   (b) The Law of Cosines formulas:

$$a^2 = b^2 + c^2 - 2bc \cos A$$

or

$$\cos A = \frac{b^2 + c^2 - a^2}{2bc}$$

$$b^2 = a^2 + c^2 - 2ac \cos B$$

or

$$\cos B = \frac{a^2 + c^2 - b^2}{2ac}$$

$$c^2 = a^2 + b^2 - 2ab \cos C$$

or

$$\cos C = \frac{a^2 + b^2 - c^2}{2ab}$$

CHAPTER TEST

T-17-1. Determine the sine of 159°.

T-17-2. Determine the cosine of 99° 27′.

T-17-3. Given the oblique triangle *ABC*, find the length of side *b*.

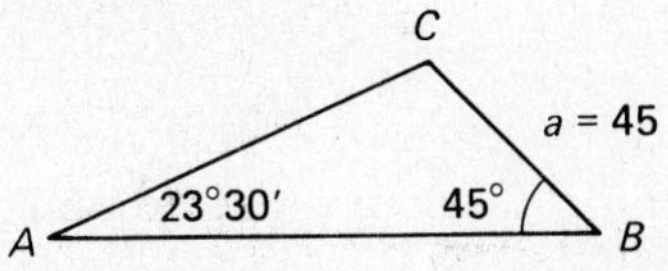

T-17-4. Given $\angle A = 65°$, side $a = 14.4$, and side $b = 12.2$, determine the degrees of $\angle B$.

T-17-5. Given the oblique triangle *GHI*, determine the degrees in $\angle I$.

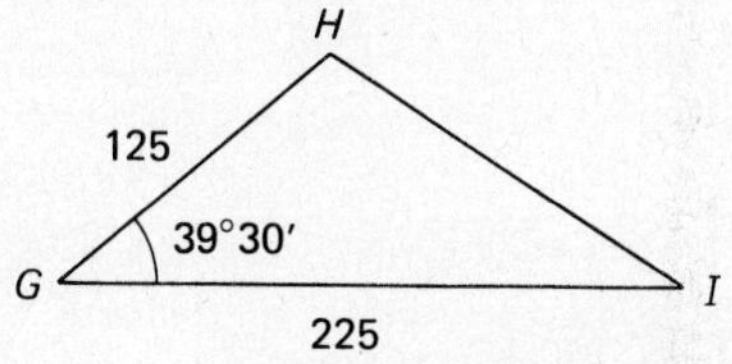

T-17-6. Given side $j = 65$, side $k = 40$, and angle $L = 37° 20'$, determine the length of side l.

T-17-7. Given triangle *MNO*, angle $O =$

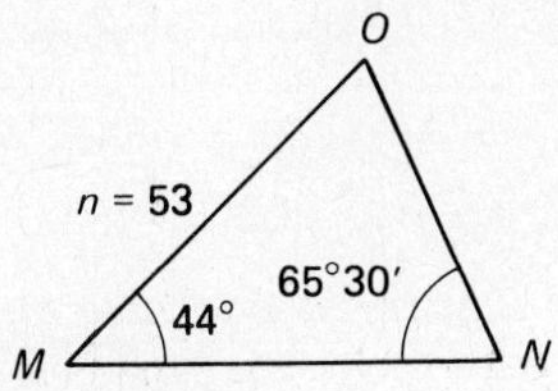

T-17-8. Side $m =$

T-17-9. Side o =

T-17-10. Two forces acting at right angles to each other produce a resultant force of 1600 pounds. One of the forces has a magnitude of 960 pounds. Determine the magnitude of the other force.

Section Five

METRICATION

CHAPTER

18

FUNDAMENTALS AND APPLICATIONS OF THE METRIC SYSTEM

OBJECTIVES

1. To identify and explain the fundamental definitions and properties of the metric system.
2. To convert one unit of measurement (length and area, capacity and volume, mass and weight) to another within the metric system.
3. To convert temperature measurements from Celsius to Fahrenheit and from Fahrenheit to Celsius.
4. To convert measurements from the metric to the English system and from the English to the metric system.
5. To solve industrial problems involving metric measurement.

SELF-TEST

This test will measure your ability to work within the metric system and to make conversions to the English system, and vice versa. You may use the conversion tables located within this chapter to solve these problems. If you are able to satisfactorily solve the problems, you will not need to further develop the metric skills as taught in this chapter. Show your work so that possible areas of difficulty may be noted.

S-18-1. 521 centimeters = ____ meters.

S-18-2. 5 dekameters = ____ meters.

S-18-3. 5280 meters = ____ kilometers.

S-18-4. 14 millimeters = ____ centimeters.

S-18-5. 625 square centimeters = ____ square inches.

S-18-6. 64,255 square meters = ____ hectares.

S-18-7. 700 miles = ____ kilometers.

S-18-8. 4 pounds 3 ounces = ____ grams.

S-18-9. If top-grade hogs weigh 220 pounds, how many metric tons would 50 top-grade hogs weigh?

S-18-10. How many kilograms does a 260-pound defensive end weigh?

S-18-11. 32.3 kilograms = ____ grams.

S-18-12. 5 pounds = ____ kilograms.

S-18-13. 2.36 metric tons = ____ pounds.

S-18-14. 5000 millimeters = ____ liters.

S-18-15. 5.55 decigrams = ____ grams.

S-18-16. How many grams does 500 milliliters of water weigh?

S-18-17. 45°C = ____°F.

S-18-18. 210°F = ____°C.

S-18-19. During a thunderstorm, the temperature dropped from 86°F to 62°F. How many degrees did the temperature change using the Celsius scale?

S-18-20. A metal shaper makes a $\frac{3}{64}$-inch cut on a piece of brass. How large is the cut in millimeters?

S-18-21. Building codes in a city require the window glass in houses to be at least $\frac{3}{16}$ inch thick. What are the code requirements in millimeters?

S-18-22. A ship loaded with 6600 tons (English) of wheat sails from Duluth, Minnesota to Yokohama, Japan. How many metric tons does the wheat weigh?

S-18-23. A classroom is 30 feet × 40 feet × 12 feet and exchanges air every 3 minutes. How many cubic meters of air are exchanged every hour?

S-18-24. A steel casting weighs 1.5 kilograms. Fifty holes are bored in the casting, and 2.75 grams of steel are removed from each hole. What is the weight of the casting after the holes have been bored?

S-18-25. An airplane can fly 1 kilometer on 12 kilograms of fuel. How many metric tons of fuel will be used to fly nonstop from New York to Minneapolis, a distance of 1760 kilometers?

INTRODUCTION

Through the centuries, man has reached out his hand to his fellowman in an attempt to come to a common understanding concerning life and the world which surrounded him. One such man, L. L. Zamenhof of Poland, observing the diversity of languages, designed an international language, which he called Esperanto. His attempt to encourage nations to adopt this universal language met little success.

Another man, Gabriel Mouton of France, saw the need for a practical decimal system of measurement. His base-10 system of measurement was introduced in 1670. People liked Mouton's system, but it was not until 1790 that the French Academy examined Mouton's system with thoroughness. Consequently, they adopted a system based on a nonchanging value. This nonchanging value was the distance from the North Pole to the equator. This distance was then divided into ten million parts which were called *meters*. The word *meter* simply means "measure." Today, because of the need for an even more precise measure for the meter, the length of the meter is based on the wavelength of the orange light given off by atoms of krypton gas. Other units of measure were derived from the meter so that length and area, capacity and volume, and mass and weight could be uniformly measured.

To insure precise measurement standards throughout the world, an international conference was called to establish standards. The Systemé Internationale, or SI, was the result of that conference. Almost all the nations of the world have since adopted the SI metric system. The United States is the only large nation which does not use the metric system. But it will be only a matter of time before the United States also adopts the system.

DEFINITIONS

There are seven base units of measure in the SI metric system. They are meter (length), kilogram (mass or weight), kelvin (temperature), second (time), ampere (electricity), candela (luminous intensity), and mole (amount of substance). Two common metric units which are not a part of the SI metric system are liter (capacity or volume) and Celsius (temperature). You will be introduced to only four basic types of metric measure: kilogram, meter, liter, and Celsius degree.

Contrary to English measurement, in the metric system an interrelationship exists between basic units of measure based on the use of water. Study the interrelationship illustrated below. This interrelationship between basic units of measure makes the metric system a most useful system of measure.

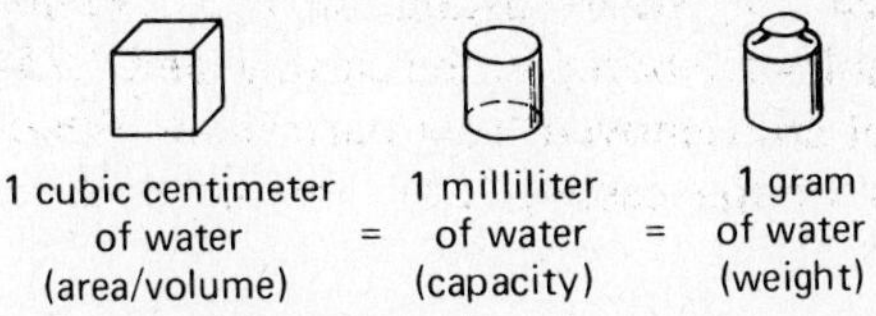

As noted before, the metric system is a base-10 or decimal measurement system. This means that each unit may be multiplied or divided by 10 to change a base unit into more workable subunits. Prefixes are added to the base units as a means of identification of the subunits. When subunits are *larger* than the base unit, Greek prefixes are used. The most common of these prefixes are *kilo* = 1000, *hecto* = 100, and *deka* = 10. When subunits are *smaller* than the base unit, Latin prefixes are used. The most common of these prefixes are *deci* = 0.1, *centi* = 0.01, and *milli* = 0.001. Thus a kilometer = 1000 meters and a millimeter = 0.001 meters. Study the chart below for a better understanding of the relationship between metric prefixes for quantities and decimal numeration for quantities.

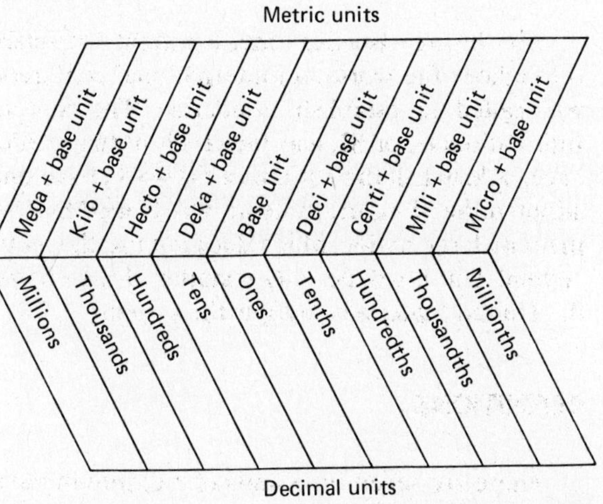

The movement of the decimal point to the left or right of the given quantity of a unit is all that is needed to change to the next higher or lower unit. Naturally, the new unit prefix should also be indicated. Thus 89 meters = 890 decimeters. The procedure for making these changes will be shown in the examples.

CONTENT

The base unit of measure for length and distance is the *meter*. A meter may be approximated for rapid measure in the following manner (see illustration). If an average adult extends his hand out from his shoulder and turns his head away from the extended hand, the distance from the tip of his fingers to the tip of his nose is approximately one meter. A millimeter is about the same length as the thickness of a dime.

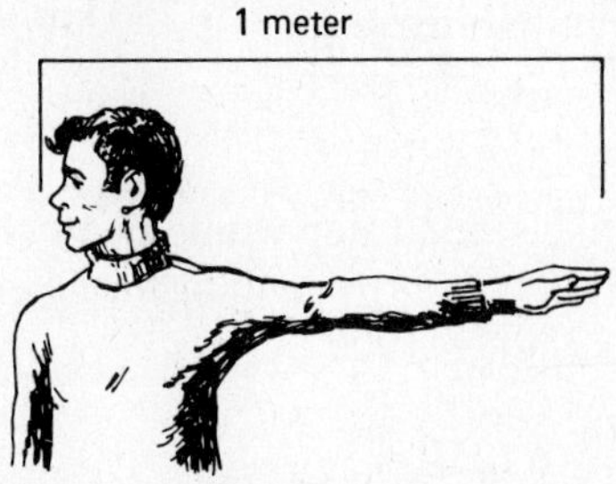

Like other base metric units, we can change the meter into more workable units by either multiplying or dividing the given quantity by 10. The following table states the meter units, their comparison in quantity to the meter, and their standard abbreviations.

METER TABLE OF EQUIVALENTS

Meter Units		*Comparison to Meter*		*Abbreviation*
1 kilometer	=	1000	meters	km
1 hectometer	=	100	meters	hm
1 dekameter	=	10	meters	dam
1 meter	=	1	meters	m
1 decimeter	=	0.1	meters	dm
1 centimeter	=	0.01	meters	cm
1 millimeter	=	0.001	meters	mm

It should be noted that the dekameter, hectometer, and decimeter are rarely used in comparison to the extensive use of the kilometer, meter, and millimeter.

Changing from one subunit to another may be accomplished simply by moving the decimal point. In order to better understand the process involved, study the examples below.

Example 18-1

PROBLEM: Change 1250 meters (m) to kilometers (km).

SOLUTION: First, using the meter chart given above, note that 1000 meters = 1 kilometer.

Second, multiply 1250 meters by unity or a fraction which is equal to 1. (Restudy Example 2-8, if necessary.) In this problem, $\frac{1 \text{ km}}{1000 \text{ m}}$ is to be used.

Note that the number, $\square$, representing unity is superimposed over the fraction for illustrative purposes.

$$\frac{\overset{1.25}{\cancel{1250}}}{1} \times \boxed{\frac{1\text{ km}}{\underset{1}{\cancel{1000}}\text{ m}}} = 1.250\text{ km}$$

An alternative second step would be to change the meters into kilometers by dividing the given number of meters (1250) by 1000.

$$\frac{1250\text{ m}}{1000\text{ m}} = 1.250\text{ km}$$

The simplest solution to the problem is to move the decimal point three places to the left.

1 km	1 hm	1 dam	1 m	1 dm	1 cm	1 mm
1000	100	10	1	0.1	0.01	0.001

(moves: 3, 2, 1)

1250 m = 1. 2 5 0 km

3 2 1

Remember: to change a smaller unit to a larger unit, move the decimal point the appropriate number of places to the left.

Example 18-2

PROBLEM: Convert 38.8 meters (m) to centimeters (cm).

SOLUTION: First, using the meter chart once again, note that 100 cm = 1 m.

Second, multiply 38.8 meters by unity, $\boxed{\frac{100\text{ cm}}{1\text{ m}}}$.

$$38.8\text{ m} \times \boxed{\frac{100\text{ cm}}{1\text{ m}}} = 3880\text{ cm}$$

An alternative second step would be to change the meters into centimeters by multiplying the given number of meters (38.8) by 100.

$$38.8\text{ m} \times 100 = 3880\text{ cm}$$

Once again, the simplest solution to the problem is to move the decimal point two places to the right.

1 km	1 hm	1 dam	1 m	1 dm	1 cm	1 mm
1000	100	10	1	0.1	0.01	0.001

(moves: 1, 2)

38.8 m = 3 8. 8 0 cm

1 2

Remember: to change a larger unit into a smaller unit, move the decimal point the appropriate number of places to the right.

EXERCISE 18-1

Convert the given quantities of one unit to the appropriate quantities for the units called for in each of the following problems.

18-1. 15 m = ____ cm.

18-2. 7 dam = ____ m.

18-3. 2 mm = ____ m.

18-4. 200 cm = ____ mm.

18-5. 53 m = ____ dam.

18-6. 3005 mm = ____ km.

18-7. 2 cm = ____ dm.

18-8. 54 hm = ____ m.

18-9. 44 cm = ____ mm.

18-10. 5280 mm = ____ m.

18-11. 12.4 km = ____ m.

18-12. 91.5 mm = ____ m.

18-13. 213.6 dm = ____ mm.

18-14. 2215 mm = ____ hm.

18-15. 302.017 cm = ____ m.

CONTENT

Recall that area is determined by finding the number of square units contained within a figure or object. Area is always expressed in square units such as square meters, square decimeters, and square centimeters. Square units may be abbreviated. For example, a square centimeter may be abbreviated as cm^2. The exponent of 2 is used to indicate square. Common abbreviations for squared units are as follows: km^2, hm^2, m^2, cm^2, mm^2.

Land measurement is denoted by a non-SI metric term, *hectare* (ha). A hectare is equal to 10,000 square meters, or a square piece of land, 100 meters × 100 meters. The following example will illustrate the use of square measurement within the metric system.

Example 18-3

PROBLEM: Determine the number of hectares (ha) in a rectangular plot of land, 1.6 km × 1.2 km.

SOLUTION: First, convert kilometers to meters in each of the measurements, since a hectare is equal to 10,000 square meters.

$$1.6\text{ km} = 1600\text{ m}$$

$$1.2\text{ km} = 1200\text{ m}$$

Second, using the formula for determining area, $A = lw$, determine the area of the plot of land.

$$A = 1600\text{ m} \times 1200\text{ m}$$

$$A = 1{,}920{,}000\text{ square meters}$$

Third, multiply 1,920,000 square meters by unity, $\frac{1\text{ ha}}{10{,}000\text{ m}^2}$ to determine the number of hectares in the plot of land.

$$\overset{192}{\cancel{1920{,}000}}\text{ m}^2 \times \frac{1\text{ ha}}{\underset{1}{\cancel{10{,}000}}\text{ m}^2} = 192\text{ ha}$$

An alternative third step would be to use the formula for converting square meters into hectares, ha = 10,000 m^2, and thus divide the square meters (1,920,000) by the number of square meters in a hectare (10,000).

$$\frac{1{,}920{,}000\text{ m}^2}{10{,}000\text{ m}^2} = 192\text{ ha}$$

EXERCISE 18-2

Determine the solutions called for in each of the following problems related to area.

18-16. 15 dam × 25 dam = ____ m^2.

18-17. 54.5 cm × 35.5 cm = ____ m^2.

18-18. .88 dm × .44 dm = ____ cm^2.

18-19. 35.1 km × .5 km = ____ ha.

18-20. $.0055 \text{ km} \times .0065 \text{ km} =$ ____ cm^2.

18-21. A shelf in a closet measures 182 cm × 30 cm. How many square meters of shelving will be needed for four shelves?

18-22. How many hectares are in a wheat farm which is 1.25 km × .35 km?

18-23. How many square meters would there be in a table top which is 214 cm × 107 cm?

18-24. What is the surface area in square centimeters of a square box which has a side length of 45 mm?

18.25. How many square meters are there in a rug which is 5 m × 3.5 m?

CONTENT

Recall that volume is determined by finding the number of cubic units within a solid. The basic unit of measure for volume in the metric system is the cubic meter (m^3). The exponent of 3 is used to indicate cubic measure. A very common usage of cubic measurement is found within the health occupations, where the term "cc" is used as an abbreviation of cubic centimeters.

The most commonly used unit to measure capacity within the metric system is the liter. Even though the liter is used extensively throughout the world, it is a non-SI unit of measure.

The words "capacity" and "volume" are often used interchangeably to mean the same thing. Technically, however, *volume* is the amount of space within a container, and *capacity* is the amount of substance a container will hold.

The relationship between volume and capacity in the metric system may best be seen in the illustration.

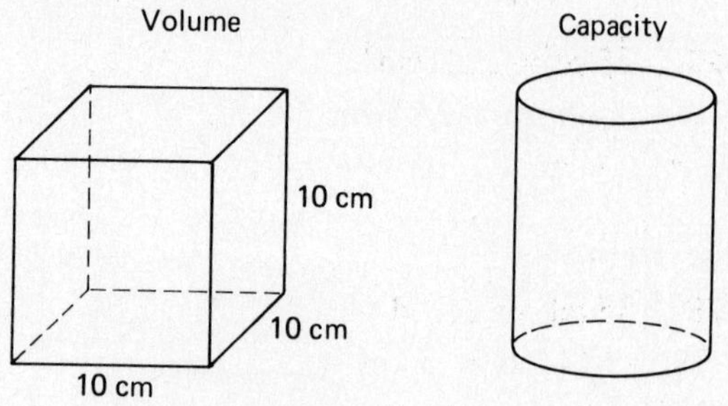

1000 cubic centimeters or 1 cubic decimeter = 1000 milliliters or 1 liter

The following table shows the relationship of 1 liter in liquid capacity to other units of liquid capacity and volume within the metric system. Note, in particular, the kiloliter, liter, and milliliter which are the most commonly used measures.

LITER TABLE OF EQUIVALENTS

Liquid Capacity			*Volume*	
1 kiloliter (kl)	= 1000 liters	= 1000 dm^3	= 1.0 m^3
1 hectoliter (hl)	= 100 liters	= 100 dm^3	= 0.1 m^3
1 dekaliter (dal)	= 10 liters	= 10 dm^3	= 0.01 m^3
1 liter (l)	= 1 liter	= 1 dm^3	= 0.001 m^3
1 deciliter (dl)	= 0.1 liter	= 100 cm^3	= 0.0001 m^3
1 centiliter (cl)	= 0.01 liter	= 10 cm^3	= 0.00001 m^3
1 milliliter (ml)	= 0.001 liter	= 1 cm^3	= 0.000001 m^3

Changing from one unit to another within the liquid capacity or liter system may be accomplished by moving the decimal point the appropriate number of spaces to the right or left. To change a larger unit (liter) to a smaller unit (deciliter), move the decimal point to the right the appropriate number of places (one). To change a smaller unit (liter) to a larger unit (dekaliter), move the decimal point to the left the appropriate number of places (one). Study the following examples which illustrate the procedures.

Example 18-4

PROBLEM: Convert 17 liters (l) to milliliters (ml).

SOLUTION: First, using the liter chart, note that 1000 ml = 1 l.

Second, multiply 17 liters by unity, $\frac{1000 \text{ ml}}{1 \text{ l}}$.

$$\frac{17 \text{ l}}{1} \times \frac{1000 \text{ ml}}{1 \text{ l}} = 17{,}000 \text{ ml}$$

An alternative second step would be to change the liters into milliliters by multiplying the given number of liters (17) by 1000.

$$17 \text{ l} \times 1000 = 17{,}000 \text{ ml}$$

The simplest solution, however, is to move the decimal point three places to the right.

1 kl	1 hl	1 dal	1 l	1 dl	1 cl	1 ml
1000	100	10	1	0.1	0.01	0.001

(1 → 2 → 3)

17 liters = 1 7. 0. 0. 0. milliliters
(1 2 3)

EXERCISE 18-3

Determine the solutions called for in each of the problems related to liter measurement. Feel free to use the liter table of equivalents.

18-26. 77 l = ____ ml.

18-28. 3.8 dm^3 = ____ cm^3.

Example 18-5

PROBLEM: Find the liquid capacity in milliliters and liters and the volume in cubic centimeters of the given aquarium.

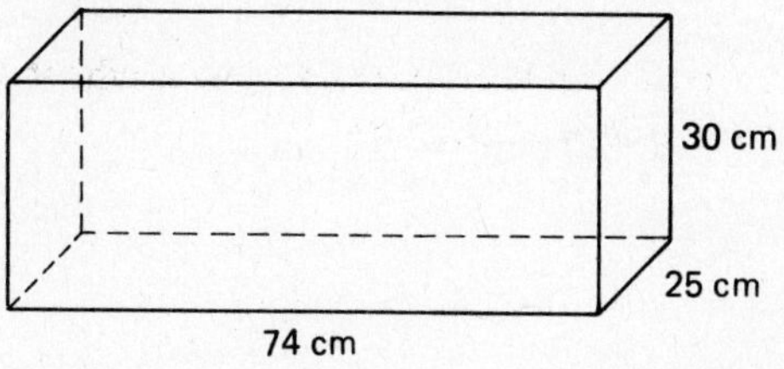

SOLUTION: First, determine the volume of the aquarium. Recall the formula for determining volume: volume (V) = Length $(l) \times$ width $(w) \times$ height (h). Thus,

$$V = lwh$$
$$= (74 \text{ cm})(25 \text{ cm})(30 \text{ cm})$$
$$= 55{,}500 \text{ cm}^3$$

Second, convert cubic centimeters to milliliters. Using the liter table of equivalents above, observe that 1 cubic centimeter = 1 milliliter. Thus,

$$55{,}500 \text{ cm} = 55{,}500 \text{ ml}$$

Third, convert 55,500 milliliters to liters. Once again, using the liter table of equivalents, note that 1000 milliliters = 1 liter. Thus,

55,500 ml = 5 5. 5 0 0 . = 55.5 l
(3 2 1)

18-27. 25 cc = ____ cm^3.

18-29. 55 ml = ____ cl.

18-30. .66 cm^3 = ____ dm^3.

18-31. 8880 ml = ____ l.

18-32. 747,720 mm^3 = ____ m^3.

18-33. 76.7 cc = ____ cm^3.

18-34. .0075 dm^3 = ____ mm^3.

18-35. 3836 dl = ____ kl.

18-36. 1.38 m^3 = ____ mm^3.

18-37. 5001 l = ____ hl.

18-38. 0.0076 cm^3 = ____ mm^3.

18-39. 3.27 ml = ____ l.

18-40. Determine the liter capacity of a small reservoir 5.5 m long, 2.5 m wide, and 2 m deep.

18-41. Determine the volume in cubic centimeters of a rectangular box 2.1 m long, 24 cm wide, and 26 mm wide.

18-42. Determine the volume in cubic meters of a cube with one side 34 cm long.

18-43. An excavation for a building is to be 55 m × 40 m × 15 m. What is the cost of excavation at $4.85 per cubic meter?

18-44. A plastic sheet is 10 km by 2 m by 4 mm. How many cubic meters of plastic are in the sheet?

18-45. A swimming pool 25 m long and 10 m wide has an average depth of 1.5 m. How many kiloliters of water will the pool hold?

18-46. How many 500-ml containers of honey could be filled from 100 hl?

18-47. How many cubic meters of blacktop are needed to make a driveway 10 m long, 5 m wide, and 20 cm thick?

18-48. If rain fell at the rate of 0.2 mm per minute, how many minutes would it take to reach 0.1 m?

18-49. Find the volume in liters of a rectangular tank 2 m by 20 dm by 28 cm.

18-50. A patient was given 2 cc of a medicine in 500 ml of water. What percent of the solution was medicine?

CONTENT

Mass is defined as the quantity of material contained in a given body. *Weight* is the measure of the force of gravity upon a given body. Although these are the technical definitions of mass and weight, in practical usage they are the same. For example, on the planet Earth, the terms "mass" and "weight" may be used interchangeably because the objects are affected by the Earth's gravity. However, on the Earth's moon, a given mass weighs about $\frac{1}{6}$ of its weight on the Earth. This is due to the moon's gravity which is about $\frac{1}{6}$ of that of the Earth. For our discussion, we will assume that mass and weight may be used interchangeably.

The basic SI metric unit for measuring mass is the kilogram. The average-sized paper clip weighs about 1 gram. Thus, a kilogram is about the weight of 1000 paper clips.

The relationship between volume, capacity, and weight in the metric system can be seen in the following illustration. It should be noted that 1 cm^3 = 1 ml = 1 g.

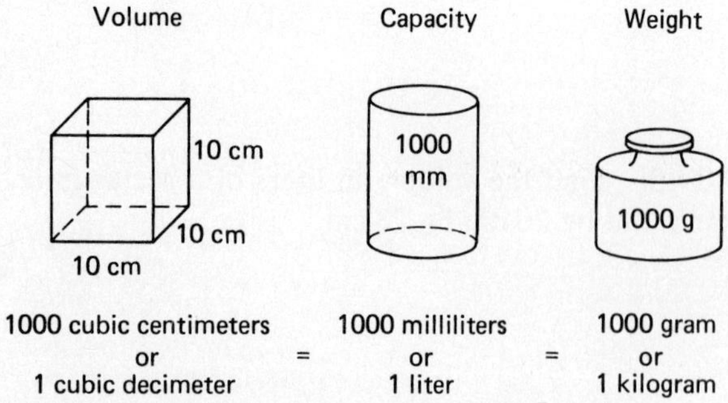

The following table shows the relationship of 1 kilogram of weight to other units of weight, volume and capacity within the metric system.

KILOGRAM TABLE OF EQUIVALENTS

Weight		*Liquid Capacity*		*Volume*
1000 kilograms (kg) = 1 metric ton (t)	=	1 kl	=	1 m^3
1 kilogram (kg) = 1000 grams	=	1 l	=	1 dm^3
1 hectogram (hg) = 100 grams	=	0.1 l	=	100 cm^3
1 dekagram (dag) = 10 grams	=	0.01 l	=	10 cm^3
1 gram (g) = 1 gram	=	1 ml	=	1 cm^3
1 decigram (dg) = 0.1 grams	=	0.1 ml	=	1 mm^3
1 centigram (cg) = 0.01 grams	=	0.01 ml	=	0.1 mm^3
1 milligram (mg) = 0.001 grams	=	0.001 ml	=	0.01 mm^3

It should be noted that 1000 kg = 1 metric ton = 1 kl = 1 m^3. The procedures for changing from one unit to another within the metric weight system and between systems are the same as those previously shown for changing other metric units. Study the examples below which illustrate the procedures.

Example 18-6

PROBLEM: Convert 2700 milligrams (mg) to grams (g).

SOLUTION: First, using the kilogram table of equivalents, note that 1000 mg = 1 g.

Second, multiply 2700 mg by unity,

$$\frac{1\ g}{1000\ mg}.$$

$$\frac{\overset{2.7}{\cancel{2700}}\ \text{mg}}{1} \times \frac{1\ \text{g}}{\cancel{1000}\ \text{mg}} = 2.7\ \text{g}$$

Instead of the above step, we can change the milligrams into grams by dividing the given number of milligrams (2700) by 1000.

$$\frac{2700\ \text{mg}}{1000} = 2.7\ \text{g}$$

Or, move the decimal point three places to the left.

1 kg	1 hg	1 dag	1 g	1 dg	1 cg	1 mg
1000	100	10	1	0.1	0.01	0.001

3 2 1

2700 mg = 2 . 7 0 0 . g

3 2 1

Example 18-7

PROBLEM: What is the weight in kilograms of gasoline in a rectangular tank 1.3 m long, 8 dm wide, and 42 cm deep? (Gasoline weighs 0.8 times as much as water.)

SOLUTION: First, change the quantities to like units.

$$1.3\text{m} = 130\ \text{cm}$$
$$8\ \text{dm} = 80\ \text{cm}$$
$$42\ \text{cm} = 42\ \text{cm}$$

Second, determine the volume of the tank using the formula, $V = lwh$.

$$V = (130\ \text{cm})(80\ \text{cm}) \times (42\ \text{cm})$$
$$V = 436{,}800\ \text{cm}^3$$

Third, change cubic centimeters into kilograms using the formula $1\ \text{kg} = 1000\ \text{cm}^3$.

$$\frac{436{,}800\ \text{cm}^3}{1000} = 436.8\ \text{kg}$$

Fourth, determine the weight of the gasoline which is 0.8 times as much as water.

$$436.8\ \text{kg} \times 0.8\ (\text{weight of gasoline}) = 349.44\ \text{kg}$$

Thus the tank holds 349.44 kg of gasoline.

EXERCISE 18-4

The following problems will give you the opportunity to develop your ability to work with gram measurement. You will need to recall some previously developed skills to solve some of the problems. Use any tables you need to solve the problems.

18-51. Convert 3.8 kilograms to grams.

18-52. Convert 520 milligrams to grams.

18-53. Convert 81.3 decigrams to milligrams.

18-54. Convert 7.8 centigrams to grams.

18-55. Convert 57.3 dekagrams to milligrams.

18-56. 7886 mg = ____ g.

18-57. 85.6 dg = ____ g.

18-58. 45,000 g = ____ kg.

18-59. 800 dag = ____ kg.

18-60. 3804 kg = ____ tons.

18-61. 84 g = ____ cg.

18-62. 81 dg = ____ mg.

18-63. 8.08 kg = ____ cg.

18-64. 5.54 tons = ____ kg.

18-65. What is the sum of the following weights: 4.3 g, 4.9 g, 78.1 g, and 5.05 g?

18-66. What is the sum in kilograms of the following: 0.83 kg, 0.38 kg, 33.3 g, 5050 mg, and 4.4 dag?

18-67. What is the weight in kg of 34 castings which weigh 38.6 dag each?

18-68. An irregular-shaped plastic container weighed 1450 g when empty. When filled with water it weighed 5.4 kg. What was the volume of the container in cm^3?

18-69. A number of fish died in polluted water because of the lack of dissolved oxygen in the water. In a new sewage treatment process, 54% of the polluting matter is removed. If there are 80 grams of polluting matter per liter of sewage, how many grams of polluting matter remain per kiloliter?

18-70. What is the weight in metric tons of 54 pieces of plastic weighing 85.6 kg each and 44 pieces of copper weighing 42.2 hg each?

18-71. An environmentalist has a beaker which contains 2 liters of aqueous solution containing 550 grams of sugar. The air evaporates 300 cubic centimeters of the water in the solution. What is the weight in grams of the remaining solution?

18-72. A brick measures 5.8 cm by 9.5 cm by 20.2 cm. What does it weigh if bricks weigh 2.4 g per cm^3?

18-73. Determine the amount of water in metric tons in a swimming pool, 25 m by 10 m by 2 m.

18-74. A beaker holds 500 ml of a salt solution containing 15 g of salt. How much salt must be added to make it a 10% solution?

18-75. Determine the weight of a solid piece of brass 95 mm thick, 5 cm wide, and 3 m long. (Brass weighs 8.2 kg per dm^3.)

CONTENT

The basic SI unit for measuring temperature is the *kelvin* (K). The measurement is used primarily by scientists and is based on a scale from absolute zero. On the kelvin scale, water freezes at 273.15°K and boils at 373.15°K. Because this scale is not very practical for popular use, the Celsius (C) or centigrade scale was devised for use within the metric system. The Celsius scale, based on the Kelvin system, lists water as freezing at 0 degrees and boiling at 100 degrees. Thus 1 degree Celsius is equal to exactly 1 degree Kelvin.

The following illustration indicates the relationships of the three types of temperature scales.

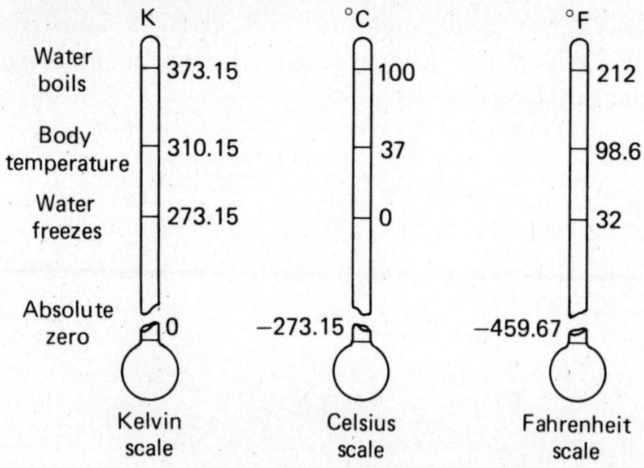

In time, the United States will adopt the Celsius system of measuring temperature as has Canada. During the process of transition from Fahrenheit to Celsius it will be necessary to be able to move freely from one system to the other. The following temperature-conversion formulas will be helpful:

From Celsius to Fahrenheit: $F = \frac{9}{5}C + 32°$

From Fahrenheit to Celsius: $C = \frac{5}{9}(F - 32°)$

The following examples will show you how the formulas are used.

Example 18-8

PROBLEM: Convert 40°C to °F.

SOLUTION: Using the formula $F = \frac{9}{5}C + 32°$, substitute 40° for C and solve the formula for the unknown, F.

$$F = \frac{9}{\cancel{5}_{1}} \times \frac{\cancel{(40°)}^{8}}{1} + 32°$$

$$= 9 \times 8° + 32°$$

$$= 72° + 32°$$

$$= 104°$$

Example 18-9

PROBLEM: A chef has an oven which has a Celsius thermostat. A recipe calls for a cake to be baked at 320°F. What would the Celsius setting be on the oven?

SOLUTION: Using the formula $C = \frac{5}{9}(F - 32°)$, substitute 320° for F and solve for the unknown, C.

$$C = \frac{5}{9}(320° - 32°)$$

$$= \frac{5}{\cancel{9}_{1}} \times \frac{\cancel{288°}^{32}}{1}$$

$$= 160°$$

Thus the setting for the Celsius thermostat would be 160°.

EXERCISE 18-5

Converting degrees from one standard of measurement to another can be simple when the correct formulas are used. Practice using the formulas of conversion as you solve the problems below. A careful look at your answer should indicate if you have used the correct formula.

18-76. 20°C = ____°F.

18-77. 28°C = ____°F.

18-78. 98.6°F = ____°C.

18-79. 100°F = ____°C.

18-80. 212°C = ____°F.

18-81. 62°F = ____°C.

18-82. 45°F = ____°C.

18-83. −30°F = ____°C.

18-84. 72°F = ____°C.

18-85. 85°F = ____°C.

18-86. The temperature in the oxyacetylene process of welding sometimes reaches 5800°F. What is the temperature on the Celsius scale?

18-87. Wrought iron melts at 1550°C. What temperature is this on the Fahrenheit scale?

18-88. Sleet occurs when the temperature is about 27°F. What is the Celsius equivalent to this temperature?

18-89. The temperature dropped from 20°F to −10°F in 5 hours. What would the change of temperature be in degrees Celsius?

18-90. At what temperature is the reading on the Celsius thermometer the same as the reading on the Fahrenheit thermometer?

CONTENT

Hopefully within a relatively short period of time, the United States will adopt the metric system as its system of measurement. When this occurs, there will be a period of time before Americans will be able to work as easily within the metric system as they do within the English system. During this period of transition there will be a need to make conversions from one system to the other. Even today, in some industries, there is a need to be able to work effectively within both systems and, when necessary, to convert measurements from one system to the other. Electronic calculators, conversion charts, and other devices have been constructed to aid in the conversion process.

When the time comes, when you find it necessary to work within both systems and to convert measurements from one system to the other, it will be important for you to learn a few common equivalents.

COMMON ENGLISH AND METRIC EQUIVALENTS

English to Metric			*Metric to English*		
1 inch	=	2.54 cm	1 cm	=	0.3937 inch
1 yard	=	0.9144 m	1 m	=	1.0936 yards
1 mile	=	1.609 km	1 km	=	0.6213 mile
1 ounce (dry)	=	28.349 g	1 g	=	0.0352 ounce
1 pound (dry)	=	0.4536 kg	1 kg	=	2.2046 pounds
1 ounce (liquid)	=	29.75 ml	1 ml	=	0.0338 ounce
1 quart (liquid)	=	0.946 l	1 l	=	1.0568 quarts

The basic conversions given above, coupled with your understanding of the units within both the metric and English systems, should enable you to move quite freely from one system to the other. You may find it to your advantage to memorize those portions of the table which are already of use to you.

If necessary, more in-depth conversion tables may be used. These are provided below.

LENGTH EQUIVALENTS

English to Metric		*Metric to English*	
1 in.	= 2.54 cm	1 mm	= 0.03937 in.
	= 25.4 mm	1 cm	= 0.3937 in.
1 ft	= 30.48 cm		= 0.0328 ft
	= 0.3048 m	1 m	= 39.37 in.
1 yd	= 91.44 cm		= 3.2808 ft
	= 0.9144 m	1 km	= 0.62136 mile
1 mile	= 1609.344 m		= 3280.8 ft

AREA EQUIVALENTS

English to metric		*Metric to English*	
1 sq in.	= 6.4516 cm^2	1 cm^2	= 0.155 sq in.
1 sq ft	= 0.09290 m^2	1 m^2	= 10.7639 sq ft
1 sq yd	= 0.83613 m^2	1 hectare	= 2.471 acres
1 acre	= 0.4047 hectare		

VOLUME EQUIVALENTS

English to Metric		*Metric to English*	
1 cu in.	= 16.387 cm^3	1 cm^3	= 0.061 cu in.
1 cu ft	= 28.317 dm^3	1 dm^3	= 61.0237 cu in.
	= 0.0283 m^3		= 0.0353 cu ft
1 cu yd	= 0.7646 m^3	1 m^3	= 35.3147 cu ft
			= 1.3079 cu yd

WEIGHT AND MASS EQUIVALENTS

English to Metric		*Metric to English*	
1 pound	= 0.4536 kilogram	1 kilogram	= 2.2406 pounds
1 short ton	= 907.2 kilograms	1 gram	= 0.0352 ounce
1 long ton	= 1016 kilograms	1 metric ton	= 2204.6 pounds

The examples which follow will give you an opportunity to follow the conversion procedures and to develop the ability to make conversions quickly.

Example 18-10

PROBLEM: How long is a 100-yard football field in meters?

SOLUTION: Using the conversion formula, 1 yard = 0.9144 meters, set up a simple equation involving unity, $\dfrac{0.9144 \text{ m}}{1 \text{ yd}}$.

$$100 \text{ yards} \times \frac{0.9144 \text{ m}}{1 \text{ yd}} = 91.44 \text{ meters}$$

Thus a 100-yard football field is 91.44 meters long.

Example 18-11

PROBLEM: In the United States, the speed limit for automobiles is 55 miles per hour. What is the speed limit in kilometers per hour?

SOLUTION: Using the conversion formula, 1 mile = 1.609 kilometers, set up an equation involving unity, $\frac{1.609 \text{ kmph}}{1 \text{ mph}}$

$$55 \text{ mph} \times \frac{1.609 \text{ kmph}}{1 \text{ mph}} = 88.495 \text{ kmph}$$

Thus 55 miles per hour is equal to 88.5 kilometers per hour.

Example 18-12

PROBLEM: The capacity of a gasoline tank in a European automobile is 42 liters. How many gallons is this to the nearest tenth of a gallon?

SOLUTION: Using the conversion formula, 1 liter = 1.0568 quarts, and the fact that 4 quarts = 1 gallon, set up a simple equation and solve for the unknown gallons.

$$\overset{10.5}{\cancel{42}} \text{ liters} \times \frac{1.0568 \text{ quarts per liter}}{\underset{1}{\cancel{4}} \text{ quarts per gallon}} = 11.0964 \text{ gallons}$$

Thus 42 liters is equal to about 11.1 gallons.

Example 18-13

PROBLEM: A chef is told by his French manager to prepare 14 servings of filet mignon for a special party. He is told that each filet should weigh about 250 grams. Without allowing for shrinkage, how many pounds of steak will the chef need to order from his supplier?

SOLUTION: First, change grams to kilograms by moving the decimal point three places to the left.

$$250 \text{ g} = .\underset{3}{2}\,\underset{2}{5}\,\underset{1}{0} \text{ kg}$$

Second, determine the total amount of meat needed in kilograms.

14 servings × .25 kg per serving = 3.5 kg of meat

Third, using the formula, 1 kilogram = 2.2046 pounds, set up an equation and solve for the unknown number of pounds.

3.5 kg × 2.2046 lbs per kg = 7.7161 lbs of meat

Thus the chef would need to order about 7.72 pounds of meat from his supplier to serve 14 people 250-gram filets.

EXERCISE 18-6

Conversion from metric to English and English to metric is not difficult when the correct formula is selected and the proper mathematical operations are applied. Practice making conversions using the correct conversion formulas.

18-91. 1 foot = ________ cm.

18-92. 1 mile = ________ m.

18-93. 8.3 km = ________ miles.

18-94. 105 mm = ________ inches.

18-95. 55 m = ________ yards.

18-96. 24 inches = ________ mm.

18-97. 66 square yards = ________ m^2.

18-98. 48 cm^2 = ________ square inches.

18-99. 66.2 mm^2 = ________ square inches.

18-100. What is the size of a piece of 12 inch × 12 inch tile in centimeters?

18-101. Find the area of the 12 inch × 12 inch tile in square centimeters.

18-102. One half gallon = ________ liters.

18-103. 25 cubic inches = ________ cm^3.

18-104. 1 liter = ________ ounces.

18-105. .35 kg = _______ ounces.

18-106. One half kilogram of nails costs $.65. What is the cost per pound?

18-107. 25,000 cubic inches = _______ cm^3.

18-108. A 12-ounce can of frozen orange juice is equal to how many liters?

18-109. How many cubic meters of earth will need to be excavated for a house 80 feet × 30 feet × 4 feet?

18-110. A contractor needs 150 cubic yards of concrete for a job. How many cubic meters is this?

18-111. 220 pounds = _______ kg.

18-112. 53 kg = _______ pounds.

18-113. A 100-pound sack of potatoes is equal to how many kilograms?

18-114. Number One spring wheat weighs 60 pounds per bushel (1.244 cubic feet). How many kilograms does the wheat weigh per dm^3?

18-115. A liter of milk equals how many ounces?

18-116. A spark plug gap is 3 mm. How many thousandths of an inch is this?

18-117. A closet shelf is 1.83 m × 61 cm. How long and wide is the shelf to the nearest inch?

18-118. A garden is 150 feet by 30 feet. How many square meters is the garden?

18-119. A jet flies from New York to San Francisco, a distance of 2560 miles, in $4\frac{1}{2}$ hours. How many kilometers per hour does the jet fly?

18-120. Determine the cost of carpeting a room 8 meters × 4.5 meters at $14.50 per square yard.

18-121. How tall is a 6 foot 5 inch man in meters?

18-122. What is the volume in kiloliters of a pool 50 feet × 25 feet × 6 feet?

18.123. Andy Marthaler drove 600 kilometers on 93.5 liters of gasoline in Spain. Determine his average miles per gallon.

18-124. Five quarts of antifreeze are added to a 22-quart cooling system. How many liters of antifreeze would be needed for a 100-quart cooling system?

18-125. Water expands 9% when becoming ice. How large a cube of ice in cubic centimeters would you get from 0.57 kilograms of water?

THINK TIME

In time, personal data will be given in metric measurement. Fill out your personal data sheet below. Be sure to estimate the measurements before you begin. This will be of value to you as you relate measurement to the metric system of measurement.

PERSONAL DATA

Measure	*Type of Unit*	*Estimate*	*Actual*
Height	Centimeters		
Weight	Kilograms		
Chest/Bust	Centimeters		
Waist	Centimeters		
Hips	Centimeters		
Foot	Centimeters		

Try measuring other common objects which are a part of your life. A few possible objects have been given. Add your own to the list.

COMMON OBJECTS

Measure	*Type of Unit*	*Estimate*	*Actual*
Height of doorway	Meters		
Weight of car	Kilograms		
Distance to work	Kilometers		
Amount of rainfall	Centimeters		
Present temperature	Celsius		

PROCEDURES TO REMEMBER

1. To change from one unit to another within the metric system, move the decimal point the appropriate number of places.
   (a) To change a smaller unit to a larger unit, move the decimal point the appropriate number of places to the left.
   (b) To change a larger unit to a smaller unit, move the decimal point the appropriate number of places to the right.
2. To change from Celsius to Fahrenheit:
   (a) Multiply the Celsius degrees by $\frac{9}{5}$.
   (b) Add 32 degrees to the product.
   (c) The resultant is the degrees Fahrenheit.
3. To change from Fahrenheit to Celsius:
   (a) Subtract 32 degrees from the given Fahrenheit degrees.
   (b) Multiply the difference by $\frac{5}{9}$.
   (c) The resultant is the degrees Celsius.
4. To convert from English to metric or metric to English:
   (a) Select the appropriate conversion formula.
   (b) Enter the given information into the formula.
   (c) Solve for the unknown.

CHAPTER SUMMARY

1. The seven basic units of measure in the SI metric system are meter (length), kilogram (mass or weight), Kelvin (temperature), second (time), ampere (electricity), candela (luminous intensity), and mole (amount of substance).
2. Two common metric units which are not a part of the SI metric system are liter (capacity) and Celsius (temperature).
3. The interrelationship between volume, capacity, and weight within the metric system is as follows:

Volume		*Capacity*		*Weight*
1 cm^3 of water	=	1 ml of water	=	1 g of water
or		or		or
1 dm^3 of water	=	1 l of water	=	1 kg of water

4. Greek prefixes identify subunits *larger* than the base unit: kilo = 1000, hecto = 100, and deka = 10.
5. Latin prefixes identify subunits *smaller* than the base unit: deci = 0.1, centi = 0.01, and milli = 0.001.

6. Subunits within the metric system and their respective values are as follows:

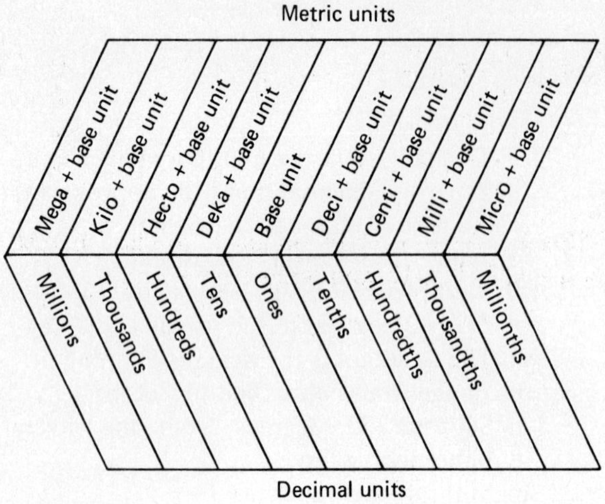

7. Although the terms "volume" and "capacity" may be used interchangeably, *volume* is the amount of space within a container and *capacity* is the amount of substance a container will hold.
8. *Mass* is the quantity of material contained in a given body.
9. *Weight* is the measure of the force of gravity upon a given body.

CHAPTER TEST

T-18-1. 33 km = ______ m.

T-18-2. 1748 m = ______ km.

T-18-3. 5280 km = ______ m.

T-18-4. 31.5 cm = ______ m.

T-18-5. 84 square inches = ______ cm^2.

T-18-6. .527 m^2 = ______ square inches.

T-18-7. 500 ml = ______ liters.

T-18-8. 31.5 gallons = ______ liters.

T-18-9. 21.5 liters = ______ gallons.

T-18-10. 1500 grams of water = ______ ml.

T-18-11. 44.2 dag = ______ kg.

T-18-12. 53,000 kg = ______ tons.

T-18-13. 5000 mg = ______ g.

T-18-14. 35.3 dg = ______ g.

T-18-15. 26,400 feet = ________ km.

T-18-16. 100 feet = ________ m.

T-18-17. 90°C = ________°F.

T-18-18. 95°F = ________°C.

T-18-19. A house contains 4000 square feet of floor space. How many m^2 is this?

T-18-20. What part of an inch, to the nearest thousandth, is a 12-mm wrench?

T-18-21. How many inches in diameter is a 155-mm howitzer?

T-18-22. A mixture of 8 g of salt and 8 g of baking soda is poured into 225 ml of water. What is the total weight of the solution?

T-18-23. A radio station broadcasts at 100 megahertz. How many kilohertz is this?

T-18-24. At a cost of $12.50 per m^2, how much would it cost to blacktop a 50-foot by 16-foot driveway?

T-18-25. Two liters of concentrated liquid fertilizer are added to water to make a total of 10 gallons of plant food. How many ml of fertilizer are added per gallon?

INDEX